Spark Ablation

Spark Ablation

Building Blocks for Nanotechnology

edited by

Andreas Schmidt-Ott

Jenny Stanford
Publishing

Published by

Jenny Stanford Publishing Pte. Ltd.
Level 34, Centennial Tower
3 Temasek Avenue
Singapore 039190

Email: editorial@jennystanford.com
Web: www.jennystanford.com

British Library Cataloguing-in-Publication Data
A catalogue record for this book is available from the British Library.

Spark Ablation: Building Blocks for Nanotechnology

Cover image inspired by Maurits F. J. Boeije, George Biskos, Bibianne E. van der Maesen, Tobias V. Pfeiffer, Aaike W. van Vugt, Bernardus Zijlstra, A. Schmidt-Ott, Attila Kohut, Gábor Galbács, Zsolt Geretovszky, Jicheng Feng, Kiwoong Lee, and Mansoo Choi

ISBN 978-981-4800-82-2 (Hardcover)
ISBN 978-0-367-81709-1 (eBook)

Contents

Preface

In the 1970s and 1980s, I witnessed and contributed to basic research in solid-state physics dealing with metal particles only a few nanometers in diameter. This size range had been identified as exciting because basic changes in the behavior of matter, "size effects," had been predicted from quantum theory and were being experimentally verified. Technological applications of these new properties were expected, and this was the dawn of nanotechnology. Today, nanotechnology is defined by the size domain below 100 nm, but it is mainly the range below 20 nm where the confinement into a small-enough particle volume makes electrons behave differently compared to the extended solid. An example of a resulting size effect is the optical absorption spectrum of semiconductors: the color of a particle changes continuously with its size. Below about 2 nm, the property–size relations change discontinuously and nonmonotonically with every atom added or removed, which opens, in principle, a myriad of possible properties with applications in catalysis, chemical sensing, electronics, and many other fields.

Surprisingly, most inorganic "nano"-products available 40 years later do not exhibit dimensions in the range of a few nanometers yet. Their structures are usually more than an order of magnitude larger, and their special properties are based on the rather trivial effects related to the surface-to-volume ratio. One of the reasons for the rather slow pace with which we are moving toward systematic exploitation of size effects below 20 nm is the difficulty of producing well-defined, stable objects and structures in this size range. Another reason is that pure sub-20-nm particles are difficult or impossible to store and transport, so a source is desirable that can easily be brought into any production process.

The present book deals with spark ablation as such a source of building blocks, which offers unprecedented flexibility regarding composition and size, opening unlimited access to novel sets of properties. New approaches to building structures by 3D additive assembly of nanoparticles are also covered.

For materials or devices exploiting size effects occurring below 20 nm, many technological applications are expected to lead to important innovations in the near future. These are emerging in important domains like energy conversion and storage, for example, the development of durable artificial leaves converting CO_2 and water into fuels for a sustainable future. Other examples making use of inorganic particles are catalysts, saving energy in the chemical industry, and avoiding or removing emissions threatening health and climate. Particle-based chemical sensors are becoming an important prerequisite for safe processes of the future, contamination control, and environmental monitoring, to name another example. These and other fields will profit especially from inorganic particles smaller than 20 nm, where size effects constitute a wealth of new properties tunable by size and composition.

The present book seeks to cover the presently most important application areas. It presents spark ablation as a research tool extremely versatile concerning particle size and composition but also in view of introducing into industry a scalable source of nanoparticulate material.

The book starts with two introductory chapters explaining the relevance of spark ablation and the basic technical features of spark discharge generators, ways of modifying the particles produced, as well as principles of immobilizing the particles for production of new materials and devices. Chapters 3 and 4 are dedicated to the basics of the ablating spark plasma and its diagnosis. Basic techniques of mixing materials, for characterizing the particles produced, and the technology for producing atomic clusters are covered by Chapters 5–7. The final seven chapters explain more technology but are mainly dedicated to application domains where spark ablation has already been applied, at least at the stage of research.

My sincere thanks go to the 28 authors of the present book, who all have many years of experience in working with spark ablation. For this book they have taken much time and effort of reviewing recent work as well as providing original contributions to this strongly developing field.

Andreas Schmidt-Ott
2020

Chapter 1

Application Domains of <20 nm Particles and the Role of the Spark Discharge Generator

Eva A. J. Rennen,[a] **Maurits F. J. Boeije,**[b] **and Andreas Schmidt-Ott**[b,c]
[a]*VSPARTICLE B.V., Molengraaffsingel 10, NL-2629 JD Delft, The Netherlands*
[b]*Faculty of Applied Sciences, Delft University of Technology, NL-2629 HZ Delft, The Netherlands*
[c]*Environment and Water Research Center, Cyprus Institute, Nicosia, Cyprus*
e.rennen@vsparticle.com

The present chapter compiles application areas of nanoparticle production by spark ablation. The spark ablation technique is best applied to produce inorganic particles in the diameter range between atomic clusters (smaller than 2 nm) and 20 nm. In this range, particle properties are very different from macroscopic (bulk) materials and strongly size dependent.

Spark Ablation: Building Blocks for Nanotechnology
Edited by Andreas Schmidt-Ott

ISBN 978-981-4800-82-2 (Hardcover), 978-0-367-81709-1 (eBook)
www.jennystanford.com

1.1 Introduction

Particles close to the atomic size obey the laws of quantum mechanics, as expressed by Schrödinger's equation, which represents the most important category of differences with respect to the bulk in the size range below 20 nm. A review by Roduner [1] explains the origin of these differences. For example, the solution of Schrödinger's equation for the valence electrons in a small metal particle (similar to the electron-in-a-box problem) demands discrete (quantized) energy levels the electrons may occupy. This change in the electronic structure of sub-20-nm particles with respect to bulk systems changes virtually *all* properties. Additional influences of the small size on particle properties also governed by quantum mechanics are associated with collective electron oscillations (plasmons), the occurrence of superparamagnetism, and more. Other size effects that specifically apply to these very small particles are related to a reduced binding energy between surface atoms due to surface curvature, such as a reduced melting point and size-dependent catalytic activity. Astonishingly, nanotechnology has, until now, hardly exploited the most interesting features of this size range, where size effects are most prominent. We claim that difficulties in producing these particles in a clean, controlled, and flexible way are one reason for holding back this development. Another feature of such small nanoparticles that has practically not been exploited is to be seen in the stronger tendency to form alloys with respect to bulk material. It has frequently been observed that the phase diagrams of mixed materials allow for new mixed phases when the dimensions shrink [2]. Atomic mixing thus receives a new quality in nanoparticles and basically opens myriads of possibilities associated with new material properties.

We believe that the development of spark ablation nanoparticle production will play an important role in moving nanotechnology into the sub-20-nm size range, where exploiting new properties for applications promises to be most rewarding. This development will be aided by the feature of spark ablation to easily mix solid materials [2], from the atomic scale (alloys) to the scale of a few nanometers. This opens an exciting spectrum of new possibilities for functional materials as well as nanodevices.

In this chapter we list the basic size effects that govern the properties of sub-20-nm particles, and link them to applications that have already been found in products and those that are in the research stage. We also speculate on possible future applications that the properties seem to open up. The reader may use this overview of what is known today also as an inspiration of what may be done tomorrow.

Subsequently, features of spark ablation are briefly outlined and compared to other nanoparticle production methods. Indications of required quantities of nanoparticulate material per item are given for industrial applications, and a selection of examples illustrates the contribution of spark ablation in the development of applications listed in the overview that make use of the size effects.

1.2 Sub-20 nm Particle Properties and Their Application Domains

1.2.1 Basic Size Effects

Table 1.1 lists basic size effects that cause properties of very small particles to change with size and that are of practical interest. It groups them into three categories. Strictly speaking, the basic categories on the left are related (e.g., binding energy is also connected to quantum mechanics), but the separation is considered as illustrative. Some effects of practical interest are listed in the right-hand column. The basic properties of category B1 deal with the rather trivial, purely geometric consequences (mainly surface-to-volumc ratio), which also apply to larger particles. The second category of surface curvature effects, B2, is based on the atomic structure of matter, which can be qualitatively explained by classical, non-quantum-mechanical arguments and that deal with important changes that occur when reducing particle size. Surface curvature in small particles reduces the coordination number of surface atoms, and that has some drastic consequences below 20 nm, as listed in the right-hand column of the second category. "Faster dissolution" is mentioned in both categories B1 and B2 because the kinetics of particle dissolution profits from a higher surface area (for the same volume), and additionally the weaker binding of surface atoms with rising surface curvature promotes dissolution. Category B3 deals with phenomena that can only be understood with the principles of quantum mechanics, even on a qualitative level.

Table 1.1 Basic size effects encountered in solids when going from larger to smaller particle sizes in the diameter range below 20 nm

Basic category	Basic size effects of practical interest
B1. Geometry or surface-to-volume ratio related	B1.1 Faster mass transport to surface of particle in medium B1.2 Faster mass transport from surface to center of particle B1.3 Faster dissolution B1.4 Reduced strain in volume expansion
B2. Reduced binding energy between atoms due to surface curvature	B2.1 Reduced melting point B2.2 Increased vapor pressure B2.3 Faster dissolution B2.4 Increased chemical reactivity B2.5 Size-dependent catalytic activity B2.6 Increased adsorptive binding of atoms/molecules B2.7 Larger variety of crystal habits B2.8 Laplace pressure inside nanoparticle B2.9 Superplasticity B2.10 Reduced surface free energy
B3. Quantum-mechanical size effects	B3.1 Size-dependent plasma resonance B3.2 Size-tunable bandgap in quantum dots B3.3 Photoelectron emission enhancement B3.4 Superparamagnetism B3.5 Formation of "superatoms" B3.6 Strong size dependence of light scattering and absorption B3.7 Surface-enhanced Raman scattering (SERS) from adsorbed molecules

Application examples related to the basic categories of Table 1.1 are listed in Table 1.2. For example, catalysis is influenced by a high surface-to-volume ratio and is additionally influenced by a "coordination number effect," that is, the changed atomic arrangement on a curved surface, leading to higher reactivity or stronger adsorptive binding with respect to the bulk. Some effects mentioned in Table 1.1 do not have any notable applications yet, which means that they are still to be found.

"Superatoms" is a term introduced in publications of the groups of Castleman and Khanna [3]. As pointed out before, valence electrons in a metal particle occupy discrete energy levels, and these particles thus mimic specific atoms in the periodic table and chemically behave like these (see also Ref. [1]). For example, Al_{13} behaves like a halogen atom, and Al_7 behaves like an alkali metal atom. Eventually, it may be possible to design stable atomic clusters that form "cluster assembled materials" with new properties or properties of more expensive materials. This vision was already expressed in 1996 in Klaus Sattler's book *Cluster Assembled Materials*, but since then there has been little progress in this field, and it remains a future perspective, which may now have become more realistic with the availability of flexible high-production-rate spark discharge generators combined with mobility classification (see Chapter 7).

1.2.2 Application Domains of sub-20 nm Particles

The basic properties described before bare the potential of many technological applications. Table 1.2 compiles applications that have been described in literature and links them to the size effects in Table 1.1. The application areas listed are examples, most of these have made use of spark ablation, and for all of them spark ablation would be an option. Chapters and sections of the present book that deal with the applications are indicated. In Section 1.4, many of these application examples are further described.

1.2.3 Requirements for Successful Applications

The effectiveness of the particles for specific applications, both in research and on an industrial scale, relies on the ability to control the size, purity, and composition, because the properties described before for sub-20-nm particles are size and composition dependent. For some applications such as model catalysts or biomedical applications of quantum dots, where the individual particle dimensions are of importance, a very narrow size distribution or size selection is usually necessary.

Table 1.2 Application examples of particles below 20 nm in diameter

Broad application domain	Specific application examples	Relation to size effect in Table 1.1	Chapter/ Section
A1 Heterogeneous catalysis	A1.1 Improved catalyst discovery by optimizing • Size (down to atomic cluster range) • Composition (atomic scale) • Composition (nanoscale, co-catalysts) A1.2 Photocatalytic degradation of airborne contaminants A1.2.1 Preserving ancient artifacts A1.2.2 Fighting the sick building syndrome A1.3 Microfluidic catalysts A1.4 Aerosol catalysis	B1 B2.5 B2.6 B2.7 B3	2 **10** 1.4.1 1.4.3 1.4.4
A2 Solar energy conversion	A2.1 Photoelectrochemical reaction acceleration A2.1.1 Water splitting A2.1.2 CO_2 reduction (artificial leaf) A2.2 Fuel cells A2.3 Photovoltaic efficiency enhancement	B1.1 B3.3	**12** 1.4.3 1.4.7 1.4.8
A3 Chemical sensors	A3.1 Metal oxide sensors A3.1.1 E-nose A3.1.2 Humidity sensing A3.2 Inert metal sensors A3.3 Plasmonic (bio)sensors A3.4 SERS-based (bio) sensors	B1.1 B1.2 B1.4 B2.5 B2.6 B3.1 B3.2 B3.7	9 **11** **12** 13 1.4.1 1.4.2 1.4.6 8.3
A4 Low-temperature welding	A4.1 Printed circuits A4.1.1 Contact lens	B2.1	1.4.2 1.4.5

Broad application domain	Specific application examples	Relation to size effect in Table 1.1	Chapter/ Section
A5 Antimicrobial, antifungal, antiviral, and cytotoxic agents	A5.1 Toxicity research A5.1.1 Inhalation toxicity A5.1.2 Organ-on-a-chip A5.2 Antibacterial surfaces A5.2.1 Medical implants	B1.3	**13** 1.4.10
A6 Magnetic materials	A6.1 Magnetic cooling A6.2 Water pollution treatment A6.3 Magnetic oil filtering A6.4 Magnetic cell separation	B3.4	1.4.15
A7. Diagnostics and drug delivery	A7.1 Diagnostics A7.1.1 Lab-on-a-chip A7.1.2 Imaging A7.2 Cancer treatment A7.2.1 Hypothermia A7.2.2 Photothermal (quantum dots)	B1.3 B3.4	**13** 1.4.10.3
A8 Atomic clusters	A8.1 Self-organized particle growth A8.2 Nanodevices A8.3 New materials composed of "superatoms"	B2.1 B3.5	**7** **1.4.4.4**
A9 Energy storage	A9.1 Batteries A9.2 Supercapacitors A9.3 Hydrogen storage materials	B1.1 B1.2 B1.4	1.4.9 1.4.14
A10. Electronics	A10.1 Magnetic cooling A10.2 LED's A10.3 Conductive paths A10.4 Welding with nanopowders	B2.1 B3.2 B3.4	8 1.4.2 1.4.5
A11. Model aerosol production	A11.1 Filter testing A11.2 Measurement device calibration		**14** 1.4.13

(Continued)

Table 1.2 (*Continued*)

Broad application domain	Specific application examples	Relation to size effect in Table 1.1	Chapter/ Section
A12. Lithography and printing	A12.1 Multimaterial printing A12.2 All-carbon electronic devices A12.3 Hybrid metal-carbon-polymer nanosystems A12.4 Soft robotics	B2.1 B3.2	2 8 **9** 1.4.2
A13 Lubricants		B2.9	8.3 1.4.11
A14 Tracers for high-speed flows		B1	1.4.12
A15 Carbon nanotube production		B2.4 B2.5	13 1.4.4.3

The possible combinations by mixing elements are endless, which provides tremendous opportunities for the discovery of novel materials. The production of arbitrary mixtures of materials will cause a revolution in catalyst discovery, when combinations are possible and simple at the atomic scale and the nanoscale and if this goes hand-in-hand with a feasible concept of catalyst production (Chapter 2). The same applies for production of chemical sensors, where, for example, a greater selectivity of mixed metal oxide sensors can be achieved by finetuning the exact material composition (see Chapter 11).

Control over the generated particles is only the first step. For successful application the particles need to be immobilized in a controlled and suitable way to produce a reproducible end product. As one application benefits from dispersed particles in a coating, another requires the ability to print porous layers of variable composition to enable the development of advanced electronic noses (e-noses). Deposition principles are described in Chapter 2, and Chapters 8 and 9 refer to specific methods.

For nanoparticle production below 20 nm our method of choice that meets the requirements above is spark ablation, the features of which are described in the following section.

1.3 Spark Ablation

1.3.1 History

Spark ablation has been used for vaporization of materials for their spectroscopical analysis for more than 100 years [5]. The reason for applying short sparks for this purpose lies in the fact that the vapor, from which the absorption spectrum is determined, has the same composition as the electrodes, which consist of the material to be analyzed [6]. When using spark ablation for nanoparticle production, this is a necessary (but not sufficient) condition for identical composition of the particles and the electrodes (see Chapter 5). In other evaporation techniques, the components with the higher vapor pressure evaporate first, that is, distillation effects occur. With the development of combustion engines, spark plugs were used, and this led to studies on the ablation rate as a parasitic effect, where the most notable publication is by Llewellyn-Jones in 1950 [7]. Burtscher and Schmidt-Ott first applied spark ablation in the gas phase in 1982 for "small particle" or "cluster" (later called "nanoparticle") production. As their focus was studying basic size effects, the authors did not dedicate more than a single sentence to the method in their first publication making use of the technique [8]. More detailed articles on spark ablation from that group followed in 1986 [93] and 1988 [9]. The effect of the vapor having the same composition as the ablated material (see Chapter 5) was exploited for production of arbitrary alloy nanoparticles by the research group of Schmidt-Ott much later [2, 10, 11], and it was one reason for several groups to dedicate much effort to the technique in the years to follow. Many of these efforts are described in the present book (Chapters 2 and 8–14). Table 2.2 of the Chapter 2 lists milestones in the technological development of the technique. Two notable commercial instruments based on the spark technique have been brought to the market, the first one at the end of the 1980s by PALAS GmbH, Karlsruhe, Germany, originally as a generator of soot-like aerosols, and the second one in 2017 by VSPARTICLE, Delft, the Netherlands, designed mainly for producing new materials and devices, presently mainly for research purposes (see Chapter 2).

1.3.1.1 Research interest

The growing interest in the spark ablation technology since publication of the first detailed article by Schwyn et al. in 1988 [9] can be demonstrated by the number of publications using or further developing that technology. Figure 1.1 shows the frequency of publications on the subject. The graphic exhibits a progressive increase with an average growth rate of ca. 15%. This suggests that spark ablation is successfully positioning itself in the field of nanoresearch, which serves a present nanotechnology market growth of 17%.

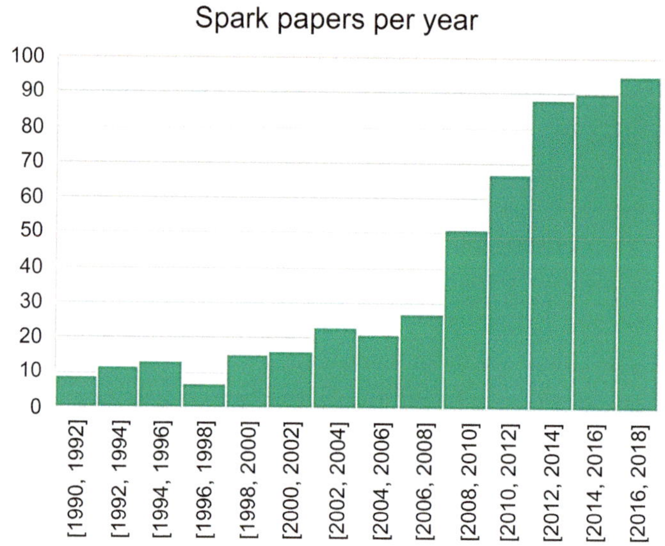

Figure 1.1 Frequency of papers published in 2-year periods from 1990.

1.3.2 Technical Features

Spark ablation covers the size range between atomic clusters and 20 nm, where size effects are important. Further technical details of the method are given in Chapter 2 and in other chapters of this book.

Spark production of nanoparticles is based on ablation of electrodes by inducing short spark discharges between them. The mean size can be tuned by varying the process parameters (see Chapter

2 and Feng et al. [13]), and the size distribution is lognormal with a geometric standard deviation around 1.35. For many applications, this is narrow enough, and where more uniformity in size is needed, size selection is easily done in the aerosol phase (see Chapter 2). The particles produced have much in common with those produced by laser ablation [14], a method that is also well established as a means of producing nanoparticles in quantities not exceeding milligrams for research purposes. The great advantage of the spark technique is its relatively low cost because no laser is required. In addition, nanoparticle production by spark ablation is scalable. The European project "Buonapart-E," which started in 2012 with 21 European partners, had scaling up of the spark method and the arc method as an objective. The project was financed mainly on the grounds of the process being environmentally friendly, as no chemical precursors are required and no waste is produced. Scale-up was demonstrated by Stein and Kruis [15]. This reference refers to arc discharge, which delivers particles larger than 20 nm (see Chapter 2), but the scale-up principle demonstrated is transferable to spark discharge. Just like laser ablation, spark ablation requires no chemicals except the substance or mixture of substances that are supposed to compose the particles. It must be mentioned, however, that laser ablation does not require conducting materials, which makes it more versatile in that respect. This disadvantage of spark ablation is partly compensated by the ease with which metal or semiconducting particles can be converted to oxides or other compounds in the aerosol phase downstream of the spark (see Chapter 2). With regard to colloidal synthesis methods, the particles produced by spark ablation are free from surfactants or precursors. Because of their purity, specific substance- and size-dependent properties can be studied without interference of surface layers. The feasibility of very high purity, enabling non-oxide particle production even for magnesium [16], is crucial when dealing with very small particles, where "each atom counts." The particles are usually crystalline, and their presence in the aerosol phase after production allows reactive modification such as oxidation and size selection. Compared to continuous arc ablation [17], an important advantage of spark ablation is that the mixing of multiple elements is possible in virtually any proportion (see

Chapter 5). The two methods, spark and arc, are complementary in the sense that arc discharge mainly covers the size domain *above* 20 nm.

The spark ablation feature of easy mixing of materials is quite unique below 20 nm. It has been demonstrated that nanoparticles made with spark ablation form crystalline mixed phases, where the components are immiscible in the bulk [2].

Liquid nanoparticle production methods require the development of a different recipe for every mixture, and contaminants are practically unavoidable when a liquid phase is used. Furthermore, reproducibility and stability of liquid methods is usually low, and continuous production is difficult. The advantage of liquid methods lies in higher controllability of the particle shapes to be produced. However, large varieties in possible controllable shapes have mainly been demonstrated in the size range above 20 nm. Mixed particle production has also been demonstrated in flame [18] and flame spray production [19], but each mixture requires different conditions, while the process is so complex that predictions on the produced particles are difficult. Mixed particle production is basically quite simple in flame spray pyrolysis [20] and spray pyrolysis, where soluble precursors such as nitrides or chlorides can be mixed in any proportion. However, as the particles have seen a liquid with its impurities and the fuel of the flame, purity is an issue if very small particles are to be produced. Modeling flame synthesis is extremely complex because the chemical reaction and flow phenomena interact with each other and the conditions are usually strongly inhomogeneous. For some applications, flame synthesis has the advantage that a particle production rate up to a few hundreds of grams per hour can be achieved at low cost.

Particles in the gas phase cannot be stored, as the particles will coagulate, aggregate, and adhere to the surface of the container over time. For some applications it would be possible to collect a nanopowder or stabilize in a colloid. However, contaminants are present as soon as liquids are involved, and particles smaller than 20 nm tend to coalesce in a powder. Even if only van der Waals forces are present, it is virtually impossible to separate particles once they are agglomerated. In addition, oxidizable nanopowders are pyrophoric and represent a safety hazard. An advantage of the spark process is that there is no need to store or transport particles.

Operating the spark discharge generator is simple; it only uses a few variable parameters, does not occupy much space, and is easily transportable. The generator can be placed right where the particles are needed for the production of materials and devices. Generating and directly depositing the particles makes storage and transport unnecessary. The generator can operate continuously and be monitored from a distance. In summary, the spark ablation method covers the requirements named in Section 1.2.3 for the particle size range below 20 nm, and it is presently the only method that does, apart from laser ablation. Compared to laser ablation, spark ablation requires a much smaller investment and is scalable for industrial use. This explains the strongly growing interest in the method in the research area and gives a perspective for industrial use.

1.3.3 Further Processing and Deposition Downstream of the Spark

Aerosol technology makes it possible to characterize, modify, and size-select the particles in-line. In particular, it is possible to perform chemical reactions of the particles with gasses added to produce compounds. Oxides have often been produced and so have hydrides [16], and other compounds like sulfides and nitrides should also be easy to produce. Chapter 2 addresses this possibility in more detail.

The particles can be directly deposited on substrates, if desired. Basically, aerosol particles that collide with a surface adhere to it due to van der Waals forces. However, deviations from this rule are possible, especially for particles in or close to atomic cluster sizes [21, 22]. Deposition can be achieved using three main methods. In diffusional deposition, particles are transported to surrounding surfaces by Brownian motion. For the sub-20-nm range, filtration is governed by diffusion and can be used to build very high-surface-area nanoparticulate structures. Here, particles are directed through a filter that is perpendicular to the gas flow. In impaction, particles are accelerated through a nozzle and impacted on a spot on the substrate. For particles smaller than 20 nm, the pressure downstream of the nozzle must be in the range of 1 mbar, and it is possible to aerodynamically focus in order to reach a spot size as small as a few micrometers [23]. Impaction can be combined with an XYZ stage making it possible to print patterns.

Aerosol particles can be captured in a liquid, for example, by bubbling (see Chapter 2). Depending on the type of deposition, it is possible to make coatings or deposit nanoporous layers. The relative standard deviation of particles from spark ablation is about 1.35. For many applications this is sufficiently narrow. If not, electrical mobility selection [24] can be used. Here another advantage of spark ablation comes into play: part of the produced particles are charged, which is a prerequisite for electrical mobility selection. An integrated system of a spark discharge generator in combination with an electrostatic mobility classifier is commercially available from VSPARTICLE. The charged state of part of the particles from spark ablation enables electrostatic classification as well as focused electrostatic deposition. Chapter 9 makes use of this effect for advanced structuring of surfaces on a nanoscale by electrostatically deposited nanoparticles from spark ablation.

1.3.4 Present State of Commercial Development

Currently, two companies have commercialized the technology in readily available generators, PALAS and VSPARTICLE.

PALAS, Karlsruhe, Germany, was the first to commercialize a spark discharge generator intended for the production of soot, for example, serving the calibration of measurement devices. It currently has two generators available, DNP digital 2000 and 3000. The difference between the two systems is the additional mass flow controller in the 3000 model. The spark generator produces agglomerates in the size range of 20–350 nm, with a mass production rate of 0.06–9 mg/h, depending on the material. The chamber is made of plastic, and easy change of materials is not intended. It requires opening up the entire system. Applications in the domain of aerosol technology such as testing of measuring instruments for particles larger than 20 nm and filter testing are the main application areas of this generator (see Chapter 14). This generator has not been built for applications where purity is necessary [25, 26], which is usually an issue for technology aiming at nanomaterials and devices.

Realizing the potential of spark ablation for the strongly growing domain of nanotechnology, VSPARTICLE, founded in 2014, is offering solutions to various branches of nanotechnology research and has chosen spark ablation as the main source of nanoparticulate building blocks for the size range below 20 nm. The spark discharge generators that have been developed by VSPARTICLE are designed in a way that any desired purity is achievable and different electrode configurations are possible depending on the application. They enable large flexibility in choosing the main parameters so that agglomerated or un-agglomerated particles with sizes from atomic clusters to 20 nm are achievable at mass production rates between 0.01 and 3g/h, depending on the material. The output particle size is adjustable by the parameters, as explained in Chapter 2. Sizes up to 100 nm can be achieved by subsequent coagulation. The generator is a device that can be connected to a variety of deposition modules, enabling flexibility in the setup. Three deposition accessories for research purposes based on electrostatic precipitation, diffusion, or impaction (see Section 1.3.3) are commercially available as well as a size selection module. A printer combining impaction with an XYZ stage has recently been developed. A powder coater and an industrial system are currently under development. The focus is a well-defined controllable, predictive, and highly variable output and particulate deposit in terms of particle size and material combinations. Using the size selection module (model VSP-S1), particle sizes with a 0.1 nm resolution can be selected.

1.4 Application Domains of Particles from Spark Ablation

As stated in Chapter 2, a mass production rate of 0.5 g/h can be assumed for a single pair of gold electrodes for a repetition frequency of 10 kHz, at which robust production was demonstrated. Ablation rates for other materials relative to gold can be derived from Table 2.1 of Chapter 2. If we allow for scaling up by increasing the number of sparks to 100 for industrial applications, we land at 50 g/h $\approx$ 1 g/min as a feasible production rate of future spark discharge generators. This limits industrial applications of spark ablation to areas that require less than 1 g per item, as a very rough estimate. Where the relative standard deviation of 1.35 is too broad, size selection reduces the maximum production rate by a factor of 10–100. Table 1.3 gives some examples of application areas grouped into three categories with different required amounts of nanoparticulate material. The first group refers to devices for which the size of humans and the objects and devices they use are of no relevance. For example, a chemical sensor could basically have any dimension above molecular, while the magnets operating the central lock of a car cannot be much smaller than human fingers. The category independent of human dimensions (1), which includes electronics and sensor production, is a subfield of nanotechnology experiencing enormous growth and in which the relevant dimensions are in the size range of nanoparticles produced by spark ablation. It is interesting to note that this technology is highly sustainable, because the exploitation of natural resources may be negligible here if one considers that a sensor may only require 1 fg (10^{-15} g) of the possibly rare active inorganic material or even less. The price of the raw material is then irrelevant, and the value added by nanostructuring and nanoparticle production is enormous.

The group of applications where human dimensions dictate the amount of material required is split into (2) surface-related and (3) bulk-related applications. Energy conversion (e.g., solar panels or electrode materials for photoelectrochemical production of fuels) is usually based on processes where thin surface layers are required. Energy storage requires minimum volumes of materials because it applies the bulk of materials to store energy (e.g., in the form of lithium or hydrogen). Considering the maximum mass production rates, spark ablation is mainly suitable for serving categories (1) and (2) in Table 1.3. These are both areas with a need for particles <20 nm.

This restriction of application areas hardly applies to research purposes, where the versatility of spark ablation is of special value, where much less material is required, and where longer collection/deposition periods can be afforded. For example, research on hydrogen storage has profited from new magnesium-based nanoparticulate phases discovered through spark ablation that have a rather high hydrogen storage capacity and that allow release of hydrogen at much lower temperatures than from pure magnesium hydride [16].

The range of material quantities required for various applications spans 18 orders of magnitude. The number of potential industrial applications of spark ablation is enormous, even without any scaling up of present spark discharge generators.

Table 1.3 Application categories and the amount of material required per item

	Nanoparticle application areas (examples)	Material quantity needed per item
1. Applications independent of human dimensions	Sensors Electronics (nanoparticulate printed circuit elements)	1 fg–1 mg
2. Surface-related applications	Solar panels Electrodes Membranes Smart windows Catalysts (automobiles) Displays	1 mg–1g
3. Bulk-related applications	Structural materials (nanocomposites) Batteries Magnetic materials Hydrogen storage materials	1 g–1 kg

1.4.1 Heterogeneous Nanoporous Layers

The application area of heterogeneous nanoporous layers is very broad and comprises the fields of catalysis, absorbers, battery electrodes, supercapacitors, and (photo)electrochemical reactors, for example, for water splitting or CO_2 reduction (artificial leaves) and more. Some of these are treated in more detail later, and chapters of this book are dedicated to the important fields of catalysts (Chapter 10) and sensors (Chapter 11). Chapter 12 deals with photoelectrochemical reactors but is focused on plasmonic particles, which could very well be embedded into porous matrices, where both the matrix and the plasmonic particles are produced by spark ablation.

Porous layers comprise a straightforward set of applications in connection with spark ablation, because deposition of aerosols onto surfaces automatically tends to form porous layers if enough material is deposited. This is due to the statistical nature of particle deposition as well as the fact that usually any collision of an aerosol particle with a surface or a deposited particle leads to sticking at the point where it hits. We have to do with a natural phenomenon of self-organization, which can be exploited for the fields mentioned before and more. Chapter 2 describes it in more detail, including approaches to alter the geometry and stability of porous layers formed this way. In co-deposition of two spark ablation–produced particle species, mixed porous structures can be produced, one component of which can be thought of as the substrate (e.g., TiO_2) and the other one as the active catalyst (e.g., Au). This example is shown in Section 2.4.2.3, Chapter 2. The great variability of this principle is easily conceived, and it is easily comprehensible that by this method of producing catalysts consisting of arbitrary mixtures of more than two materials are made possible. This constitutes a new generation of catalysts with respect to the conventional structure, where an active catalyst sits on a substrate. With the new approach, co-catalysts (requiring phase boundaries of two or more materials) can easily be made. The feasibility of this method of building catalysts or other products is presently a subject of research projects [27]. In connection with spark ablation, the additional aspect of straightforward production of alloys (Chapters 2 and 5) or mixed oxide particles is of special interest and basically opens a universe of new possibilities.

Of course, the particles can be modified in the aerosol phase before deposition, for example, by surface coating, for which a number of options exist in aerosols [28, 29]. A technique that further expands the range of possibilities here is described in the article by Pfeiffer et al. [30] and that was named "spark coating." The particles to be coated are passed by the vicinity of the spark, where they are exposed to a high concentration of vapor and atomic cluster–sized material, which rapidly coagulates with the particles, forming a surface layer or islands of the sparked material on virtually any kind of "substrate particles"; see the example in Fig. 2.9, Chapter 2. It is clear that various applications requiring porous layers can benefit from the simple way in which aerosol deposition enables fabrication of nanoscale heterogeneous structures, as well as the ease of atomic mixing (alloying) made possible by spark ablation.

1.4.2 Additive Manufacturing

Three-dimensional (3D) nano- or microstructures created by high-resolution deposition of nanoparticles from spark ablation opens up new possibilities for nanodevices such as sensors and electronics, as well as for energy conversion systems. Charged patterns on the substrate attract nanoparticles of the opposite charge, which selectively deposit on the desired spots. An early example using an electron beam–charged substrate is explained in Section 2.4.2.1, Chapter 2 [31], and Chapter 9 reviews a more advanced nanoscale patterning technology called electric-field-assisted aerosol lithography (EAAL), which uses nanosized electrostatic lenses to increase resolution. Spark ablation is the most suitable nanoparticle source for this application because of its versatility and because a large portion of the particles is charged.

The first attempts of using a cluster beam to print 3D nanostructures were published in 2000 by the Milani group [32]. A nanoparticle or atomic cluster source such as a spark discharge generator produces nanoparticles that are released into vacuum and focused to a narrow beam (see Chapter 8). In supersonic cluster beam deposition (SCBD), the beam hits a substrate, where the nanoparticles fuse together to form lines of arbitrary thickness or other structures. SCBD has been shown to be suitable for 3D printing purposes by the same group [32]. The combination with a state-of-the-art technique of polymer additive manufacturing (rapid prototyping) called fused filament fabrication can be combined with SCBD [33]. The authors show that a 3D structure containing electronic elements like embedded electrical conducting lines and resistors with tailored conductivity can be produced this way.

Aerosol methods allowing focused deposition of particles are an alternative to vacuum beam–based methods. They do not require the high vacuum necessary for cluster beams but are also applicable for 3D printing. Until now, they have mainly been restricted to particles/droplets larger than the size range of interest here. These methods make use of particle inertia, when an aerosol stream exiting a nozzle impinges upon a substrate, so that the particles maintain their motion toward the substrate and impact with it [33]. This works well with particles larger than 1 μm under near-atmospheric pressure, and the method called "aerosol jet printing" uses this effect. It applies droplets in which solid species like nanoparticles may be suspended [34]. In connection with spark ablation, direct impaction of 5 nm particles is the more elegant approach, though. It requires reduced pressures downstream of the nozzle in the range of 1 mbar. The work of Fernandez de la Mora and his group on low-pressure impaction of very small particles [35] basically paved the way for nanoparticle impaction printing, but it was aimed at size analysis in a variable pressure setup. Aerodynamic focusing was developed around the same time for the same purpose [36]. Demonstrations of impacting very small particles (down to 2 nm) using subsonic and supersonic flows had also been made [37, 38], and feasibility of additive manufacturing using this impaction and focusing technology has mainly been demonstrated with particles far larger than 20 nm (e.g., [39]). However, impaction of very small particles is also of interest and has theoretically been covered [40]. Figure 1.2 shows the result of impacting 5 nm gold particles from spark ablation (VSP-G1 combined with impaction and XYZ stage modules) onto a contact lens by VSPARTICLE. Here the liquid-like behavior of 5 nm metal particles and the impaction energy lead to fusing of the particles on the substrate to form a solid gold layer. This is indicated by the gold color created by the surface plasma absorption of bulk gold. Agglomeration or powder formation would broaden the absorption range to spread over the visible spectrum so that it would appear black. Indeed, diffusional deposition, where diffusion dominates the deposition process, the impact energy is

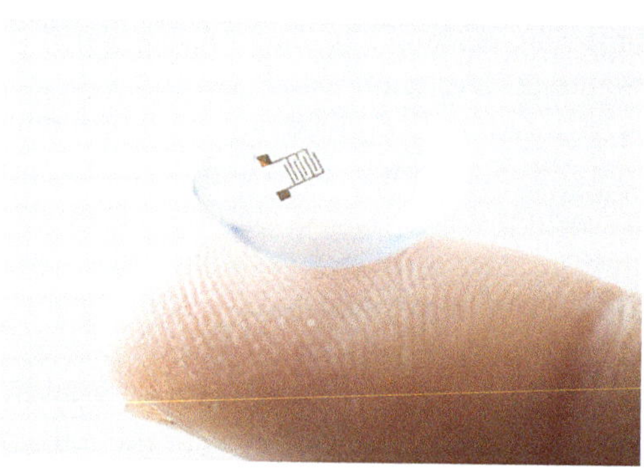

Figure 1.2 Soft contact lens with gold conducting lines printed on it by focused impaction of 5 nm particles. Picture from VSPARTICLE.

low and black deposits are obtained with metal nanoparticles. The remarkable thing is that the impaction-based 3D printing/coating method works at low temperature so that the polymer substrate is not damaged, just like in vacuum cluster beam deposition (see Chapter 8).

1.4.3 Applications of Plasmonic Particles

Collective electron oscillations in metal nanoparticles, called plasmons, interact with electromagnetic radiation, and resonance occurs (localized surface plasmon resonance, LSPR) if the size- and composition-dependent plasma frequency of the particle coincides with the frequency $\upsilon = c/\lambda$ (c: speed of light) of the electromagnetic radiation with wavelengths λ in the ultraviolet, visible, or infrared part of the spectrum. Upon illumination near the resonance, such particles concentrate the electric field of incoming radiation; they generate hot electrons or heat up the particle, representing a nanoscale heat source. Chapter 12 treats these phenomena in more detail and explains the advantage of spark ablation as it enables flexibility in defining the composition. As further shown in that chapter, hot electrons can be extracted for solar cell applications or to drive chemical reactions (e.g., CO_2 reduction) in an electrochemical cell. Chapter 13 describes, among others, medical applications (diagnostics and cancer treatment) as well as the opportunities these phenomena represent for analysis of (bio)molecules, with sensitivities ranging up to single-molecule detection. Environmental monitoring and forensics are further application domains. Approaches of producing 3D structures that can be used for applications making use of the plasmonic effect are treated in Chapters 2 and 9.

1.4.4 Heterogeneous Catalysis

Heterogeneous catalysts require high-surface-area structures in order to reduce the transport limitation. Catalyst properties referring to the temperature range, turnover rate, selectivity, surface stability, and resistance to poisoning depend on the chemical surface structure, which is determined by the particle composition as well as its size. In addition, contact with a substrate or contact with particles of a different composition (co-catalysts) may be beneficial for catalyst optimization for specific purposes. The perspective of easy production of virtually any material combination by spark ablation with arbitrary length scales explained in Section 1.4.1 and further in Section 2.4.2.3, Chapter 2, is believed to be a great asset for the future of heterogeneous catalysis. Chapter 5 focuses further on the subject of mixing, and Chapter 10 is dedicated to catalysis.

Of course, it will not be trivial to find stable combinations such as high-entropy alloys, optimizing performance and earth abundance of the materials used. However, the enormous space of possibilities as well as the rather easy way in which these nanostructures can be varied in composition from nanostructure to the atomic scale will allow screening experiments involving a great number of variations to find the right catalyst. The size range of atomic clusters should be of special interest here (see Section 1.4.4.2 and Chapter 7). The problem of immobilization of atomic clusters while they retain their integrity must be solved for this purpose. Metal organic frameworks have been used to protect single-metal atoms or clusters [41] by encapsulation, while allowing reactant molecules to pass, and this is a possible way forward. The approach of co-depositing atomic clusters with support particles and spacer particles described in Section 2.4.2.3, Chapter 2, to embed them in porous structures is also promising and would be connected to a very simple process.

The catalytic properties of nanoparticles depend on the electronic structure on their surface, which may induce reactions of adsorbed molecules. These electronic configurations also govern the adsorption strength of these educt molecules as well as their products, and both should be in an adequate range. Surface curvature lowers the coordination number of the catalyst surface atoms, which can be imagined to increase the number of dangling bonds at the surface and to alter the electronic orbital configuration. This picture illustrates that particle size plays a crucial role.

In connection with the ease of building porous structures with arbitrary components, including metals, semiconductors, and insulators, it is worth mentioning that plasma resonance in metal particles was discovered by Christopher et al. in 2012 as an effect that can promote photocatalysis [42]. Using spark ablation, the particle size and composition of the plasmonic particles could be chosen to optimize their response with respect to the incoming light.

1.4.4.1 Aerosol catalysis as a screening method

Aerosol catalysis, introduced by the group of Weber (1999), refers to experiments where the particles are suspended in a gas without any contact with a substrate [43]. Such studies determine the pure catalytic activity of substrate-free nanoparticles. Using this method, the influence of strong electric fields on the catalyst surface for a gas reaction was demonstrated for the first time. This was simply done by varying the particle charge state by Peineke et al. [44]. Spark ablation was used for these experiments, and they demonstrate the potential of aerosol catalysis as a screening method for catalyst activity or for catalyst discovery. Such experiments could be extended to mixed particles, where the particles are agglomerates containing the "substrate material" or co-catalysts. See Chapter 10 for details on aerosol catalysis.

1.4.4.2 Photocatalytic degradation of contaminants

Photocatalysts such as TiO_2 and ZnO are known to decompose organic contaminants, in particular in water or in the air. Volatile organic carbon (VOC) indoor airborne contaminants are brought in connection with the sick building syndrome, and it would mean great progress if an inexpensive method to oxidize them to H_2O and CO_2 would be found. TiO_2 and ZnO are nontoxic, low-priced, and chemically stable photocatalysts. However, the large bandgap requires UV lamps for this purpose. A shift of the light required to the visible and a higher efficiency under given radiation is achieved by doping the catalyst [45]. Very small (i.e., high-surface-area) doped particles could easily be produced by spark ablation of mixed electrodes [2, 10] or by combining a Ti electrode with an electrode of the dopant (see Chapter 5) with subsequent controlled oxidation in the aerosol phase (see Chapter 2). To our knowledge, spark ablation has never been applied for this purpose. Recent preliminary results of our own research indicate another possible path of producing TiO_2-based photocatalysts operating in the visible range: They have revealed that controlled oxidation of 5 nm Ti particles from spark ablation in the aerosol phase at low enough temperature produces black TiO_2. The black color indicates oxygen deficiency, from which a narrow bandgap results, which should allow photoactivation in the visible. This process of producing black TiO_2 is much simpler than other approaches that have been published, for example, by Wang et al. [46], but stability issues have to be solved.

It has also been demonstrated that porous nanocomposites show advantageous performance as photocatalysts [47]. Here, the high flexibility of spark ablation in mixing at any scale (Chapter 5) and the option of forming porous layers (Section 1.4.1 and Section 2.4.2, Chapter 2) has great potential in finding photocatalysts superior to the existing ones, where decomposition of VOCs is just one example.

1.4.4.3 Carbon nanotube production

A state-of-the-art method of carbon nanotube production uses catalyst nanoparticles deposited on a surface as nuclei, from where the tubes grow, in a chemical vapor deposition reactor. The carbon source can be an organic vapor or CO. Alternatively, high-quality individual, that is, nonbundled single-walled carbon nanotubes (SWCNTs), have been produced at ambient pressure in a continuous-flow aerosol reactor in a substrate-free process by using 3–4 nm iron nanoparticles produced by spark ablation as the catalyst [48]. Deposition of these SWCNTs onto a surface produced transparent conducting films with exceptionally low sheet resistances. Process parameters, catalyst particle composition, and size have an influence on chirality and structure of the SWCNTs, where the flexibility of spark ablation should be useful in obtaining the desired properties for various applications. An earlier study has shown that the tube diameter is closely related to the catalyst particle diameter under constant conditions [49].

1.4.4.4 Atomic cluster catalysis as a future perspective

The domain of atomic clusters is defined here as the "nonscalable" range (diameter smaller than about 2 nm), where properties change with size in a discontinuous and nonmonotonic way (see Chapter 7). Research on particles in this range has mainly been restricted to vacuum-based experiments with cluster beams (see Chapter 8), and transition of the interesting properties discovered into applications has hardly taken place. Note that the term "cluster" is not consistently used in literature and that some publications (also Chapter 8) on cluster beam research deal with particles up to 50 nm. Only few examples of materials composed of atomic clusters in the nonscalable sense have been reported. C60 is an example of a very stable atomic cluster that forms a crystalline material, in which the clusters have the function of atoms. However, practical applications of such cluster-assembled materials have hardly emerged, although theory predicts very interesting catalytic properties for specific clusters, for example, in splitting H_2O [50]. As all the cluster properties change if only one atom is removed, added, or replaced, a simple calculation reveals that myriads of combinations are possible, representing myriads of possible catalysts or other useful materials. One reason for the stagnation of this field from the view of applications is probably due to the difficulty of immobilizing the clusters, while retaining their properties, as already addressed before in Sections 1.4.4 and 1.4.1 with the approach of integrating clusters into the type of porous structures described as a possible solution. Furthermore, it has been demonstrated that single atoms can be immobilized on specific sites on a surface (see, for example, Ref. [51]), in which C atoms form a 2D lattice on a substrate, well separated from each other. Reacting such immobilized single atoms with clusters could lead to a new concept of producing catalysts with very specific properties. Apart from trial and error, using spark ablation and size classification, the discovery process can be given a direction by calculations, for example, using density functional theory (e.g., [50]). Another reason for the slow pace with which atomic clusters are reaching the domain of applications is the difficulty of producing large enough amounts of cluster material, as the vacuum-based experiments only deal with very small quantities. Here the approach of combining spark ablation with mobility analysis described in Chapter 7 may lead the way to exploitation of inorganic atomic clusters in the future.

1.4.5 Electronics

Printing of conductive paths (see Fig. 1.2) as well as resistive elements (Chapter 8 and Ref. [33]) can make use of the basic size effect of melting point reduction when particles around 5 nm in diameter are used. The same effect makes welding at low temperature possible using nanoparticulate powder. This could be useful for metal–metal contacts in electronics. Figure 1.3 shows a proof of principle using

5 nm gold particles from spark ablation. The 1-mm-diameter spot of 5 nm Au particles first appears black, indicating nanogold. After annealing to 200°C, the characteristic yellowish color of gold occurs, indicating the formation of a continuous gold layer. Welding of gold to gold with nanogold is thus possible at 200°C, while the melting point of gold is above 1000°C. Moving from gold to copper is desirable for cost reasons. Here solutions have to be found to protect the particles from oxidation.

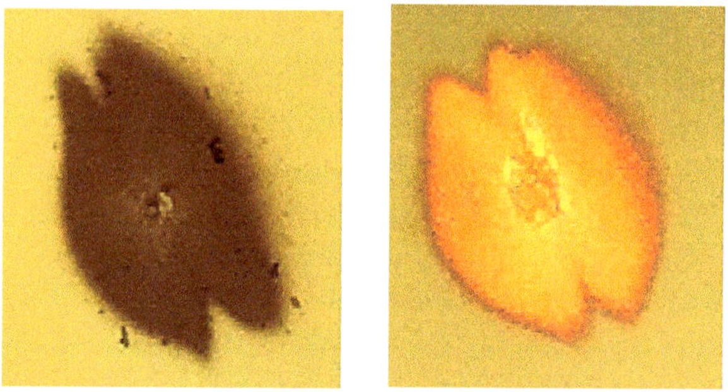

Figure 1.3 Micrograph of (softly) impacted nanogold layer on gold-coated glass before and after sintering at ca. 200°C (spot diameter ca. 1 mm). Picture contributed by Joost van Ginkel, Delft University of Technology.

Such applications of nanoparticles from spark ablation profit from advances in additive manufacturing technology involving nanoparticles. The technique of impaction (see Section 2.4.2.2, Chapter 2) enables inertial focusing and does not rely on electrostatic charge. For electrostatic focusing, the large portion of charged particles produced by spark ablation is a great advantage with respect to other production methods (Chapter 9). This is true especially for the size range of atomic clusters (see Chapters 7 and 8) but also made use of for focusing larger particles (see Chapter 9). Additive manufacturing by aerosol focusing, applied to nanoparticles from spark ablation, has a huge potential for integrating multiple sensors into electronic circuits for production of e-noses. These can consist of hundreds of or even much more small electroresistive sensors (cf. Section 1.4.6 and Chapter 11), each of which possesses a different composition or nanomixture.

1.4.6 Chemical Sensors

Chemical sensors are usually based on high-surface-area materials with which specific molecules in the surrounding gas or liquid reversibly interact, changing their properties. Easily accessible properties like conductivity, impedance, or color can be recorded as the sensor signal. Different particle sizes and compositions show sensitivities to different molecules, where small variations of the composition can already lead to significant changes. Besides noble metals and carbon materials, oxides are often applied as the sensor material [52]. For detection of molecules with specific functional groups, including biomolecules, molecules linking to these can be attached to the sensing surface, and numerous variations of this sensing principle have been introduced [53]. Due to its versatility, spark ablation is a unique tool for discovery of inorganic sensor materials. Nanoparticles from spark ablation can easily be brought to surfaces by diffusion, impaction, or, if charged, electrostatic precipitation to form a spot of porous particulate deposit, as described in Section 1.4.1 and in Chapters 2 and 11. A small particle size is usually preferred because surface changes lead to large effects on properties. Due to the small amount of material a sensor requires, which could very well lie in the picogram range or even far below that, spark ablation in connection with additive manufacturing has the potential of fast industrial production.

Sensor arrays, with each sensor element being sensitive to a different spectrum of molecules, can be used as e-noses if connected to multiplexing electronics, capable of recognizing specific molecules or mixtures of molecules by means of pattern recognition [54]. This publication uses laser ablation to produce SnO_2 nanoparticles and "doping" by adding thin layers of different oxides between ZnO layers for different sensor elements of the array. The dopants were also produced by laser ablation so that each sensor element required many successive runs. A process offering much higher variability for such combinatorial methods would consist in mixing aerosols containing different particle materials and sizes in arbitrary proportions before deposition, as illustrated for the simple "nanomixing" example in Fig. 2.7d, Chapter 2. Such a process would be feasible by applying multiple spark ablation units. See Chapter 8 for an example of a sensor array produced by a combined spark ablation and particle beam technique.

Gold particles on graphene can lead to injection of n or p carriers in the graphene, depending on the size and density of the particles [55]. This effect of converting n-conduction to p-conduction is of special interest for chemical sensors because the conductive properties of the graphene layer are also expected to be very sensitive to adsorption of electropositive or electronegative molecules on the gold particles, for similar reasons. There is ongoing research at TU Delft, in which equally sized Au particles from spark ablation are being used to optimize the sensor performance for NO_2.

For more sensor applications based on spark ablation, see Chapter 11 dedicated to this subject.

1.4.7 Photoelectrochemical Conversion

Photoelectrochemical conversion uses light-absorbing semiconductors to convert solar radiation into a chemical fuel (e.g., hydrogen). It is one of the most promising technologies in connection with sustainable energy production. The research goal is to obtain high conversion efficiencies with abundant materials. Porous semiconductor materials represent an advantage with respect to flat surfaces, and much is still to be done to optimize the electrode structure, where the spark ablation–based principles described in Section 1.4.1 will certainly play a significant role. In addition, LSPR in metallic nanoparticles deposited onto the semiconductor surface has been shown to enhance light absorption and to increase conversion efficiency (see Chapter 12 and Refs. [56, 57]). Apart from varying the size and composition of the deposited particles, it has been demonstrated that the degree of agglomeration plays an important role, as it creates more plasmon resonances [58].

1.4.8 Photovoltaic Efficiency Enhancement

The best results in photovoltaic energy conversion (solar cells) have actually been obtained with plasmonic particles larger than 20 nm [59], which forms an exception with respect to the other application examples compiled here. To obtain particles up to 100 nm in diameter, the small particles leaving the spark ablation process were given time to agglomerate by diffusional coagulation, and subsequently the aerosol was passed through a tube furnace, where they coalesced into sphere-like particles, from which the desired size was extracted by mobility classification. These uniformly sized charged particles were then deposited onto the surface by applying an electric field. It had earlier been demonstrated that such a process may lead to rather dense particle deposits, while agglomeration on the substrate is suppressed by Coulomb repulsion of the charged depositing spheres [31, 49]. Avoidance of particle contact leads to rather narrow plasmon absorption peaks, whereas agglomeration leads to multiple resonances, so the deposit absorbs a larger range of wavelengths (see Section 1.4.7). Depending on the specific application, one or the other is preferred. The spark method was preferred with respect to other methods due to the higher purity of the particles.

1.4.9 Supercapacitors

The porous aerogel-type structures [60] that can be formed by spark ablation in combination with diffusional or impaction dominated deposition are described in Section 1.4.1 and treated further in Chapter 2. If these nanoporous deposits are conducting as well as the substrate, the electrical contact with the substrate enables their application as supercapacitors [61]. As described on the example of surface-contacted carbon nanotubes [62], such structures can be soaked with an electrolyte, the composition of which determines the surface charge double layer on the nanostructure surface, which represents a large capacitance. Redox reactions may form an addition to the charge stored [63].

1.4.10 Biomedical Applications

Biomedical applications of particles from spark ablation are extensively treated in Chapter 13. We add important points in the following sections referring to antimicrobial applications on the one hand, where toxic properties of specific nanoparticles are exploited for the benefit of humans, and nanoparticle toxicity threatening human health on the other hand.

1.4.10.1 Antimicrobial applications

Spark ablation has proven to be a handy method of producing nanoparticles for testing antimicrobial activity. Chapter 13 treats this topic in detail.

The experiment reported by Kourmouli et al. [64] refers to nanosilver as an effective antimicrobial agent and supports the model according to which the silver dissolves (probably as Ag^+). As studies suggest that the antimicrobial effect rises with decreasing size [65], a preliminary study was carried out by Bhamidipati et al. [66, 67] with Ag particles that are much smaller than those used in almost all the studies reported until now. Agglomerated 5 nm Ag particles produced by spark ablation were collected on a quartz fiber filter. The surprising result was that agglomerated 5 nm Ag particles dissolve in phosphate-buffered saline (PBS), as Fig. 1.4 demonstrates. The small discs punched out of the large disk had been submerged in the solution and the silver on the large filter had not been. Substantial dissolution can be concluded from the fact that the characteristic black color of the silver nanoparticles became much fainter. The antibacterial activity (on *Escherichia coli* strain DH5α) as a function of time is compared to a blank sample in Fig. 1.5. Before deposition on the filter, the 5 nm silver particles were allowed to form agglomerates in the aerosol. In an additional run, the aerosol was passed through a tube oven prior to deposition, which coalesced them to larger particles, reducing their surface area. The blue curve clearly shows an effect of this surface reduction, which explains a lower rate of dissolution and therewith antibacterial activity. The 1.08 μg of filter-deposited Ag submerged per mL solution was comparable to $AgNO_3$ of the same molar Ag concentration. Further studies are necessary to confirm this result, but it is clearly demonstrated that very small nanoparticles produced by spark ablation have a great potential as antimicrobial agents. Silver has shown to be more effective for specific bacteria [68].

For broad-band antibiotic applications it will therefore be interesting to mix materials, for which spark ablation is certainly a good method of choice. Mixing may also lead to synergetic effects, so rather expensive materials may become replaceable by mixtures of abundant materials.

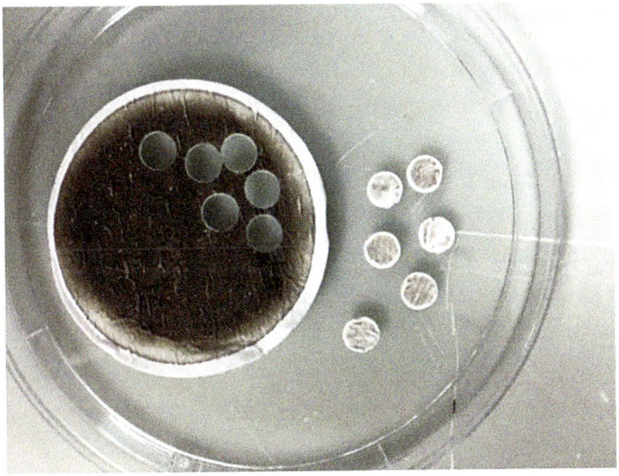

Figure 1.4 Ag particles (5 nm) deposited on a filter (left), with punched-out samples (right), after being submerged in phosphate-buffered saline.

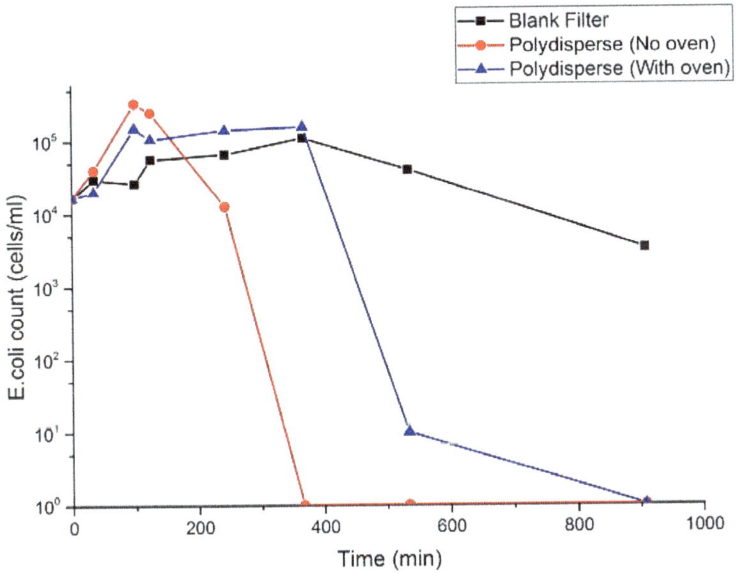

Figure 1.5 Time–kill curve of *E. coli* against silver nanoparticles (mean primary particle diameter 5 nm). "Polydisperse (No oven)" refers to agglomerated, high-surface-area particles. "Polydisperse (With oven)" refers to coalesced agglomerates.

1.4.10.1.1 *Textile nanofinishing*

An attractive application of the antimicrobial effect of silver nanoparticles is finishing of textiles, where the particles kill odor-causing bacteria. Silver nanoparticles ca. 3 nm in diameter from spark ablation have been deposited onto the textile fibers in a dry aerosol process, in which the aerosol is passed through the textile [69]. Due to their smallness, the particles are deposited very efficiently on the textile fibers by diffusion. The adhesion forces of particles of that size are usually stronger than any forces that tend to remove them from the fibers, so after washing, the antimicrobial activity is hardly diminished. The small particle size also leads to an antimicrobial activity comparable to conventional finishing processes involving silver, while only 1/10 of the metal was used (10 ppm; that is, 10 mg Ag per kg of textile). Even for this reduction in the necessary amount of active material, silver may be replaceable by other inorganic materials that are more abundant and where there are no environmental concerns. Of course the chance is high that spark ablation can deliver other effective materials equally well. At the same time, it is an ideal method to discover them. Here as in other cases, a mixture will probably be advantageous.

1.4.10.2 Toxicity research

Reliable and reproducible testing of the toxicity of pure nanoparticles smaller than 20 nm has been challenging. The spark discharge generator enables us to investigate systematically their size-dependent toxicity properties. There are roughly three categories of nanoparticles to which individuals can be exposed: particles derived from combustion sources, engineered nanoparticles, and nanoparticles that occur in nature. Risks of exposure during the life of these nanoparticles, from production to use and disposal, and consequently adverse effects on health can occur. The main exposure route is by inhalation. The role of both chemical composition and particle size in exposure effects has been demonstrated [70], for example, in the translocation from the lung to the liver [71] and the brain [72]. Cells, tissues, and test animals can be exposed to sub-20-nm airborne particles using nanoparticles generated by spark ablation, for example, in combination with a liquid–air interface [73]. An important advantage of using spark ablation, besides the control over the size range, is the purity of the test particles. Even minor impurities in particles can change their properties and make the comparison of studies challenging, if not impossible. Mimicking real-life environments without the presence of surfactants and precursors enables a benchmark for risk assessment (see Chapter 14).

Spark ablation would enable studying the biodistribution of sub-20-nm particles. A spark discharge generator can process radioactive electrodes to produce particles that are traceable in extremely small quantities for in vivo experiments to study biodistribution, which would not be possible with other methods. As the particles rapidly grow downstream of the spark, high concentrations of sub-5-nm particles have to be produced in very close proximity of the exposed subject, which is easily done with a spark discharge generator.

1.4.10.3 Diagnostics and drug delivery

There are numerous applications of inorganic particles that can be produced by spark ablation in the biomedical field, especially at the stage of research, and Chapter 13 treats this topic in detail. Medical diagnosis and targeted drug delivery gain great advantages through superparamagnetic particles, as described in Section 1.4.15. In tumor therapy, the magnetic material is attached to the drug; it can magnetically be drawn to the tumor and heated by an external radiofrequency magnetic field, which induces drug release and/or destruction of tumor tissue by the raised temperature (hyperthermia). For diagnosis, superparamagnetic material can be applied as a contrast agent in magnetic resonance imaging (MRI) tomography (see Chapter 13). For these applications it is very important that the material be purely superparamagnetic, because any remnant magnetism would lead to clustering of the particles and clogging of the blood vessels.

1.4.11 Lubricants

Nanoparticles may have a lubricant effect, that is, they reduce friction between solids that are in contact. An example of such an application of particles from spark ablation is shown in Section 8.3, Chapter 8. For particles in the size range below 20 nm the characteristic roughness length scale of the shearing surfaces is much larger than the radius of the nanoparticles. An important effect leading to lubrication can then be seen in filling of the valleys between asperities of the surfaces [74]. The shearing surfaces are then smoother; thus the tribological properties are improved.

1.4.12 Tracers for High-Speed Flows

Diagnostics of flow fields is done by measurement techniques, including particle image velocimetry (PIV), Doppler global velocimetry (DGV), and laser Doppler anemometry (LDA). The flow is visualized by adding tracer particles to it that give detectable laser scattering signals, while inertial effects must be small so that they perfectly follow the flow. For high-speed flows, the criterion for the latter condition is a small enough relaxation time given by

$$\tau = Bm. \tag{1.1}$$

Here m is the particle mass and B is its mechanical mobility. B depends on the geometry of the particle, and for small primary particles composing the agglomerate, it is inversely proportional to the agglomerate's surface area. It has been shown by Ghaemi et al. [75] that fractal-like agglomerates with a fractal dimension smaller than 2, as they automatically form when noncoalescing particles coagulate, represent a geometry that has the smallest possible τ for a given agglomerate mass and primary particle size. Reducing the primary particle size further reduces τ. The recipe of producing particles that optimally follow the flow is thus: Produce the smallest possible primary particles and let them agglomerate in an aerosol. This was done by producing small Al particles by spark ablation in high concentration, causing rapid agglomeration. The agglomerates were oxidized to Al_2O_3, and the oxidation process created stiff necks between the primary particles. Giving the agglomerates a large enough size, they were optically detectable and exhibited the smallest relaxation time ever obtained for tracer particles. This enables visualizing high-speed turbulent flows as they occur, for example, around a space capsule under simulated conditions of diving into the earth's atmosphere.

1.4.13 Model Aerosol Production for Research, Instrument Testing, and Calibration

Carbonaceous aerosols from combustion, for example, from cars, belong to the main polluting species in ambient air and the atmosphere as a whole. Great efforts are therefore undertaken to understand the effects that these particles have on health, atmospheric chemistry, and climate to develop measuring instruments and efficient air filters. Spark ablation has widely been used for production of soot-like model aerosols in this important research field and it still is today. Chapter 14 deals with this topic. Pure carbon electrodes are then applied in the spark discharge generator. Spark ablation serves as a simple and reproducible source, and as shown in Chapter 2, the mass production rate is much higher for carbon than for metals. In Germany, the method has been standardized in the technical rule VDI 3491 [76]. The technique has the advantage that the particles consist of an analytically pure material if care is taken that the high plasma temperature involved does not heat any material in the vicinity of the spark (see Chapter 2). An important application could be calibration of measurement devices for combustion engine measurements, where presently there is no standard that prescribes calibration material. Other fields requiring model aerosols are studies on inhalation in view of health effects and simulation of the atmospheric aerosol (see Chapter 14).

As a convenient source of small nanoparticles, spark ablation has broadly been applied in testing of aerosol filters of all kinds. See Ref. [77] for the case of testing the filtration and scavenging efficiency for particles between 6 and 16 nm in diameter. Filter testing is the main application area of the commercial spark generators offered by Palas GmbH, Karlsruhe. The design of these instruments is treated in Chapter 14.

A source that delivers particles smaller than 20 nm is required in the development of modern measuring instruments, as the importance of particles down to 1 nm in size for atmospheric processes and climate has recently been recognized [78]. Counters for particles that small have been developed by Airmodus Oy, Helsinki, Finland [79], and for testing them spark ablation is applied [80]. The impact of particles in the atmosphere on atmospheric chemistry has been known for a long time and has attracted more interest in connection with the ozone hole, when the Crutzen–Arnold model showed that an aerosol process played a key role here (polar stratospheric clouds). An early study on direct catalytic ozone destruction using spark ablation [81] revealed for the first time that carbon particles lead to a high rate of ozone destruction. Later, an experiment in a smog chamber enabling more realistic conditions confirmed ozone destruction by a carbon aerosol, but the rates measured were much lower [82]. Unfortunately, in both cases a commercial spark discharge generator type was applied for which Roth et al. [83] showed later that it delivers significant concentrations of impurities due to heating of material in the vicinity of the spark (see Chapter 2). It was thus the wrong source for studying a catalytic surface phenomenon extremely sensitive to any contamination, and these results have to be regarded with skepticism.

As mentioned, measuring instruments for particles smaller than 20 nm and down to the atomic cluster range are of great importance for monitoring and understanding the role of pollutants in the atmospheric environment, which calls for a reliable calibration source. The rapidly growing need for building blocks in nanotechnology, part of which can be generated in the gas phase, manifests the same requirement. Calibration procedures for sizing instruments [84] make use of standard particles in liquid suspension where polystyrene latex spheres are the most popular materials besides silica. The liquid is spray-dried to bring these particles into the aerosol state. This method is not feasible for particles smaller than 10 nm, though, because also the very purest liquids available contain dissolved impurities that adhere to the aerosolized particles. Using electrospray eliminates this problem, and Fernandez de la Mora et al. [85] demonstrated how to produce ionic clusters of the general form $(A_n^{+} + B_{n\pm1}^{-})^{\pm1}$ of precisely defined composition in the size range between a single ion to 4 nm to be used as a calibration means. As the series of ions appears simultaneously, the resolving power of the size analyzer defines the size of the largest ion that can be distinguished from the rest. A similar series of peaks appears in the mobility spectrum of metal atomic clusters produced by spark ablation (see, for example, Fig. 7.3 in Chapter 7). Under clean conditions and with inert metals, these should also be well defined. The same source, a spark discharge generator could produce the calibrating species and subsequently the desired particles or clusters.

A calibrated source consisting of a spark discharge generator and a mobility analyzer could then be used as a size standard for the range between atomic single atoms and 20 nm particles. The flexible design of the VSPARTICLE G1 allows for configurations that deliver unagglomerated particles below 5 nm, and if larger sphere-like particles are required, coalescence can be guaranteed beyond that size by heating the aerosol stream (see Chapter 2).

1.4.14 Basic Research on Hydrogen Storage Materials

The mass production rates feasible with spark ablation will probably remain too low for industrial production of hydrogen storage materials (see Table 1.3), but the method has successfully been applied to the search for materials that have large hydrogen storage capacities but form bonds with the hydrogen that are weak enough to release the gas at low temperatures. Such materials are of special interest for mobile applications. Vons et al. [86] produced 10 nm Mg particles and showed that thermal hydrogen desorption already starts at temperatures of 350 K and extends to 700 K, while bulk Mg releases hydrogen at much higher temperatures. Mixing with small amounts of Pd further promotes hydrogen release. A further study using spark mixing revealed that Mg–Ti composite nanomaterials are promising for achieving favorable hydrogen storage properties, because the presence of Ti destabilizes MgH_2 [87]. This led to preparation of thin-film multilayer stacks by vapor deposition, showing similar properties [88], and such a technique may enable larger mass production rates of destabilized MgH_2 in the future.

1.4.15 Magnetic Materials

Particles of ferro- or ferrimagnetic materials smaller than 20 nm exhibit a phenomenon at room temperature called superparamagnetism. Each particle comprises a single magnetic domain, wherein the elementary magnetic dipoles are parallel, as in a ferromagnet. However, they have lost their fixed orientations with respect to the crystal lattice, and the material is unmagnetic in the absence of an external field. The condition for superparamagnetism is

$$kT \geq KV, \tag{1.2}$$

where k is the Boltzmann constant, T is the temperature, K is the magnetic anisotropy constant of the material, and V is the particle volume. Equation 1.2 states that for any value of K and particle size (defined by V), there is a temperature, called blocking temperature, above which the particle becomes superparamagnetic. Superparamagnetic materials have no hysteresis in the magnetization as a function of the external magnetic field as ferromagnetic materials do. Their coercivity and remanence is zero, which means they have no permanent magnetization. However, they have a high magnetic susceptibility, which means that their magnetization rises steeply with an external magnetic field. To make use of superparamagnetism, a composite material basically has to be made consisting of small enough magnetic particles. These particles should only have weak magnetic interaction with each other so that the size effect is retained. This can be achieved by coating them with an unmagnetic material such as a polymer [29].

Maghemite (γ-Fe_2O_3) aggregates consisting of 7 nm primary particles were produced by spark ablating pure Fe, letting the aerosol coagulate and subsequently oxidizing the agglomerates by adding O_2 to the aerosol [88]. As already shown by Ghaemi et al. [74], oxidation of agglomerates causes neck formation between the primary particles, making the agglomerates stiff, in which case they should more correctly be referred to as aggregates. These aggregates were collected as a powder, and the magnetization curve was measured. The superparamagnetic state was clearly identified, and coating of the particles was not necessary to retain this size effect. It can be concluded that the stiffness of the aggregate structure guarantees a certain mean distance between superparamagnetic units, avoiding their magnetic interaction to a sufficient degree. The powder had a saturation magnetization of 22 emu/g and a blocking temperature of 65 K.

Another application of superparamagnetic particles is cell separation [89]. The particles are functionalized with cell-specific proteins, which makes them attach selectively to specific cells in solution. Subsequently, they are precipitated with the aid of an inhomogeneous magnetic field. Similarly, toxic components can be removed from waste streams if the magnetic particulate material is covered by suitable adsorbents [90].

For cooling purposes, the magnetocaloric effect can be used. If a strongly magnetized material is demagnetized in an adiabatic process, the conservation of entropy causes a reduction of temperature. This works very well with superparamagnetic material, which loses its magnetic ordering the moment the external field is removed. Magnetocaloric cooling is a realistic future perspective if a low-cost mixed material can be found that has its blocking temperature around room temperature. Spark ablation is strong in mixing on a nanoscale, and an important step in this search has been taken by demonstrating production of a ternary alloy with interesting magnetic properties by spark ablation [91].

1.4.16 Medical Therapy and Diagnostics

Inorganic nanoparticles are widely used in biomedical research, and spark ablation can play an important role here. These particles are required for therapy and diagnostics, including biosensing and drugs, as well as biomolecule delivery. The research group of Byeon has frequently applied spark ablation for their research in these areas, and in Chapter 13 they report about these studies as well as areas where spark ablation can potentially be applied and put them into a broad perspective.

1.5 Conclusion

Although the most pronounced size effects, drastically changing material properties and also creating completely new ones, occur in the size range below 20 nm, nanotechnology is only just beginning to exploit the new opportunities arising here. Many activities are in the research phase, and here spark ablation is increasingly being chosen as a versatile tool of producing inorganic particles, the composition and size of which can easily be varied. We anticipate that it will play a much larger role in the years to come. As the technique is scalable, there is also a broad range of possible applications in industrial processes, especially in connection with emerging developments in materials and devices, for example, in the broad domains of catalysts and sensors. Feasible production rates can certainly satisfy the application category, where devices can be made much smaller than human dimensions (electronics and sensors), as well as most applications that can be described as "surface related" (Table 1.3). Other methods such as laser ablation, flame synthesis, low-pressure plasma synthesis, and chemical vapor synthesis may, from their principle, also cover most of this size range. In particular, flame synthesis has the potential of a much higher mass production rate than spark ablation, which makes it applicable also for the third category in Table 1.3 of "bulk related" materials on an industrial production scale. However, the methods mentioned all have the drawbacks regarding a less predictable output for each new material. In addition, spark ablation has the advantage of not requiring precursors and purity. Unlimited variation of the material composition is probably the most important feature. Singlet particle production ("unagglomerated" particles) is no problem in spark ablation, because vapor production and coalescence are decoupled and can be controlled independently. Where narrow size distributions are necessary, electrostatic size classification is easily applied, because the spark plasma delivers a high fraction of charged particles. This feature is also exploited for additive manufacturing, as referred to in Chapter 13.

References

1. Roduner, E. (2006). Size matters: why nanomaterials are different, *Chem. Soc. Rev.*, **35**, pp. 583–592.
2. Tabrizi, N. S., Xu, Q., Van Der Pers, N. M. and Schmidt-Ott, A. (2010). Generation of mixed metallic nanoparticles from immiscible metals by spark discharge, *J. Nanopart. Res.*, **12**, pp. 247–259.
3. Castleman, A. W. and Khanna, S. N. (2009). Clusters, superatoms, and building blocks of new materials, *J. Phys. Chem. C*, **113**, pp. 2664–2675.
4. Sattler, K. (ed.) (1996). *Cluster Assembled Materials*, Vol. 232, Trans Tech Publications, Switzerland.
5. Hemlasech, G. A. and de Watteville, M. C. (1907). Sur une nouvelle méthode de production des spectres de flamme des corps mètalliques, *C.R. Hebd. Seances Acad. Sci.*, **144**, p. 1338.
6. Scheeline, A. and Coleman, D. M. (1987). Direct solids elemental analysis: pulsed plasma sources, *Analyt. Chem.*, **59**, pp. 1185A–1196A.
7. Llewellyn-Jones, F. (1960). Electrode erosion by spark discharges, *Br. J. Appl. Phys.*, **1**, pp. 60–65.

8. Burtscher, H. and Schmidt-Ott, A. (1982). Enormous enhancement of van der Waals forces between small silver particles, *Phys. Rev. Lett.*, **48**, pp. 1734–1737.

9. Schwyn, S., Garwin, E. and Schmidt-Ott, A. (1988). Aerosol generation by spark discharge, *J. Aerosol Sci.*, **19**, pp. 639–642.

10. Tabrizi, N. S., Xu, Q., van der Pers, N. M., Lafont, U. and Schmidt-Ott, A. (2008). Synthesis of mixed metallic nanoparticles by spark discharge, *J. Nanopart. Res.*, **11**, p. 1209.

11. Lafont, U., Simonin, L., Tabrizi, N. S., Schmidt-Ott, A. and Kelder, E. M. (2009). Synthesis of nanoparticles of Cu, Sb, Sn, Sb and Cu_2Sb by densification and atomization process, *J. Nanosci. Nanotechnol.*, **9**, pp. 2546–2552.

12. https://www.reportlinker.com/p03472838/Global-Nanotechnology-Market-Outlook.html] Global Nanotechnology Market Outlook 2024 (2018).

13. Feng, J., Biskos, G. and Schmidt-Ott, A. (2015). Toward industrial scale synthesis of ultrapure singlet nanoparticles with controllable sizes in a continuous gas-phase process, *Sci. Rep.*, **5**, 15788.

14. Ullmann, M., Friedlander, S. K. and Schmidt-Ott, A. (2002). Nanoparticles formation by laser ablation, *J. Nanopart. Res.*, **4**, pp. 499–509.

15. Stein, M. and Kruis, F. E. (2018). Scaling-up metal nanoparticle production by transferred arc discharge, *Adv. Powder Technol.*, **29**, pp. 3138–3144.

16. Vons, V. A., Anastasopol, A., Legerstee, W. J., Mulder, F. M., Eijt, S. W. H. and Schmidt-Ott, A. (2011). Low-temperature hydrogen desorption and the structural properties of spark discharge generated Mg nanoparticles, *Acta Mater.*, **59**, pp. 3070–3080.

17. Stein, M. and Kruis, F. E. (2016). Optimization of a transferred arc reactor for metal nanoparticle synthesis, *J. Nanopart. Res.*, **18**, p. 258.

18. Roth, P. (2007). Particle synthesis in flames, *Proc. Combust. Inst.*, **31**, pp. 1773–1788.

19. Strobel, R., Alfons, A. and Pratsinis, S. E. (2006). Aerosol flame synthesis of catalysts, *Adv. Powder Technol.*, **17**, pp. 457–480.

20. Teoh, W. Y., Amal, R. and Mädler, L. (2010). Flame spray pyrolysis: an enabling technology for nanoparticles design and fabrication, *Nanoscale*, **2**, pp. 1324–1347.

21. Wang, H. C. and Kasper, G. (1991). Filtration efficiency of nanometer-size aerosol particles, *J. Aerosol Sci.*, **22**, pp. 31–41.

22. Van Gulijk, C., Bal, E. and Schmidt-Ott, A. (2009). Experimental evidence of reduced sticking of nanoparticles on a metal grid, *J. Aerosol Sci.*, **40**, pp. 362–369.

23. Wang, X., Kruis, F. E. and McMurry, P. H. (2005). Aerodynamic focusing of nanoparticles: I. guidelines for designing aerodynamic lenses for nanoparticles, *Aerosol Sci. Technol.*, **39**, p. 611.

24. Knutson, E. O. and Whitby, K. T. (1975). Aerosol classification by electric mobility: apparatus, theory, and applications, *J. Aerosol Sci.*, **6**, pp. 443–451.

25. Meuller, B. O., Messing, M. E., Engberg, D. J. L, Jansson, M. A., Johansson, L. I. M., Norlén, S. M., Tureson, N. and Deppert, K. (2012). Review of spark discharge generators for production of nanoparticle aerosols, *J. Aerosol Sci. Technol.*, **46**, pp. 1256–1270.

26. Roth, C., Ferron, G. A., Karg, E., Lentner, B., Schumann, G., Takenaka, S. and Heyder, J. (2004). Generation of ultrafine particles by spark discharging, *J. Aerosol Sci. Technol.*, **38**, pp. 228–235.

27. Braams, P. (2015). A clean and flexible catalyst synthesis method (Unpublished master's thesis), Delft University of Technology, Delft, Netherlands.

28. Harra, J., Juuti, P., Haapanen, J., Sorvali, M., Roumeli, E., Honkanen, M., Vippola, M., Yli-Ojanperä, J. and Mäkelä, J. M. (2015). Coating of silica and titania aerosol nanoparticles by silver vapor condensation, *Aerosol Sci. Technol.*, **49**, pp. 767–776.

29. Vollath, D., Szabo, D. V. J. Fuchs, J. (1999). Synthesis and properties of ceramic – polymer composites, *NanoStruct. Mater.*, **12**, pp. 433–438.

30. Pfeiffer, T. V., Kedia, P., Messing, M. E., Valvo, M. and Schmidt-Ott, A. (2015). Precursor-less coating of nanoparticles in the gas phase, *Materials*, **8**, pp. 1027–1042.

31. Zonnevylle, A. C., Hagen, C. W., Kruit, P. and Schmidt-Ott, A. (2009). Directed assembly of nano-particles with the help of charge patterns created with scanning electron microscope, *Microelectron. Eng.*, **86**, pp. 803–805.

32. Barborini, E., Piseri, P., Podesta, A. and Milani, P. (2000). Cluster beam microfabrication of patterns of three-dimensional nanostructured objects, *Appl. Phys. Lett.*, **77**, pp. 1059–1061.

33. Bellacicca, A., Santaniello, T. and Milani, P. (2018). Embedding electronics in 3D printed structures by combining fused filament fabrication and supersonic cluster beam deposition, *Addit. Manuf.*, **24**, pp. 60–66.

34. Binder, S., Glatthaar, M. and Rädlein, E. (2014). Analytical investigation of aerosol jet printing, *J. Aerosol Sci. Technol.*, **48**, pp. 924–929.

35. de Juan, L. and Fernández de la Mora, J. (1998). Sizing nanoparticles with a focusing impactor: effect of the collector size, *J. Aerosol Sci.*, **29**, pp. 589–599.

36. Fernandez de la Mora, J. (1996). Drastic improvement of the resolution of aerosol size spectrometers via aerodynamic focusing: the case of variable pressure impactors, *J. Chem. Eng. Commun.*, **151**, pp. 101–124.

37. Fernandez de la Mora, J. and Schmidt-Ott, A. (1993). Performance of a hypersonic impactor with silver particles in the 2 nm range, *J. Aerosol Sci.*, **24**, pp. 409–415.

38. Fernández de la Mora, J., De Juan, L., Liedtke, K. and Schmidt-Ott, A. (2003). Mass and size determination of nanometer particles by means of mobility analysis and focused impaction, *J. Aerosol Sci.*, **34**, pp. 79–98.

39. Park, J. J., Lee, J. G., Kim, D. Y., Lee, J. H., Yun, J. H., Gwak, J., Eo, Y. J., Cho, A., Swihart, M. T., Al-Deyab, S. S., Ahn, S., Kim, D. H. and Yoon, S. S. (2017), Rapid supersonic spraying of $Cu(In,Ga)(S,Se)_2$ nanoparticles to fabricate a solar cell with 5.49% conversion efficiency, *Acta Mater.*, **123**, pp. 44–54.

40. Li, C., Singh, N., Andrews, A., Olson, B. A., Schwartzentruber, T. E. and Hogan, C. J. (2019). Mass, momentum, and energy transfer in supersonic aerosol deposition processes, *Int. J. Heat Mass Transfer*, **129**, pp. 1161–1171.

41. Liu, Y. and Tang, Z. (2013). Multifunctional nanoparticle @ MOF core–shell nanostructures, *J. Adv. Mater.*, **25**, pp. 5819–5825.

42. Christopher, P., Xin, H., Marimuthu, A. and Linic, S. (2012). *Singular characteristics and unique chemical bond activation mechanisms of photocatalytic reactions on plasmonic nanostructures*, Nat. Mater., **11**, pp. 1044–1050.

43. Weber, A. P., Seipenbusch, M., Thanner, C. and Kasper, G. (1999). Aerosol catalysis on nickel nanoparticles, *J. Nanopart. Res.*, **1**, pp. 253–265.

44. Peineke, C., Davoodi, P., Seipenbusch, M., Weber, A. and Schmidt-Ott, A. (2011). Charge dependent catalytic activity of gasborne nanoparticles, *J. Nanosci. Nanotechnol.*, **11**, pp. 8628–8633.

45. Shayegan, Z., Lee, C. and Haghighat, F. (2018). TiO2 photocatalyst for removal of volatile organic compounds in gas phase: a review, *Chem. Eng. J.*, **334**, pp. 2408–2439.

46. Wang, X., Fu, R., Yin, Q., et al. (2018). Black TiO_2 synthesized via magnesiothermic reduction for enhanced photocatalytic activity, *J. Nanopart. Res.*, **20**, p. 245.

47. Min, X., Li, H., Jian-Qiang, W., Ying, W., Ling, G., Jian-hua, Z. and Zhi-Gang, Z. (2008). The direct synthesis of mesoporous structured MnO2/TiO2 nanocomposite: a novel visible-light active photocatalyst with large pore size, *J. Nanotechnol.*, **19**, p. 8.

48. Mustonen, K., Laiho, P., Kaskela, A., Zhu, Z., Reynaud, O., Houbenov, N., Tian, Y., Susi, T., Jiang, H., Nasibulin, A. G. and Kauppinen, E. I. (2015). Gas phase synthesis of non-bundled, small diameter single-walled carbon nanotubes with near-armchair chiralities, *J. Appl. Phys. Lett.*, **107**, art. no. 013106.

49. Zonnevylle, A. C., Hagen, C. W., Kruit, P., Valenti, M. and Schmidt-Ott, A. (2009). Positioning Pd catalyst particles for carbon nanotube growth using charge patterns created with a scanning electron microscope, *J. Vac. Sci. Technol., B*, **27**, pp. 3048–3050.

50. Roach, P. J., Woodward, W. H. Castleman Jr., A. W., Reber, A. C. and Khanna, S. N. (2009). Complementary active sites cause size-selective reactivity of aluminum cluster anions with water, *Science*, **323**, pp. 492–495.

51. Streletskii, O. A., Khvostov, V. V., Novikov, N. D., Guseva, M. B. and Aleksandrov, A. F. (2012). Secondary emission properties of the 2D ordered linear chain carbon films, *J. Commun. Technol. Electron.*, **57**, pp. 424–428.

52. George, J. M., Antony, A. and Mathew, B. (2018). Metal oxide nanoparticles in electrochemical sensing and biosensing: a review, *Microchim. Acta*, **185**, pp. 319–326.

53. Krasteva, N., Besnard, I., Guse, B., Bauer, R. E., Klaus Mullen, K., Yasuda, A. and Vossmeyer, T. (2002). Self-assembled gold nanoparticle/dendrimer composite films for vapor sensing applications, *Nano Lett.*, **2**, pp. 551–555.

54. Aronova, M. A., Chang, K. S. and Takeuchi, I. (2003). Combinatorial libraries of semiconductor gas sensors as inorganic electronic noses, *J. Appl. Phys. Lett.*, **83**, p. 1255.

55. Wu, Y., Jiang, W., Ren, Y., Cai, W., Lee, W. H., Li, H., Piner, R. D., Pope, C. W., Hao, Y., Ji, H., Kang, J. and Ruoff, R. S. (2012). Tuning the doping type and level of graphene with different gold configurations, *J. Small*, **8**, pp. 3129–3136.

56. Valenti, M., Jonsson, M. P., Biskos, G., Schmidt-Ott, A. and Smith, W. A. (2016). Plasmonic nanoparticle-semiconductor composites for efficient solar water splitting, *J. Mater. Chem. A*, **4**, pp. 17891–17912.

57. Valenti, M., Dolat, D., Biskos, G., Schmidt-Ott, A. and Smith, W. A. (2014). Enhancement of the photoelectrochemical performance of CuWO4 thin films for solar water splitting by plasmonic nanoparticle functionalization, *J. Phys. Chem. C*, **119**, pp. 2096–2104.

58. Valenti, M., Kontoleta, E., Digdaya, I. A., Jonsson, M. P., Biskos, G., Schmidt-Ott, A. and Smith, W. A. (2016). The role of size and dimerization of decorating plasmonic silver nanoparticles on the photoelectrochemical solar water splitting performance of BiVO4 photoanodes, *ChemNanoMat*, **2**, pp. 739–747.

59. Pfeiffer, T. V., Ortiz-Gonzalez, J., Santbergen, R., Tan, H., Schmidt-Ott, A., Zeman, M. and Smets, A. H. M. (2014). Plasmonic nanoparticle films for solar cell applications fabricated by size-selective aerosol deposition, *Energy Procedia*, **60**, pp. 3–12.

60. Lindquist, G. J., Pui, D. Y. H. and Hogan Jr., C. J. (2014). Porous particulate film deposition in the transition regime, *J. Aerosol Sci.*, **74**, pp. 42–51.

61. González, A., Goikolea, E., Barrena, J. A. and Mysyk, R. (2016). Review on supercapacitors: technologies and materials, *Renew. Sustain. Energy Rev.*, **58**, pp. 1189–1206.

62. Kanninen, P., Luong, N. D., Sinh le, H., Anoshkin, I. V., Tsapenko, A., Seppälä, J., Nasibulin, A. G. and Kallio, T. (2016). Transparent and flexible high-performance supercapacitors based on single-walled carbon nanotube films, *Nanotechnology*, **27**, pp. 1–6.

63. Jain, D., Kanungo, J. and Tripathi, S. K. (2018). Enhanced performance of ultracapacitors using redox additive-based electrolytes, *J. Appl. Phys. A*, **124**, p. 14.

64. Kourmouli, A., Valenti, M., van Rijn, E., Beaumont, H. J. E., Kalantzi, O. and Schmidt-Ott, A. (2018). Can disc diffusion susceptibility tests assess the antimicrobial activity of engineered nanoparticles?, *J. Nanopart. Res.*, **20**, p. 62.

65. Morones, J. R., Elechiguerra, J. L., Camacho, A., Holt, K., Kouri, J. B., Ramírez, J. T. and Yacaman, M. J. (2005). The bactericidal effect of silver nanoparticles, *Nanotechnology*, **16**, pp. 2346–2353.

66. Bhamidipati, M., Adam, S., Valenti, M., Kooyman, P. and Schmidt-Ott, A. (2017). Water disinfection by antimicrobial fibrous material, European Aerosol Conference, Zurich.

67. Bhamidipati, A. S. M. (2017). Industrial integration of aerosol route to nanofinishing of highly effective antibacterial textiles, Master thesis, TU Delft (unpublished).

68. Deshmukh, S. P., Patil, S. M., Mullani, S. B. and Delekar, S. D. (2019). Silver nanoparticles as an effective disinfectant: a review, *J. Mater. Sci. Eng. C*, **97**, pp. 954–965.

69. Feng, J., Hontañón, E., Blanes, M., Meyer, J., Guo, X., Santos, L., Paltrinieri, L., Ramlawi, N., Smet, L. C. P. M. D., Nirschl, H., Kruis, F. E., Schmidt-Ott, A. and Biskos, G. (2016). Scalable and environmentally benign process for smart textile nanofinishing, *ACS Appi. Mater. Interface*, **8**, pp. 14756–14765.

70. Maynard, A. D. and Baron, P. A. (2004). Aerosols in the industrial environment, in Ruzer, L. S. and Harley, N. H. (eds.) *Aerosols Handbook: Measurement, Dosimetry and Health Effects*, CRC Press, Boca Raton, pp. 225–264.

71. Kreyling, W. G., Mayer, P., Takenaka, S., Schulz, H., Oberdörster, G. and Ziesenis, A. (2002). Translocation of ultrafine insoluble iridium particles from lung epithelium to extrapulmonary organs is size dependent but very low, *J. Toxicol. Environ. Health Part A*, **65**, pp. 1513–1530.

72. Oberdörster, G., Sharp, Z., Atudorei, V., Elder, A., Gelein, R., Kreyling, W. and Cox, C. (2004). Translocation of inhaled ultrafine particles to the brain, *Inhalation Toxicol.*, **16**, pp. 437–445.

73. Brossell, D., Troller, S., Dziurowitz, N., Plitzko, S., Linsel, G., Asbach, C., Azong-Wara, N., Fissan, H. and Schmidt-Ott, A. (2014). A thermal precipitator for the deposition of airborne nanoparticles onto living cells: rationale and development, *J. Aerosol Sci.*, **63**, pp. 75–86.

74. Akbulut, M. (2012). Nanoparticle-based lubrication systems, *J. Powder Metall. Min.*, **1**, p. e101.

75. Ghaemi, S., Schmidt-Ott, A. and Scarano, F. (2010). Nanostructured tracers for laser-based diagnostics in high-speed flows, *Meas. Sci. Technol.*, **21**, art. no. 105403.

76. VDI 3491 Blatt 4:2018-03 Technical rule: Measurement of particles - Methods for generating test aerosols - Condensation methods, Publication date: 2018-03.

77. Förster, H., Thajudeen, T., Funk, C. and Peukert, W. (2016). Separation of nanoparticles: filtration and scavenging from waste incineration plants, *Waste Manage.*, **52**, pp. 346–352.

78. Tröstl, J., et al. (2016). The role of low-volatility organic compounds in initial particle growth in the atmosphere, *Nature*, **533**, p. 527.

79. Lehtipalo, K., Leppa, J., Kontkanen, J., Kangasluoma, J., Franchin, A., Wimnner, D., Schobesberger, S., Junninen, H., Petaja, T., Sipila, M., Mikkila, J., Vanhanen, J., Worsnop, D. R. and Kulmala, M. (2014). Methods for determining particle size distribution and growth rates between 1 and 3 nm using the Particle Size Magnifier, *Boreal Envrion. Res.*, **19**, pp. 215–236.

80. Kangasluoma, J., Attoui, M., Junninen, H., Lehtipalo, K., Samodurov, A., Korhonen, F., Sarnela, N., Schmidt-Ott, A., Worsnop, D., Kulmala, M. and Petäjä, T. (2015). Sizing of neutral sub 3nm tungsten oxide clusters using Airmodus Particle Size Magnifier, *J. Aerosol Sci.*, **87**, pp. 53–62.

81. Fendel, W., Matter, D., Burtscher, H. and Schmidt-Ott, A. (1995). Interaction between carbon or iron aerosol particles and ozone, *Atmos. Environ.*, **29**, pp. 967–973.

82. Kamm, S., Möhler, O., Naumann, K. H., Saathoff, H. and Schurath, U. (1999). The heterogeneous reaction of ozone with soot aerosol, *Atmos. Environ.*, **33**, pp. 4651–4661.

83. Roth, C., Ferron, G. A., Karg, E., Lentner, B., Schumann, G., Takenaka, S. and Heyder, J. (2004). Generation of ultrafine particles by spark discharging, *J. Aerosol Sci. Technol.*, **38**, pp. 228–235.

84. Wiedensohler, A., Wiesner, A., Weinhold, K., Birmili, W., Hermann, M., Merkel, M., Müller, T., Pfeifer, S., Schmidt, A., Tuch, T., Velarde, F., Quincey, P., Seeger, S. and Nowak, A. (2018). Mobility particle size spectrometers: calibration procedures and measurement uncertainties, *Aerosol Sci. Technol.*, **52**, pp. 146–164.

85. Fernandez de la Mora, J. and Barrios-Collado, C. (2017). A bipolar electrospray source of singly charged salt clusters of precisely controlled composition, *Aerosol Sci. Technol.*, **51**, pp. 778–786.

86. Vons, V. A., Anastasopol, A., Legerstee, W. J., Mulder, F. M., Eijt, S. W. H. and Schmidt-Ott, A. (2011). Low-temperature hydrogen desorption and the structural properties of spark discharge generated Mg nanoparticles, *Acta Mater.*, **59**, pp. 3070–3080.

87. Anastasopol, A., Pfeiffer, T. V., Middelkoop, J., Lafont, U., Canales-Perez, R. J., Schmidt-Ott, A., Mulder, F. M. and Eijt, S. W. H. (2013). Reduced enthalpy of metal hydride formation for Mg-Ti nanocomposites produced by spark discharge generation, *J. Am. Chem. Soc.*, **135**, pp. 7891–7900.

88. Asano, K., Westerwaal, R. J., Anastasopol, A., Mooij, L. P. A., Boelsma, C., Ngene, P., Schreuders, H., Eijt, S. W. H. and Dam, B. (2015). Destabilization of Mg hydride by self-organized nanoclusters in the immiscible Mg–Ti system, *J. Phys. Chem. C*, **119**, pp. 12157–12164.

89. van Benten, J. (2018). Synthesis of superparamagnetic Fe_xO_y nanoparticles by spark ablation (Unpublished master's thesis) Delft University of Technology, Delft, Netherlands.

90. Kyeong, S., Jeong, C., Kang, H., Cho, H.-J., Park, S.-J., Yang, J.-K., Kim, S., Kim, H.-M., Jun, B.-H. and Lee, Y.-S. (2015). Double-layer magnetic nanoparticle-embedded silica particles for efficient bio-separation, *PLoS One*, **10**, art. no. e0143727.

91. Jiang, C., Sun, Y., Yu, X., Zhang, L., Sun, X., Gao, Y., Zhang, H. and Song, D. (2012). Removal of sudan dyes from water with C18-functional ultrafine magnetic silica nanoparticles, *Talanta*, **89**, pp. 38–46.

92. Feng, J., Geutjens, R., Thang, N. V., Li, J., Guo, X., Kéri, A., Basak, S., Galbács, G., Biskos, G., Nirschl, H., Zandbergen, H. W., Brück, E. and Schmidt-Ott, A. (2018), Magnetic phase transition in spark-produced ternary LaFeSi nanoalloys, *ACS Appl. Mater. Interfaces*, **10**, pp. 6073–6078.

93. Burtscher, H. and Schmidt-Ott, A. (1986). In situ measurement of adsorption and condensation of a polyaromatic hydrocarbon on ultrafine C particles by means of photoemission, *J. Aerosol Sci.*, **17**(3), p. 456.

Chapter 2

Nanoparticle Production by Spark Ablation: Principle, Configurations, and Basic Steps toward Application

Maurits F. J. Boeije,[a] George Biskos,[b,c] Bibianne E. van der Maesen,[a] Tobias V. Pfeiffer,[a] Aaike W. van Vugt,[a] Bernardus Zijlstra,[a] and A. Schmidt-Ott[b,d]

[a]*VSPARTICLE B.V., Molengraaffsingel 10, NL-2629 JD Delft, The Netherlands*
[b]*Environment and Water Research Center, Cyprus Institute, Nicosia, Cyprus*
[c]*Faculty of Civil Engineering and Geosciences, Delft University of Technology, NL-2628 CN Delft, The Netherlands*
[d]*Faculty of Applied Sciences, Delft University of Technology, NL-2629 HZ Delft, The Netherlands*
m.boeije@vsparticle.com

This chapter reviews basic configurations of the spark discharge generator (SDG). It briefly describes SDG devices that are offered on the market. Furthermore, it describes the options aerosol technology offers to modify the produced particles while they are in gas suspension. The basic features of spark ablation are elucidated, focusing on sizes below 20 nm, and it is shown how particle size

Spark Ablation: Building Blocks for Nanotechnology
Edited by Andreas Schmidt-Ott

ISBN 978-981-4800-82-2 (Hardcover), 978-0-367-81709-1 (eBook)
www.jennystanford.com

distribution can be predicted as a function of the SDG operating parameters. Then, different methods of immobilization of the particles are explained for producing nanoscaled materials and devices. Finally, a list of milestones in producing nanoparticles by spark ablation technology is given with the respective references.

2.1 Introduction

Spark ablation (SA) refers to electrical discharges that are induced in the gap between two electrodes. In an inert gas atmosphere, the spark plasma ablates material from the electrodes (Chapters 3 and 4), from which nanoparticles form by condensation. This principle was introduced in detail by Schwyn, Garwin, and Schmidt-Ott in 1988 [1], after having been developed for basic studies on very small particles by Burtscher and Schmidt-Ott in 1982 [2].

The present chapter, as well as the entire book, mainly refers to ablation using sparks that have a duration between 1 and 10 μs. The spark setup includes a housing and means for introducing a carrier gas that flushes the region around the electrode gap and carries the growing particles to an outlet. Such a setup is usually referred to as a spark discharge generator (SDG; also used for "spark discharge generation"), although the terms "spark ablation generator" and "spark generator" are also in use. Downstream of the SDG outlet, the particles can be modified in many ways and immobilized on a substrate or in a liquid (see Section 2.4.1), depending on the application. See Chapter 1 as well as Chapters 9–14 for specific applications.

Restriction to short sparks serves two important purposes, namely the production of particles smaller than ca. 20 nm in diameter and the ability of producing internally mixed particles, the composition of which is well defined, as it is the composition of the electrodes. Obviously, the ablation process would approach the state of the continuous arc if the spark duration was increased. This means that at some spark duration above 10 μs, distillation effects would occur just like they do in continuous arcs, meaning that the more volatile material in mixed electrodes evaporates with a higher rate. The fact that such effects do not occur with short enough sparks makes SDG a unique tool for producing mixed particles with controlled composition. Chapter 5 is dedicated to the production of internally mixed particles. The second reason for restricting the discharge time is the ease with which particles smaller than 20 nm in diameter and down to atomic cluster sizes can then be produced. SDGs have been shown to cover this important size range (see Chapter 1), where most other methods have difficulties.

Repetitive sparks have shown to be a reliable controllable source of vapor, which can be easily and reproducibly quenched by providing an appropriate room temperature gas flow, and spark durations of a few microseconds are easily reached in the RCL circuit (resistance [R], capacitance [C], and inductance [L]); in Fig. 2.1, L is usually the inductance of wiring. R is predominantly the resistance of the spark gap, unless a series resistance is added, for example, to influence the mixing ratio in spark mixing (see Chapter 5). If the inductance is increased, the spark duration is also increased, as explained in Chapter 5. SA circuits that apply inductors, as described by Horvath et al. [3] are not considered here, because of their increased L, which leads to much longer spark durations.

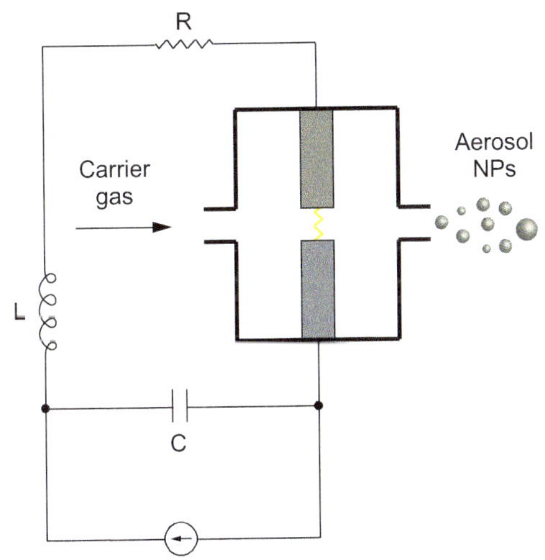

Figure 2.1 Simple-circuit spark discharge generator (SDG) containing a current source, a capacitance (C), an inductance (L), a resistance (R), and the electrodes between which sparks are induced.

In contrast to laser ablation, which produces similar particles as SA, SA is scalable, as many sparks can be operated in parallel. This is usually not relevant for research activities, but it is for applications, where production of more than 1 g/h is required (see Chapter 1). The basic principles of scaling up are treated in the study by Stein and Kruis [4]. They show the feasibility for the case of ablation by continuous arcs, but their main conclusions can be taken over for SA.

2.2 Basic Configurations

2.2.1 Different Geometries

Each spark produces a small vapor cloud, typically a millimeter in diameter (cf. Chapter 3), which is very rapidly mixed into an inert gas flow. The cooling associated with the mixing leads to extremely high supersaturation of the vapor and spontaneous condensation of the vapor to produce atomic clusters, which further coagulate to form larger particles. Figure 2.2 shows two frequently used geometries. A number of other electrode geometries, such as hole-to-plate, are reported in Chapter 9 as well as in the review by Meuller et al. [5]. Quite recently, a concentric cylindrical electrode arrangement for lab use was introduced by Ahmad et al. [6], which enables a rather constant, low-concentration output over a long time, but results in a slowly changing geometry, which cannot be compensated for by electrode adjustment. As the volume between the electrodes should be flushed continuously to avoid (re)deposition of vapor or the forming particles on the electrodes, the configurations shown in Fig. 2.2 are usually favored. Another important feature of these configurations is that the electrode position can be continuously controlled, compensating for ablation and keeping the geometry constant. Commercial SDGs are required to provide a strictly constant output, which has practically restricted them to the options in Fig. 2.2.

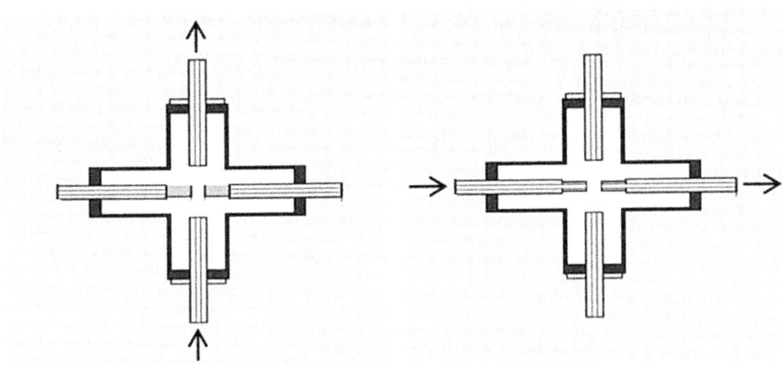

Figure 2.2 Two basic configurations of the spark discharge generator (SDG; proportions of the VSPPARTICLE model G1). Left: cross-flow; right: flow through the electrodes. Gas flow direction is indicated by arrows.

2.2.2 Simple Circuit

In the most simple case, the SDG electronics (Fig. 2.1) consists of a high-voltage source, usually delivering a constant current and charging the capacitor until the field between the electrodes induces gas breakdown and spark formation (Chapters 3 and 4). The energy to be released in the spark is stored in this capacitor parallel to the spark gap. The process restarts after discharge of the capacitor, and a series of sparks occurs with a repetition frequency f given by the charging current I, capacitance C, and mean voltage across the spark, $\bar{U}$ (see Eqs. 2.1 and 2.2 in Section 2.2.5.2).

The standard circuit becomes less useful at higher frequencies, when a high spark-to-spark variation starts to occur as subsequent sparks are influenced by each other. Better flushing of the spark gaps combined with a large series resistance reduces the spark-to-spark variation [7], but the large series resistance introduces a significant energy loss and waste heat.

2.2.3 Switching Circuit

New developments of fast, high-power electronic switches have opened possibilities to obtain better control over the discharge circuit and thus expand the possibilities of SDGs. In the example shown in Fig. 2.3 [8], the SDG electronics produces a low-current continuous discharge (glow current) in the spark gap, too low in energy to induce any ablation, and a pulsing network, including the parallel capacitor, to superimpose a series of short, high-current spark discharges onto the glow current, which are responsible for the ablation. Advantages of this concept are that the spark energy E_s (Eq. 2.2) can be controlled over a wide range by varying the voltage up to which the capacitor is charged. This need not be the break-down voltage, because the glow discharge enables pulsed discharges within a wide range of initial voltages. Direct control over the mass production rate is gained this way, which is roughly proportional to the power transferred to the spark, $P = fE_s$ (see Section 2.3.1). In addition, a more regular operation is possible and the repetition frequency can be increased up to $f = 20$ kHz. Such high frequencies, associated with high production rates, are not possible if the spark itself is the switch, unless a high series resistor is added [7], because the discharge tends to go into a continuous mode above a certain frequency [8]. More details are provided in Section 2.3.4.

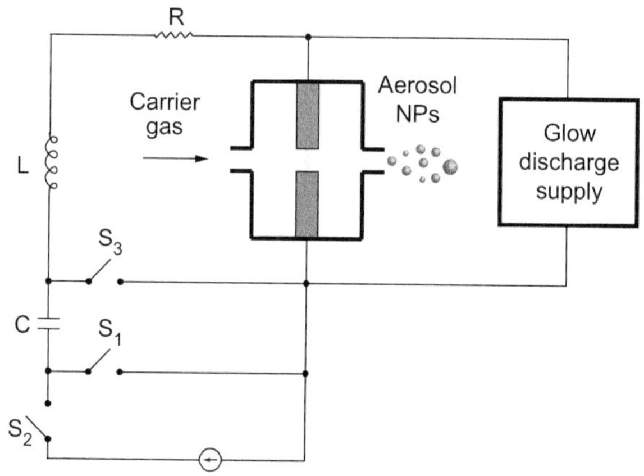

Figure 2.3 Switching circuit (principle), containing switches and a glow discharge supply, in addition to the elements of the simple circuit shown in Fig. 2.1. When switches S_2 and S_3 are closed and S_1 is open, the capacitance C is charged. For a spark discharge to occur, S_1 is closed, with S_2 and S_3 open. A continuous glow discharge enables spark ignition in a wide range of voltages.

2.2.4 Particle Modification and Selection Downstream of the Spark

Aerosol technology offers a large number of possibilities for influencing particle growth and the final size, shape, and composition of the particles. Round singlet particles can be produced by keeping the temperature in the growth zone high (Fig. 2.4) [9], and large-surface-area fractal-like agglomerates form when the temperature is low enough to avoid coalescence and enough time is provided for agglomeration (Fig. 2.5). Alternatively, nonspherical agglomerates can be sintered into round (singlet) structures after they have formed [10, 11]. Furthermore, letting the particles react with a gas to form oxides or other compounds [12] is easily done by the concept outlined in Fig. 2.6. If the particles are as small as 5 nm, the reaction often proceeds rapidly at room temperature.

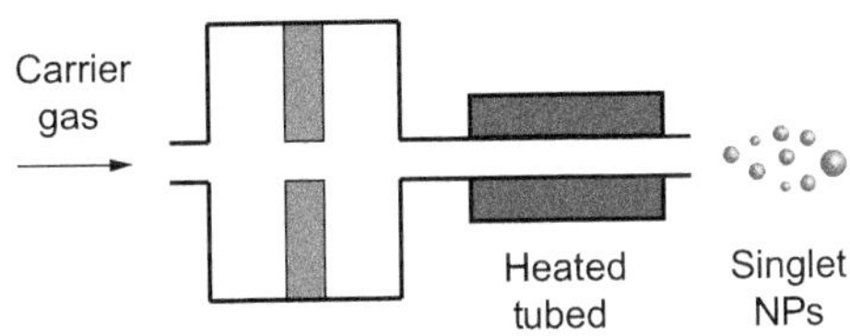

Figure 2.4 Production of singlet particles by heating the growth zone.

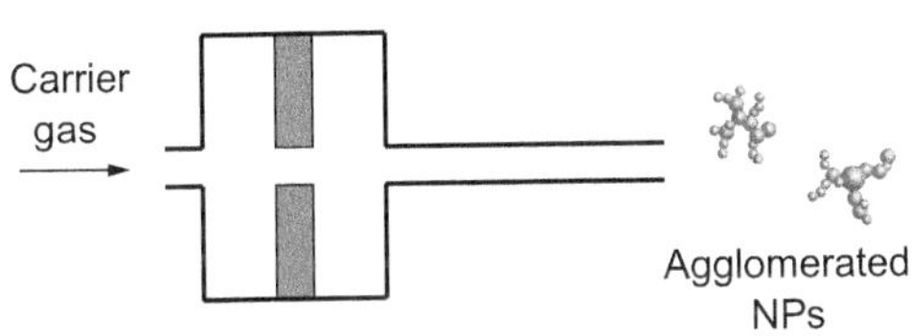

Figure 2.5 Production of agglomerates.

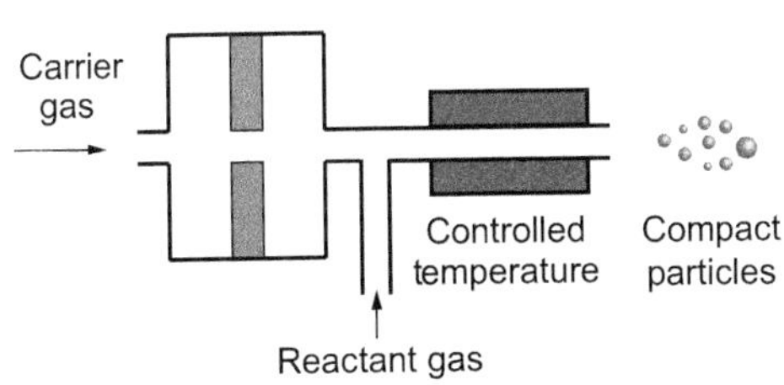

Figure 2.6 Production of compound nanoparticles by gas reactions.

Various combinations of the principles shown in Figs. 2.4–2.6 are possible. For example, Al agglomerates can be produced and subsequently exposed to low concentrations of O_2. The oxidation process then forms necks between the primary particles, yielding rigid aggregates [13]. That can serve purposes where high-surface-area nanostructures are required (Chapter 1). An interesting application of these specific particles was a significant improvement in the diagnostics of flow fields (Section 1.4.12). Careful dosing of the O_2 concentration under a controlled temperature can be used to influence the oxidation state and has recently allowed production of "black TiO_2," for which interesting properties have been predicted in connection with photocatalysis under the visible spectrum (see Chapter 1).

Coating of particles in the aerosol phase (forming core-shell particles) is possible by different techniques. Vapors of volatile materials can be condensed onto the particles [14] or precursor vapor can be added, which decomposes under the influence of radiation or heating [15]. "Spark coating" is a new technique described in Section 2.2.4.2.

2.2.4.1 Spark ablation for mixed nanomaterial production

A significant feature of SA is its unique capability of producing nanoparticulate mixtures of virtually any materials [16–20]. If mixed electrodes are used, the particles obtain the same composition (cf. Section 2.1 and Chapter 5). Two different electrodes produce alloyed particles consisting of a predictable mean composition with a predictable distribution (cf. Chapter 5). Figure 2.7 shows different configurations of SA that can yield mixed nanoparticles. The technique can be used for mixing materials at the atomic scale to produce well-defined alloy nanoparticles (cf. Fig. 2.7a,b). Recently, it was demonstrated that ternary mixtures can also be produced by SA (just as, in principle, mixtures with any number of components) [21], expanding the capabilities of the technique. Apart from alloy nanoparticles, SA can be used to mix nanoparticles at the nanoscale [8, 22]. The latter can yield core-shell nanoparticles (cf. Fig. 2.7c), a method described in more detail in Section 2.2.4.2 below. Alternatively, agglomerates consisting of two (or more) primary particle species (cf. Fig. 2.7d) can be produced. Because other methods of nanoparticle production do not possess these mixing options, this feature is probably the most valuable that SA has to offer, and a strong reason to keep the discharge time short (cf. Section 2.1 and Chapter 5). Optimized materials for sensors (cf. Chapter 11) or catalysts (cf. Chapter 10) can hardly be expected to consist of a pure (one element) metal or metal oxide. It can therefore be anticipated that these and other fields will be revolutionized by SA.

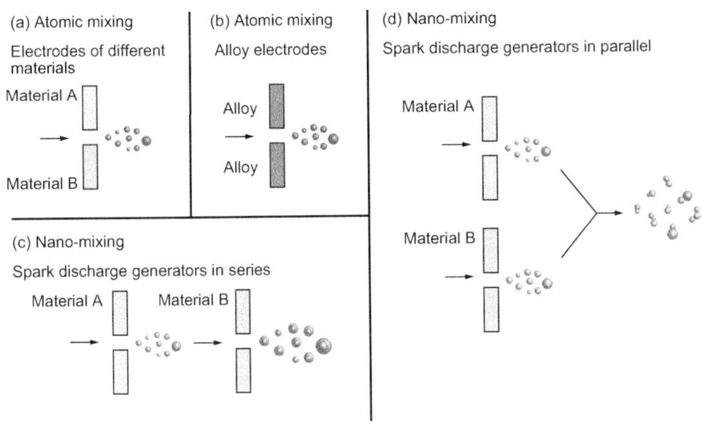

Figure 2.7 Configurations of spark discharge generators (SDGs) for producing nanoparticles that are mixed at the atomic (a, b) and the nanometer (c, d) scales.

2.2.4.2 Spark coating

SA itself has also been demonstrated as a technique for coating particles [23]. This is of special interest because no precursors are necessary to achieve such structures. The hollow-electrode configuration shown in Fig. 2.2 (right) was used for this purpose. The aerosol with the particles to be coated flows through the electrodes, and a small inert gas flow through the gap rapidly brings vapor from SA or growing atomic clusters into the vicinity of the particles, as illustrated in Fig. 2.8. As the temperature in the zone where the vapor mixes with the aerosol has rapidly decreased toward room temperature, temperature-sensitive particles can be coated with high-boiling-point materials. Figure 2.9 shows polystyrol latex (PSL) particles covered with smaller gold nanoparticles. Some of the PSL spheres were doublets or triplets, and the micrograph in Fig. 2.9 indicates that slight melting near the spark may have occurred. Whether a smooth/uniform coating is formed or the resulting particles are decorated with smaller nanoparticles depends on the surface free energies of the two materials, with the resulting structure being the one corresponding to the minimum surface free energy of the system.

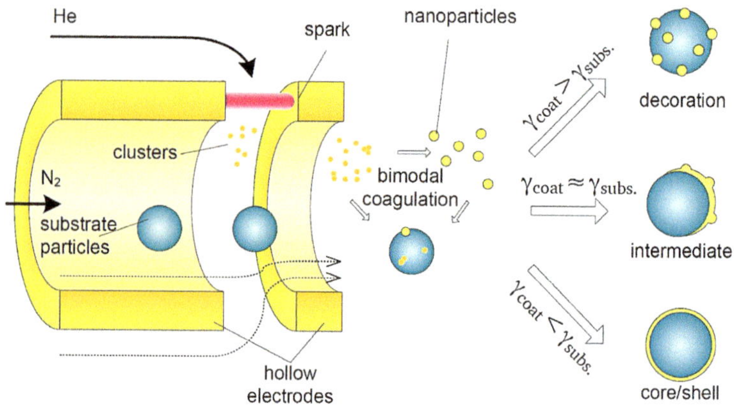

Figure 2.8 Spark coating [23]. The hollow electrode configuration (Fig. 2.2, right) enables rapid mixing of the vapor and growing atomic clusters from spark ablation with particles to be coated. Whether decoration or core-shell coating occurs depends on the relationship between the surface free energy of the coating, γ_{coat}, and of the particulate substrate, γ_{subs}.

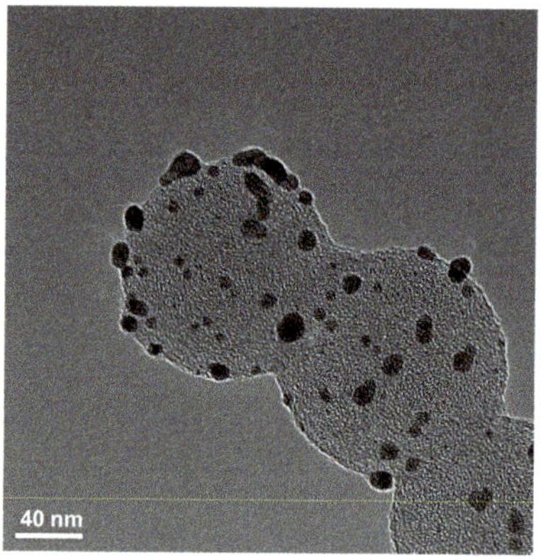

Figure 2.9 Spark-coated PSL particles (from Ref. [23]).

2.2.4.3 Combination with size classification

As discussed in Section 2.3.3.3, the size distributions of the particles produced by SA typically have a geometric standard deviation of about 1.35. If particles with narrower size distributions are required, one can use a differential mobility analyzer (DMA; [24]), which has become a standard aerosol tool over the years, downstream of the SDG to produce nearly monodisperse nanoparticles (based on their electrical mobility, which strongly depends on their size). An advantage of SA is that a good fraction of particles are electrically charged and can therefore be classified directly through a DMA avoiding any charging stage, thereby achieving a relatively high throughput of nearly monodisperse samples. Considering that existing DMA designs can efficiently size-select particles having sizes down to those of atomic clusters [25], one has a wide range of options for producing highly monodisperse nanoparticles for building nanomaterials. The fact that 100% of the particles exiting a DMA are charged may be beneficial for the following immobilization step, necessary to produce a product, as illustrated in Fig. 2.10. Immobilization principles are treated in Section 2.4 and Chapter 9.

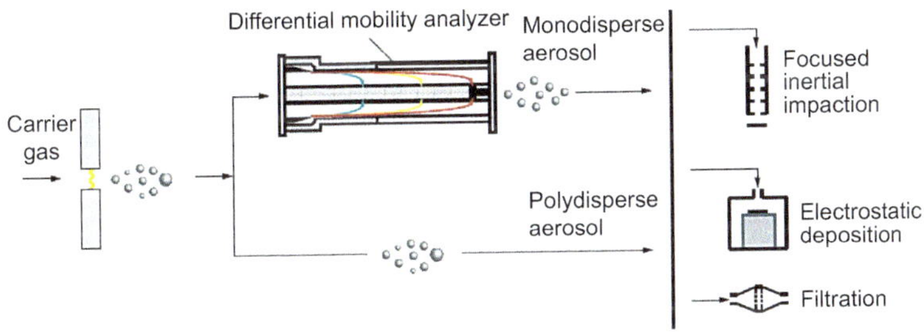

Figure 2.10 Alternative options for producing nanomaterials using either monodisperse (through the DMA) or polydisperse nanoparticle building blocks generated by spark ablation.

The DMA could, in principle, be replaced by an aerosol particle mass (APM) analyzer (see Chapter 6 and Ref. [26]) that selects a narrow size interval based on the mass-to-charge ratio. Typically, APMs can classify particles with masses down to half an attogram, which correspond to unit-density spherical particles having diameters of ca. 10 nm [27]. Recent efforts have shown that the technique can also distinguish particles having masses down to 20 zeptograms, corresponding to spherical unit-density particles with diameters of ca. 3.5 nm [28]. The resolution of APMs is not as high as that of DMAs, and thus the latter are more commonly used to produce nearly monodisperse nanoparticles.

2.2.5 Commercial Spark Discharge Generators

2.2.5.1 PALAS spark generator

Based on the original paper [1], PALAS GmbH, Karlsruhe, Germany, developed the first commercial spark generator GFG 1000 in the late 1980s [30], followed by the more recent models DNP digital 2000 and DNP digital 3000. These generators were originally developed as a source of soot-like aerosols in order to simulate combustion sources (see Refs. [12, 30] and Chapter 14) and are excellently applicable to filter testing. They produce agglomerates larger than 20 nm [29], the size of which can be influenced by admixing dilution air to the aerosol stream and the spark frequency [31]. Neither versatility with regard to particle composition or structure nor purity was the objective for this development. Figure 2.11 shows a cross section of the PALAS model GFG 1000 spark chamber, the design of which has largely been maintained for the later models. In this design, the electrodes are guided by an insulating block originally consisting of polyamide, and a channel in this material guarantees very efficient flushing of the space between the electrodes. The newer generators produce constant reproducible outputs of 9 mg/h for carbon. The desired efficient flushing of the spark carries the disadvantage with it that materials other than the desired particle constituents are close to the spark and are thus heated. In addition, the spark chamber is not completely sealed with respect to the external atmosphere, which may introduce further contaminants [5]. Roth et al. [12] state that the concentration of contaminants has been reduced from 25% to about 6% by replacing the polymers by ceramic material in the GFG 1000, which is a standard option in the more recent models DNP digital 2000 and 3000 [5]. For applications like filter testing, a 6% contamination is of little importance, and this is the main domain where the PALAS SDGs are used. See Chapter 14 for more details and application areas of this SDG.

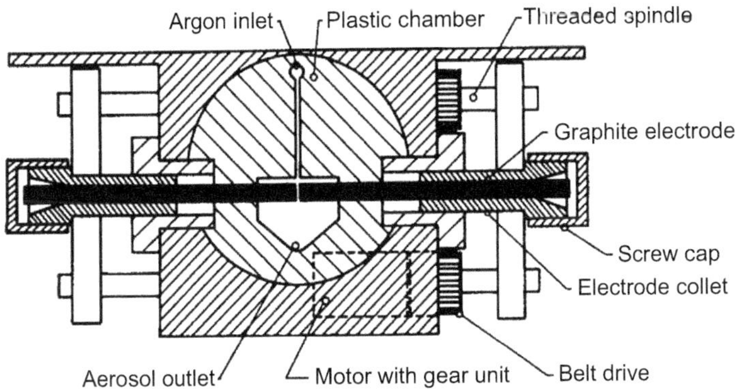

Figure 2.11 Cross section of the PALAS GFG100 spark generator. Reprinted from Ref. [30], Copyright (1993), with permission from Elsevier.

2.2.5.2 VSPARTICLE spark generator model G1

The VSPARTICLE model VSP-G1 has been designed mainly for research purposes and possess a high degree of flexibility. It is applicable for the production of nanoparticles from any conducting and semiconducting material, agglomerated or in the form of singlet particles. One of the major advantages of the SDG principle with regard to other evaporation-condensation methods for nanoparticle production consists in the fact that it can be operated in a contamination-free way. The spark chamber for the G1 avoids vicinity of any material with the spark other than the material to be evaporated. It is thus designed for generation of any particles, including uncontaminated species for high-tech applications, and can, for example, produce unoxidized nonnoble metals. The cross sections in Fig. 2.2 refer to the G1, whereas Fig. 2.12 shows a picture of the apparatus. The spark chamber is mounted on top of the power supply. This flexible design has made it a popular instrument for research purposes. The concept permits gas and product aerosol transport through the electrodes (Fig. 2.2, right) or in the cross-flow configuration (Fig. 2.2, left). Modifications of these cases are also possible, for example, if reactive gasses are to be added. An inert gas can then be introduced through both electrodes, protecting them from reacting, while the reactant gas mixture is added through one of the vertical inlets, with the other one then acting as the outlet. The open design also makes it possible to perform, for example, optical measurements within the chamber. The spark gap automatically adjusts to the chosen voltage $\bar{U}$ (see below), and the two tubes shown vertically in Fig. 2.1 are adjustable in their position. For production of atomic clusters or small singlet particles, the particle growth time can be kept small enough by using a hollow electrode as the outlet, behind which rapid dilution or deposition may take place. Alternatively, a heated tube can be attached, where further growth under full coalescence guarantees sphere-like particles. The generator produces a constant, reproducible output up to 3 mg/h for Au or 20 mg/h for C. See the table in Section 2.3.1 to derive the production rates concerning other materials from these numbers. The adjustable parameters of the G1 are the current I that continuously charges the capacitor in Fig. 2.1, which defines the repetition frequency f according to

$$f = \frac{I}{2C\bar{U}} \tag{2.1}$$

and the mean voltage $\bar{U}$, approximately determining the energy per spark E_s through

$$E_\mathrm{s} \approx 2C\bar{U}^2 . \tag{2.2}$$

The energy storage capacitance has the standard value $C = 20$ nF, but modifications are possible.

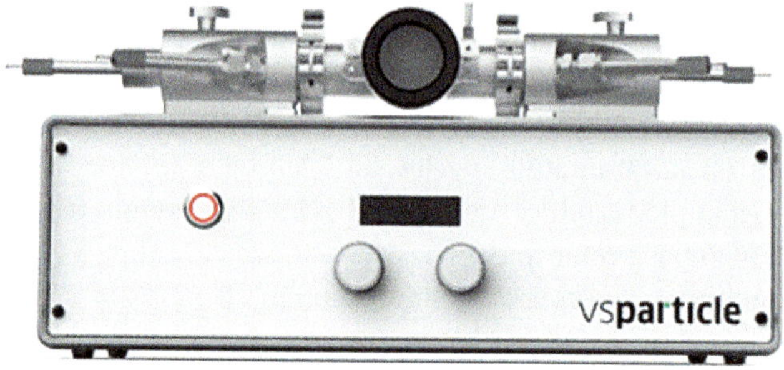

Figure 2.12 VSP-G1 spark discharge generator.

Both E_s and f determine the mass production rate (see Eq. 2.5 in the following section). The volume flow rate Q is flexible and can be controlled externally. It is shown below that the size distribution can be estimated by a simple formula as a function of the operation parameters (Section 2.3.3.2) or predicted more precisely by software developed by the supplier (Section 2.3.3.1).

The focus of the design on purity is of special importance for nanostructured products like sensors or catalysts, and it gains importance the smaller the particles get. For atomic clusters (see Chapters 7 and 8), the partial pressure of any contaminant gases or vapors must approach "high to ultrahigh vacuum conditions" within the suspending noble gas, which typically has a pressure of 1 bar. It has been demonstrated earlier that the spark method can even produce nanoparticles of metallic Mg in this way [31].

VSPARTICLE also offers an SDG-DMA combination for monodisperse particle production (VSP-S1), deposition accessories according to Fig. 2.10 (VSP-A), as well as a nanoparticle printer (VSP-P1) based on impaction. In addition, the company is developing a spark generator that uses the switching concept described in Section 2.2.3, tentatively called G2, for more control and flexibility. It enables a production rate of about 0.5 g/h at f = 10 kHz for Au. For stationary production, 10 kHz should not be exceeded without special cooling.

All the data presented below has been produced with the VSPARTICLE G1 or G2 model or very similar configurations if not otherwise stated.

2.3 Basic Features of Spark Discharge Generators

2.3.1 Mass Production Rate and Energy Efficiency

In 1950 Llewellyn-Jones [32] set up an energy balance equation containing material-dependent terms, which mainly had the purpose of estimating the ablation of spark plugs in combustion engines. This approach leads to the relation between the energy E_S transferred to a single spark and the electrode mass Δm evaporated by it:

$$\alpha E_s - \pi r^2 \sigma_S \tau T_b^4 - 2\pi r \tau k (T_b - T) = \Delta m c_{ps} (T_m - T) + \Delta m c_{pl} (T_b - T_m) + \Delta m H_m + \Delta m H_v \quad (2.3)$$

Equation 2.3 has been found useful for rough estimates of relative ablation rates for different materials [33] and for certain alloys (cf. Chapter 5). The approach assumes that the portion αE_s (with $\alpha < 1$) of the spark energy E_s is transferred to the "hot spot," where the spark hits the electrode, the rest being dissipated in the spark plasma. The second term represents energy loss of the hot spot by radiation, given by the spark radius r, the hot spot's temperature T_b, and the Stefan–Boltzmann constant σ_S. The next term covers heat dissipated by conduction through the electrodes, given by the spark radius r, the thermal conductivity k of the electrode material, and T_b. On the right-hand side of the equation we find the energy required for evaporation of the mass Δm, given by the melting and boiling temperatures T_m and T_b as well as the specific heats for the solid and liquid phase, c_{ps} and c_{pl} and the melting and evaporation enthalpies H_m and H_v. Note that this model assumes that the temperature does not exceed the boiling temperature of the electrode materials, T_b, so it excludes the phenomenon of short-term formation of a superheated phase. It is presently not clear, if superheating plays any significant role in SA for microsecond discharges.

A publication by Domaschke et al. [34] has pointed out that the electrode cross section significantly influences heat conduction away from the hot spot, which should lead to apparent variations in α from one setup to the next, because the formula does not account for effects of electrode geometry. In Table 2.1 the electrode diameter dependence according to Ref. [34] is indicated. In Chapter 5 it is further suggested that the ionization energy of the target material in the vapor phase may play a role for the mass production rate. Another phenomenon not considered by the formula is ejection of liquefied droplets, which is referred to as "splashing," further explained in Section 2.3.2. Accordingly, some metals show significant deviations from the Llewellyn-Jones formula in terms of relative ablation rates [33], which can, at least in part, be explained by the points mentioned above.

It has empirically been shown by Feng et al. [9] and the supplementary information of that publication [35] that the particle mass produced per spark, Δm, is connected to the energy E_s per spark through

$$\Delta m = \alpha\, C_{mat}\,(E_s - E_0) \quad (E_s > E_0). \tag{2.4}$$

E_0 is a material-dependent barrier below which material ablation is negligible. In practice, a proportional relation between Δm and E_s

leads to tolerable errors in the fit with experiments (see Fig. 2.13) [9] so that the energy efficiency, that is, the mass produced per unit energy consumed or the mass production rate $\dot{m}$ per unit of power $P = E_s f$ is

$$\eta = \frac{\Delta m}{E_\mathrm{s}} = \frac{\dot{m}}{E_\mathrm{s} f} \cong \alpha C_\mathrm{mat}, \tag{2.5}$$

with the repetition frequency f. Comparison of Eq. 2.5 with Eq. 2.3 shows that this is equivalent to neglecting the radiation loss term as well as the heat conduction loss term, and

$$C_\mathrm{mat} \cong \frac{1}{c_\mathrm{ps}(T_\mathrm{m} - T) + c_\mathrm{pl}(T_\mathrm{b} - T_\mathrm{m}) + H_\mathrm{m} + H_\mathrm{v}} \tag{2.6}$$

Radiation loss is usually negligible with respect to heat conduction, and the latter can, in principle, also be neglected for any materials and physical configurations if the spark duration, τ, is short enough. Another reason for proportionality between Δm and E_S (when varying only E_S under otherwise constant conditions) could be that the heat loss term in Eq. 2.3 is not independent of but rather analogous to the energy input E_S, suggesting that the assumptions of Lewellyn Jones are too crude in this respect. This would explain the significant contribution of heat conductivity that some studies reveal (cf. Chapter 5).

Equation 2.5 indicates that the proportionality constant between energy input and produced mass can be split into the material-dependent term C_mat (referred to as ablatability; see Chapter 5) and α, which depends on physical process parameters such as the characteristics of the spark. This concept has been found particularly useful for predicting and controlling the relative ablation of the electrodes in oscillating sparks if the anode and the cathode consist of different materials and mixed particles are the product. See also Ref. [36] for details.

Pfeiffer et al. [8] determined that the value of α to be inserted into Eq. 2.4 in order to match experimental production rates is $\alpha = 0.0015$. This value, however, can only be used as a rough estimate for setups similar to the cross-flow arrangement according to Fig. 2.14, and similar conditions, if we consider the reservations with respect to Eq. 2.3 mentioned above. Furthermore, the spark discharge can be influenced by gas flow, and so can re-deposition of evaporated material on the electrodes. $\alpha = 0.0015$ means that 0.15% of the consumed spark energy is transferred to the hot spot, and the rest is dissipated in the spark plasma and lost to the associated shock wave and radiation of the plasma. The spark energy consumed per unit of mass ablated (and thus converted to nanoparticles) is the energy efficiency, $\eta_{Au,Ar} = \alpha C_{mat} = 0.82$ μg/J for the case presented in Fig. 2.13. This corresponds to $\eta_{Au,Ar} = 3$ g/kWh for Au in Ar. Note that this value is basically independent of the particle sizes produced, because it is based on evaporated mass. The resulting particle size distribution depends on the process parameters governing particle growth downstream of the spark (see below). It should be noted also that particle losses are not considered in these estimations, which of course depend very much on the processes to follow downstream.

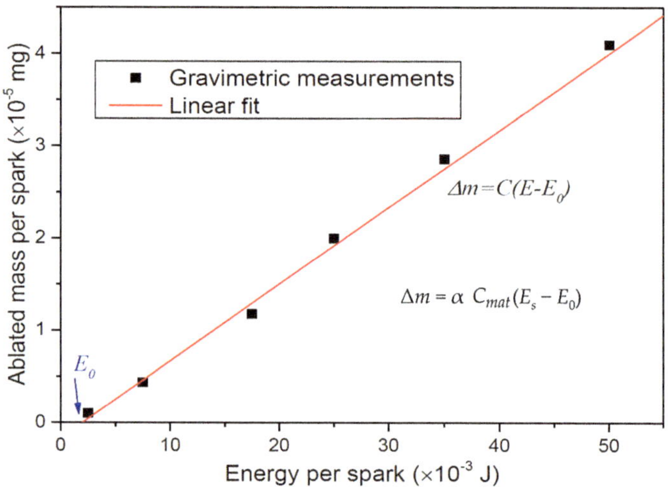

Figure 2.13 Ablated mass per spark as function of energy per spark for Au electrodes, Ar carrier gas, electrode diameter 3 mm, electrode gap 1 mm. The measurement has been carried out with the configuration of Fig. 2.14 [40].

The most frequently used carrier gasses in SDGs are Ar and N_2. Tabrizi et al. [33] showed that using the simple circuit shown in Fig. 2.1, the mass production rate per spark with the same circuit, and the same electrode distance is about 3 times higher in N_2, while the breakdown voltage is 2.8 times higher. According to Eq. 2.2, the energy per spark should therefore be ca. 7.8 times higher, but the mass production rate is not 7.8 times higher, as Eq. 2.5 predicts. We can conclude from this data that *for the same energy per spark*, the production rate in Ar would be ca. 2.6 times higher than in N_2. This means that α must be higher by this factor in Ar and so must the energy efficiency. This can be explained by larger energy losses in the N_2 plasma, where the molecule is split at a temperature in the order of 20,000 K (see Chapters 3 and 4). This is important to note because the higher mass production rate in N_2 is often pointed out in studies that use the simple circuit, but it is associated with a 2.6-fold energy consumption per unit mass.

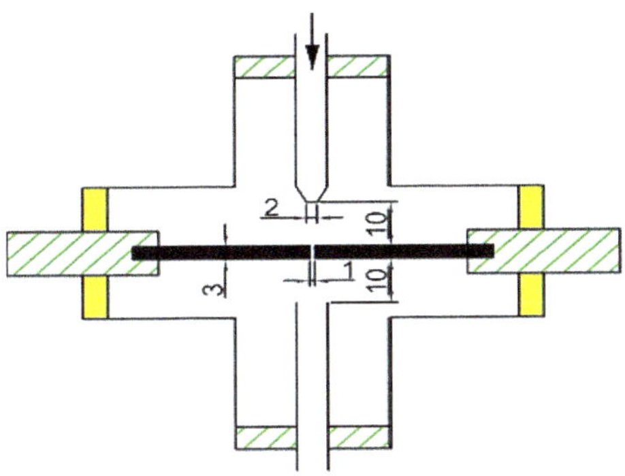

Figure 2.14 Spark generator configuration applied for determining the relation between the energy per spark and the ablated mass. The numbers indicate millimeters [40].

Relative production rates for different materials under equivalent conditions have experimentally been determined in various studies [20, 33] and in Chapter 5, and new values determined for the VSP-G1 are shown in Table 2.1 below. These values can be regarded as more reliable than the estimations that the Llewellyn-Jones equation (Eq. 2.3) delivers. From the energy efficiency for Au in Ar, $\eta_{\mathrm{Au,Ar}}$, we can, for example, estimate the energy efficiency $\eta_{\mathrm{Cu,N2}}$ for Cu in N_2 by applying Eq. 2.5 to different materials, yielding

$$\eta_{\mathrm{Cu,N2}} = \left(\frac{\dot{m}}{P}\right)_{\mathrm{Cu}} \approx \eta_{\mathrm{Au,Ar}} \frac{\alpha_{\mathrm{N2}}}{\alpha_{\mathrm{Ar}}} \frac{C_{\mathrm{mat(Cu)}}}{C_{\mathrm{mat(Au)}}}. \tag{2.7}$$

Tabrizi et al. [33] showed that the ablatability ratio Cu over Au is ca. 1:6.5 in Ar, and according to the arguments above, the α

value for ablation in N_2 is a factor of 2.6 smaller than for ablation in Ar, yielding $\eta_{Cu,N2}$ = 0.13 g/kWh, which is a factor of 20 smaller compared to $\eta_{Au,Ar}$. Table 2.1 in Section 2.3.1.1 lists ablatability ratios with respect to gold for different metals.

The estimated value of $\eta_{Cu,N2}$ = 0.13 g/kWh for SA allows direct comparison with the process of Cu particle production from a continuous arc, which can be deducted from Stein and Kruis [37]. That value of $\eta_{Cu,N2}$ = 1.5 g/kWh is an order of magnitude higher. However, particles having sizes in the range between 0.5 and 10 nm cannot be produced with arc methods, and no physical production process has been shown to have higher energy efficiency than SA in this size range.

The following data has been produced using a switching circuit based on the principle of Fig. 2.3 with a generator equivalent to a demonstration unit of VSPARTICLE (VSP-G2). This option offers a large range of energies per spark and a broad frequency range up to 20 kHz. In principle, this frequency should allow a production rate up to 2.9 g/h for Au [35]. Energy is also consumed by the electronics (including transformation to high direct current (DC) voltage and the switching circuit) of a modern SDG unit. For this VSP-G2 prototype, the total power consumed was determined to be

$$P = 40 \text{ mJ} f + 90 \text{ W}, \tag{2.8}$$

where f is the spark repetition frequency chosen. Here the frequency-proportional term represents the energy per unit time contributing to the spark at E_s = 40 mJ, and the constant term is consumed by electronic circuitry, including a transformer in the DC power supply. As already indicated above, the mass production rates and energy efficiencies reported in various papers vary considerably, and discrepancies of a factor of 2 for similar conditions are not uncommon. The reasons for these differences are manifold. The electrode structure as well as gas impurities may sensitively influence the state of the discharge plasma, and slight oxidation of the electrodes certainly has an influence on ablation behavior. According to Ref. [34], the mass production rate is influenced by the heat flux away from the hot spot, which significantly depends on the electrode diameter (see Table 2.1). In addition, it is known that high particle losses to the electrodes occur, and these are very sensitive to the degree to which the volume between the electrodes is flushed between two discharges by the gas inflow. Furthermore, residues of the ablated material will influence the successive discharges and so will the inert gas velocity distribution in the discharge volume. A higher mean temperature of the electrodes increases mass production.

The mass production numbers above are based on weight loss of the electrodes, which should be a realistic measure if the particles are immobilized with a small time delay, that is, near the point of production, and transport tubes are kept short. This assumption is based on small electrostatic losses in the spark chamber. Deposition of particles to the tubes or reactor walls is certainly to be considered in addition, if the particles are to be modified in any way by aerosol processes. Diffusion is the mechanism that governs wall deposition, as the particles are initially very small. Where a nanopowder is the desired material, the material deposited on the walls must not be regarded as lost. After the deposition of thick-enough layers on the walls, resuspension should be in equilibrium with deposition.

2.3.1.1 Ablatabilities of different materials

The ablatabilities C_{mat}, as defined by Eq. 2.4, are different for each material (see also Chapter 5). Table 2.1 lists the relative ablation rates for a number of metals and carbon. They have been determined under identical conditions with VSP-G1 using the cross-flow configuration. As the generator is automatically controlled to keep the conditions constant, even after exchanging the electrode material, these ratios can be regarded as reliable. It is remarkable how far apart the ablatabilities of different materials may be. In terms of the amount of material in moles, SA produces 116 times more nanoparticulate carbon than copper. Most of the reasons for apparent variabilities mentioned earlier in the ablatabilities reported are eliminated by reporting *relative* values determined under identical conditions. The electrode diameters have not always been kept the same for the data in Table 2.1, and this has been compensated for by applying a correction according to Ref. [34] (see Table 2.1).

Table 2.1 Relative ablatabilities of different materials (defining gold as 1) determined with VSP-G1 in cross-flow mode I = 8 mA, $\overline{U}$ = 1.1 kV, Q = 3 lpm in Ar. All values are for electrode diameters of 3 mm. The values for Al, C, Co, Cu, Fe, Mo, Ni, Sn, and Ti have been determined using 6 mm electrodes and multiplied by $(6\ \text{mm}/3\ \text{mm})^{0.91} = 1.88$ according to Ref. [34] (derived from data measured by Isabel Postmes at VSPARTICLE).

Element	Ablatability ratio with Au [moles/mole Au]	Ablatability ratio with Au [g/g Au]
Cu	1.05	0.34
Au	1.00	1.00
Ni	2.92	0.87
Mo	2.36	1.15
Sn	3.74	2.25
C	116.2	7.08
Fe	2.60	0.74
Pd	2.76	1.49
Ag	1.34	0.73

2.3.2 Splashing

In addition to the desired evaporation-condensation effect in SDGs, the ejection of solidified droplets has been observed and is referred to as "splashing" [8]. It leads to the formation of large spherical particles having sizes from 50 nm up to a few micrometers. Crater-shaped structures have been observed on the electrodes and associated with this phenomenon [38]. Ejection of droplets from micron-sized liquid pools can possibly be explained by the acoustic shock wave associated with each spark (Chapter 4). Redeposition of such material has also been observed [33, 39]. A recent study by VSPARTICLE has revealed peaks and wires on the electrodes, and much of this material can be explained by deposition of "splashing" particles at the counterelectrode. The splashing particles are certainly negligible number-wise, and they typically represent a small portion of the produced particle mass [40], if at all present. Depending on the application, splashing particles can reduce the quality of the product. Means of reducing their formation have therefore been searched for.

A promising approach avoids that consecutive sparks hit the same electrode spot to limit local heating. The above-mentioned study by VSPARTICLE showed that wire growth by splashing is eliminated at far enough electrode distances. A magnetic field has been used to deflect the glow discharge between sparks in a switching circuit (cf. Section 2.2.3), and the results strongly suggest a reduction of particles produced by the splashing effect [40]. More research is clearly necessary here.

The tendency for splashing rises with rising electrode temperature in high-frequency arrangements, where a high average power is transferred to them. At a power of 500 W, the inert gas flow should be at least 10 L/min and focused to the spark for sufficient cooling. Arrangements allowing 1000 W and more should apply additional cooling means. Reports on energy efficiencies significantly higher than the ones mentioned in Section 2.3.1 for gold and corrected according to Table 2.1 for other materials should raise caution, if they do not comment on the portion of splashing particles they contain.

2.3.3 Particle Size Distribution

In SA, the vapor produced by each spark is drastically quenched down to a near-ambient temperature, inducing an enormously high supersaturation. This shifts the critical nucleus size for particle condensation into the atomic size range. We assume that under these conditions the vapor atoms that collide with other atoms or atomic clusters stick at every collision, and Smoluchowski-type coagulation [41] describes the growth process from the very beginning, without the need of employing nucleation theory. In Section 2.3.3.1 a numerical model is described and in Section 2.3.3.2 a simple analytical one. Both models further assume that the vapor coming from the spark is instantaneously distributed homogeneously into the flow so that the starting particle size is atomic size [9]. It can be seen as a useful feature of SA that it allows for these assumptions, which imply, in particular, a complete separation of vapor production and particle growth. This makes modeling of particle growth in SDGs simple with respect to other aerosol production processes and leads to useful predictions of general value.

2.3.3.1 Numerical model

The following numerical model has recently been developed by VSPARTICLE. It calculates the evolution of the size distribution starting from the vapor phase using Smoluchowski theory [41]. The population balance equation according to that theory is extended by inserting a loss term, and the resulting differential equation is

$$\frac{\partial n(v,t)}{\partial t} = \frac{1}{2}\int_0^v \beta_D(v',v-v')\,n(v',t)\,n(v-v',t)\,dv'$$
$$-\,n(v,t)\int_0^\infty \beta_D(v,v')\,n(v',t)\,dv' + \frac{\partial n_{\text{sink}}(v,t)}{\partial t}). \qquad (2.9)$$

This equation expresses the rate of change of the particle concentration for each particle size represented by the particle volume v, and $n(v,t)$ represents the time-dependent size distribution. β_D is the coagulation kernel, and we apply the one according to Dahneke [42, 43] as a function of the sizes of the colliding particles. We use the loss term on the right-hand side of the equation according to Park et al. [44], assuming a cylindrical tube. It is given by

$$\frac{\partial n_{\text{sink}}(v,t)}{\partial t} = 0.440\frac{kT}{\eta}\frac{C(D_p)}{R^2 D_p}n(v,t)\,, \qquad (2.10)$$

where R is the tube radius, η the gas viscosity, k the Boltzmann constant, and T the absolute temperature. The Cunningham slip correction factor $C(D_p)$ is given below (Eq. 2.14). Dahneke's coagulation kernel is expressed through the kernel β_{Co} for the continuum regime ($D_p >> \lambda$) and the kernel β_F for the free molecular regime ($D_p << \lambda$) by

$$\beta_D = \beta_F\frac{(\beta_{Co}+\beta_F)^2-\beta_F{}^2}{(\beta_{Co}+\beta_F)^2+\beta_F{}^2}\,. \qquad (2.11)$$

The coagulation kernel in the continuum limit is given by

$$\beta_{Co}(v,v') = \frac{2kT}{3\eta}\left(v^{\frac{1}{3}}+v'^{\frac{1}{3}}\right)\left(\frac{C(v)}{v^{\frac{1}{3}}}+\frac{C(v')}{v'^{\frac{1}{3}}}\right), \qquad (2.12)$$

and the coagulation kernel for the free molecular limit is

$$\beta_{\mathrm{F}}(v,v')=\left(\frac{6kT}{\rho_{\mathrm{g}}}\right)^{\frac{1}{2}}\left(\frac{3}{4\pi}\right)^{\frac{1}{6}}\left(v^{\frac{1}{3}}+v'^{\frac{1}{3}}\right)^{2}\left(\frac{1}{v}+\frac{1}{v'}\right)^{\frac{1}{2}}, \tag{2.13}$$

where k is the Boltzmann constant, T the absolute temperature, ρ_{g} the gas density, and η the gas viscosity, and the Cunningham slip correction is given by

$$C(D_{\mathrm{p}})=1+\frac{2\lambda_{\mathrm{g}}}{D_{\mathrm{p}}}\left(\alpha+\beta\exp\left(-\gamma\frac{D_{\mathrm{p}}}{2\lambda}\right)\right) \tag{2.14}$$

with the gas molecule mean free path λ_{g}. For argon, the empirical constants are $\alpha = 1.227$, $\beta = 0.42$, and $\gamma = 0.85$. As spherical particles are assumed, the particle volume can be expressed by the particle diameter through $v=\frac{\pi}{6}D_{\mathrm{p}}^{3}$. This value of β_{D} does not consider van der Waals forces between the particles. For metals, this attractive force is significant, and this requires multiplication of β_{F} and β_{Co} with an enhancement factor, as given by Ouyang et al. [45]. The supplementary material of that article [46] contains a table, with which the enhancement factor can be determined for any value of the Hamaker constant. Hamaker constants for the elements and many compounds are available from many sources. A Hamaker constant for gold of $A_{\mathrm{H}} = 3.5\ 10^{-19}$ J is adapted from Ref. [47], which leads to a van der Waals enhancement factor of 3.1 and 1.48 for β_{F} and β_{Co}, respectively.

The equation is solved by means of the method of moments described by Yu et al. [44] and Otto et al. [48]. The model calculates the final particle size distribution for the coagulation time given by V/Q, where Q is the SDG volume flow and V is the volume that the flow occupies between the spark and the point, where coagulation is stopped by deposition or dilution, or where measurement takes place.

If the model is applied for the case where there is no heating downstream of the spark, fractal-like agglomerates form above a critical particle size (the primary particle size in the agglomerates), typically above 5 nm at room temperature. The coagulation coefficient can be corrected to account for the fractal shape. As we can derive from Rogak and Flagan [49], the coagulation coefficients versus the volume equivalent diameters of coalesced spheres and agglomerates of small fractal dimensions in the range below 20 nm do not differ much from each other. Such a correction is therefore not applied.

The geometric mean that the model predicts is shown in Section 2.3.3.3, where it is compared to experimental values. The size distributions are lognormal and the geometric standard deviation σ_g approaches a self-preserving value, which is a weak function of the final geometric mean diameter and lies between 1.29 and 1.33, as shown in Fig. 2.15. A former study by Lee et al. [50] determined a value of 1.32 for σ_g for the free molecule regime and 1.33 for the continuum regime, and in between these extremes, in the transition range, the values drop down to about 1.29. This is in agreement with the range of values determined by this model. Another important conclusion from the numerical model is that for coagulation times $t = V/Q < 1$ s diffusional losses are negligible for our spark process.

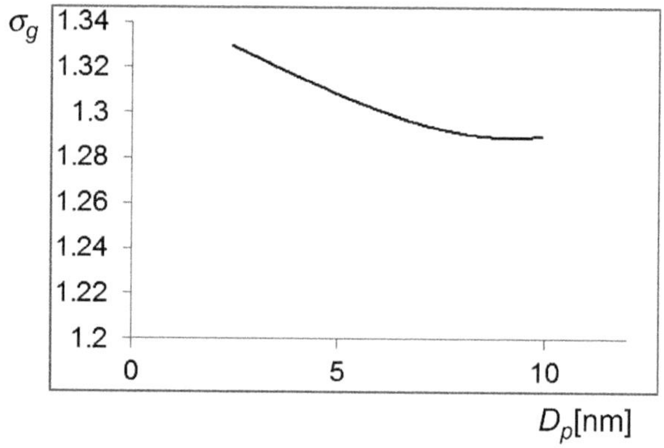

Figure 2.15 Geometric standard deviation as a function of the geometric mean diameter calculated by the VSPARTICLE model.

2.3.3.2 Analytical model

A simple model for estimating the size distribution was published by Feng et al. [9]. This model is described next and is based on pure Smoluchowski-type growth, just like the VSPARTICLE numerical model above, and is valid for spherical particles. It has been experimentally verified for Au, which coalesces to round particles at room temperature up to the size of ca. 5 nm. If larger spherical particles are to be produced, the temperature in the particle growth zone has to be increased to ensure coalescence. To describe the process, the formula of monodisperse coagulation is used, namely

$$\frac{dN}{dt} = -\frac{1}{2}KN^2. \tag{2.16}$$

It expresses the rate of change of concentration N, and K is the monodisperse coagulation coefficient. Losses by diffusion to the wall are neglected, as justified by the numerical model above. The analytical model is restricted to coagulation limited processes. In these, the initial concentration N_0 is very high and decreases by several orders of magnitude by coagulation. This is normally the case in SA, where N_0 is typically between 10^{18} m^{-3} and 10^{20} m^{-3}, with the resulting aerosols having concentrations many orders of magnitude lower. For this case the solution of Eq. 2.16 can be simplified as

$$N(t) = \frac{2}{Kt}, \tag{2.17}$$

where t is the coagulation time. The particle concentration is thus independent of the initial concentration in an aerosol from spark discharge. Another feature of coagulation-limited processes is that the final particle size is practically determined during the last reduction in the particle concentration by 1 order of magnitude or the last growth in the particle size by a factor of 2. Applying this feature, we ignore the fact that during growth from atomic size to, say, 20 nm, K may vary by a factor of 5, and a constant value of K corresponding to the final particle size (the geometric mean to be precise) is used.

As derived by Feng et al., Eq. 2.17, together with conservation of mass per unit volume, leads to the estimation of the geometric mean particle sizc

$$\overline{D_\mathrm{p}} \cong \left(\frac{3\dot{m}K(\overline{D_\mathrm{p}})V}{\pi\rho Q^2} \right)^{\frac{1}{3}}. \tag{2.18}$$

Here ρ is the density of the electrode material (assumed also for the particles), and V is the volume being occupied by the flow downstream of the spark to the "exit," which is the place where the size is measured or coagulation is stopped by immobilization or dilution. The coagulation time in Eq. 2.18 has been expressed by V/Q, with Q standing for the volume flow rate. We derive from Eqs. 2.11–2.14 the monodisperse coagulation coefficient from Dahneke's theory as

$$K(\overline{D_\mathrm{p}}) = \beta_\mathrm{D}(v, v') \qquad (v = v' = \frac{\pi}{6} D_\mathrm{p}{}^3). \tag{2.19}$$

With the gas viscosity η = 2.25 × 10^{-5} Pa s, the mean free path λ = 64 nm, and the density ρ_g = 1.78 kg/m^3 for argon at 1 bar and 294 K, we get the size-dependent monodisperse coagulation coefficient, as shown in Fig. 2.16.

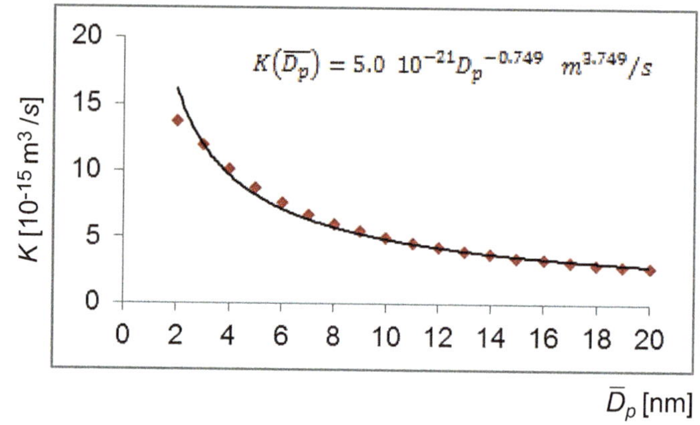

Figure 2.16 Monodisperse coagulation coefficient according to Dahneke (brown diamonds) approximated by the power law indicated between 2 and 20 nm. Coefficient of determination r^2 = 0.99 .

Between 2 nm and 20 nm, the function $K(\overline{D_\mathrm{p}})$ can be fitted by the power law

$$K(\overline{D_\mathrm{p}}) = 5.0 \times 10^{-21} D_\mathrm{p}{}^{-0.749} \frac{m^{3.749}}{s}, \tag{2.20}$$

with a coefficient of determination of r^2 = 0.99, as shown in Fig. 2.16. Together with Eq. 2.18 we then get a useful approximation for the particle diameter as a function of the mass production rate, the volume flow rate, and the coagulation volume V expressed by

$$\overline{D_\mathrm{p}} \cong 3.75 \times 10^{-6} \left(\frac{\dot{m}V}{\rho Q^2} \right)^{0.267} m/s^{0.267} (2nm \le D_\mathrm{p} \le 20nm). \tag{2.21}$$

For VSP-G1 in the cross-flow configuration, Fig. 2.17 shows a comparison of experimental values for $\overline{D_p}$ with the VSPARTICLE numerical model and the approximation according to Eq. 2.21.

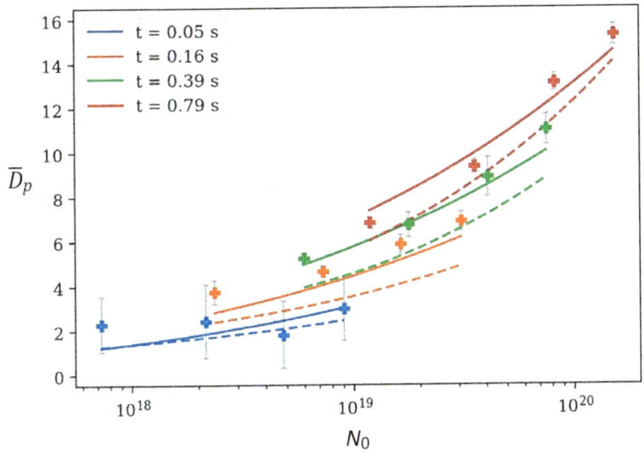

Figure 2.17 Geometric mean particle diameter $\overline{D_p}$ as a function of initial concentration $N_0 = \dfrac{\dot{m}N_A}{M_A Q}$, where $\dot{m}$ is the mass production rate, Q the volume flow rate, M_A the atomic mass, and N_A Avogadro's number. Coagulation time $t = V/Q$ contains the coagulation volume V. Solid lines: numerical model; dashed lines: analytical model (Eq. 2.21); crosses: measurement.

The model assumes a monodisperse aerosol with spherical particles, no diffusional losses, and no van der Waals attraction. As the study by Feng et al. already showed, the size distributions obtained from SDGs are close to lognormal with a relative standard deviation similar to the self-preserving value (1.32–1.36) [9]. Furthermore, K is assumed to be constant in time and determined by the final particle size. As in the numerical model, a single coagulation time is assumed, although even if the flow is laminar, different volume elements require different times to reach the exit.

It is worth pointing out that the assumption of Smoluchowski-type coagulation with sticking at every collision from the very beginning, as also assumed for the numerical model in Section 2.3.3.1, is not valid yet at the first collisions of vapor atoms and atomic clusters. This is evident through the fact that different cluster sizes occur in different concentrations, depending on their stability (see Chapter 7 and Ref. [51]). This indicates that collisions that do not irreversibly stick do occur and that corresponds to a smaller value of K than in the case of 100% sticking probability. However, this should not introduce any large error if the product particles are larger than 2 nm, because the final size is very insensitive to the initial value of K, as pointed out above. The dominant value of K corresponds to the final size.

2.3.3.3 Application of the model results to VSP-G1

For a comparison of modeling results with measurements, the setup in Fig. 2.18 is used. The spark generator (VSP-G1) is operated in the cross-flow mode (Fig. 2.2, left) with an inlet and outlet inner diameter of 4 mm and a distance of the tubes from the electrodes of 7 mm. It was connected to a mobility analyzer (TSI 3086) in connection with a fA sensitivity aerosol electrometer to determine the size distribution. The size distributions are near-lognormal, and we represent them by the geometric mean diameter $\overline{D_p}$ and the relative standard deviation σ_g.

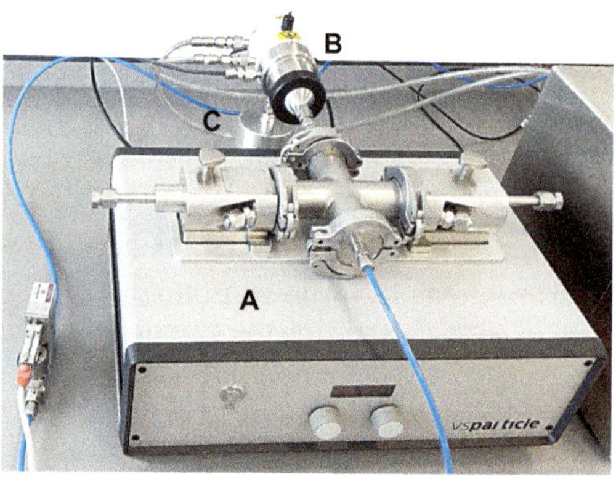

Figure 2.18 Experimental setup showing (A) VSP-G1, (B) TSI mobility analyzer model TSI 3086, and (C) aerosol electrometer with fA sensitivity.

We must account for the fact that the mobility diameter is measured, but the models work with the volume equivalent one. These are identical, where complete coalescence takes place, but the mobility equivalent diameter $D_{p(me)}$ has to be corrected to give the volume equivalent diameter$D_{p(ve)}$, where fractal-like shapes form. This is done by using the transformation from Schmidt-Ott et al. (1990) [52]:

$$D_{\mathrm{p(ve)}} = D_{\mathrm{p(p)}} \left(\frac{D_{\mathrm{p(me)}}}{D_{\mathrm{p(p)}}} \right)^{\frac{D_{\mathrm{f}}}{3}} \quad (D_{\mathrm{p(me)}} \geq D_{\mathrm{p(p)}};\, D_{\mathrm{f}} \geq 2) \tag{2.22}$$

An estimate for the primary particle size $D_{\mathrm{p(p)}}$ is usually available (here: 8.5 nm), and the fractal dimension is assumed to have the value of 2, which is a good estimate for normal coagulation, also referred to as cluster–cluster aggregation.

The model results were compared with a measurement using VSP-G1. For this comparison, the coagulation time, t, and the initial concentration, N_0, are chosen to represent the SDG parameters through

$$t = V/Q \tag{2.23}$$

and

$$N_0 = \frac{\dot{m} N_{\mathrm{A}}}{M_{\mathrm{A}} Q}, \tag{2.24}$$

where V is the coagulation volume, Q the volume flow rate, $\dot{m}$ the mass production rate, M_{A} the atomic mass of the material ablated, and N_{A} Avogadro's constant. Equation 2.24 follows from conservation of mass per unit volume, for which the numerical model gave the justification (negligible losses). The mass production rate was measured by mass loss measurements of the electrodes. Figure 2.17 shows a comparison between the measured geometric mean particle diameter and the model predictions. The numerical model comes quite close to the measured values, and the analytical model is correct within about 1 nm and can be used as a crude approximation. The coagulation volume V (13.2 cm^3) includes part of the volume around the spark, the outlet tube, and the mobility analyzer inlet. We must consider the fact that the literature on coagulation coefficients reveals quite some ambiguities and that the mobility diameter measurement assumes an equilibrium bipolar charge distribution of the particles, which has been confirmed by the data of Tabrizi et al. within an uncertainty that could amount to a factor of 3 [33]. Considering this and other assumptions mentioned above, the rather good match of theory and experiment can be regarded as fortuitous. As the approximative formula (Eq. 2.18) shows, part of the explanation for this good fit may be seen in the fact that K only enters as the cube root. Uncertainties regarding the particles' charge state are expected to influence primarily the concentration values measured but to have a minor influence on the sizes determined. The gravimetric measurements were affected by a rather large error, explaining the error bars.

We prefer to express the expected particle size by $\dot{m}$ rather than the SDG power because of the concerns expressed in Section 2.3.1 about regarding α in Eq. 1.5, Chapter 1, as a constant. It will have a different value for each configuration, and it is safe to determine m $\dot{m}$ (fE_S) from case to case. Predicting the final size in terms of $\dot{m}$, as in Fig. 2.17, should be of a general value for the different configurations possible with VSP-G1. This does not include situations, where the spark is unstable (see criteria in Section 2.3.4) or where the losses are too large, for example, at coagulation times much larger than 1 s.

The geometric standard deviations measured were significantly larger than the predictions by the numerical model (cf. Fig. 2.15). They showed a correlation with the volume flow rate Q, as shown in Fig. 2.19. This may be due to a buoyancy effect near the spark, which spreads more of the aerosol into the spark chamber before the outlet tube is reached when the flow is slow.

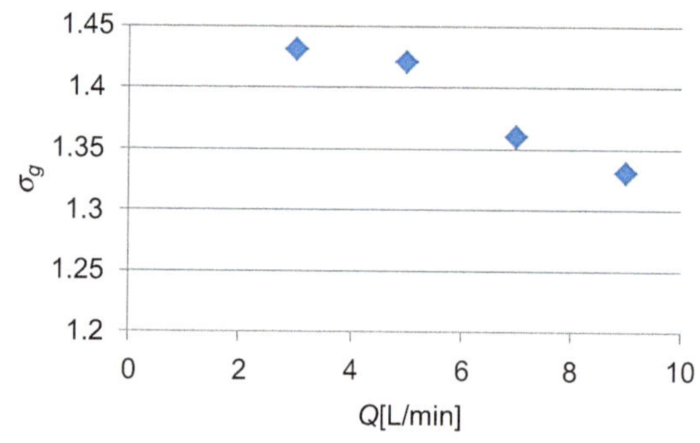

Figure 2.19 Measured relative standard deviation σ_g, averaged over different power settings $P = \overline{UI}$ of VSP-G1 as a function of flow rate Q.

In summary, the output size distribution of the VSP-G1 model SDG is lognormal, and the geometric mean size can be predicted by a numerical model. Between geometric mean diameters of 2 and 20 nm, the relative standard deviation σ_g is between 1.29 and 1.33 according to the model (Fig. 2.15), and measured values (Fig. 2.19) are systematically higher, between 1.33 and 1.43. The explanation is the assumption of a single coagulation time rather than a distribution of coagulation times under laminar flow conditions. The sensitivity of σ_g to the flow situation is indicated by the slight narrowing of the distribution with rising volume flow (Fig. 2.19). A simple analytical model gives a useful estimate of the particle size.

2.3.4 Stability Range of the Spark Mode

As pointed out in Section 2.2.2, the spark discharge in the simple circuit becomes unstable for high frequencies. This phenomenon has been studied systematically with the VSP-G1 model. In Figs. 2.20 and 2.21, the limit, where the spark mode transfers to a glow discharge mode, is indicated as it has been determined for each combination of the adjustable parameters mean voltage $\bar{U}$ and current I. The flow rate showed negligible influence and the data is averaged for measurements at 1 to 4 L/min. The cross-flow mode corresponding to Fig. 2.2 (left) was chosen. The carrier gases used were nitrogen and argon with purities of 99.9% and 99.999%, respectively. If I, and therewith f, is too large for a given voltage, the spark mode is transferred into the glow mode, where no ablation of the electrodes takes place. Close to the spark discharge limit, the spark tends to be unstable.

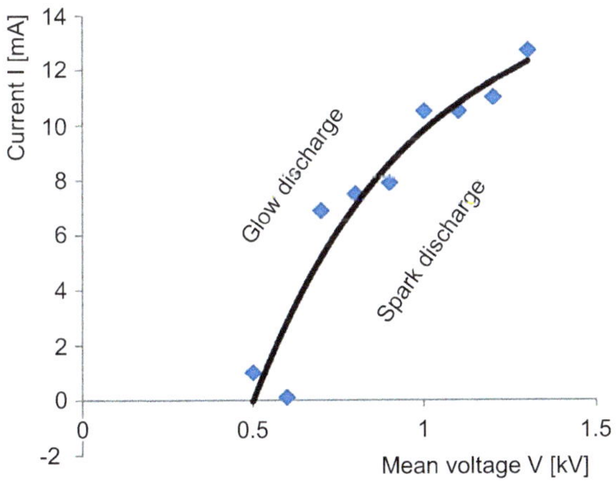

Figure 2.20 Spark discharge limit in argon with VSP-G1.

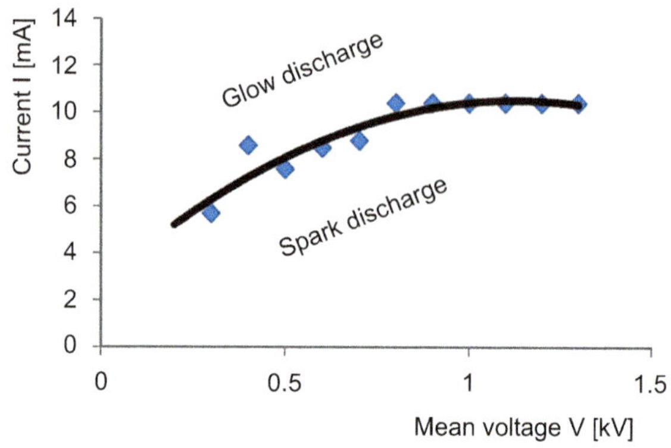

Figure 2.21 Spark discharge limit in nitrogen with VSPARTICLE G1.

As pointed out in Section 1.2.3, Chapter 1, the switching circuit avoids transition of the spark mode to the glow mode.

2.4 Immobilization for Synthesis of Nanomaterials and Devices

2.4.1 Transfer of Particles into a Liquid

For a number of applications, for example, in the biomedical field, as described in Chapter 13, the particles have to be transferred into a water-based liquid. Since it is practically impossible to purify liquids as can be done with gases, this cancels out an important feature of SA, the ability of producing ultrapure particles. Nevertheless, there are advantages of SA that may be dominant, especially in the domain of research, such as the flexibility in producing arbitrary mixtures or the simple way in which superparamagnetic high-surface-area structures can be produced (cf. Section 1.4.15, Chapter 1, and Chapter 13) for biomedical applications.

VSPARTICLE has developed a high-efficiency bubbler prototype consisting of a perforated disc at the base of a glass tube 2.6 cm in diameter. For testing it, VSP-G1 was followed by a heated tube to obtain full coalescence and thus singlet particles. The settings of the apparatus were adjusted to produce gold particles with a geometric mean diameter of 17 nm, which were passed through the bubbler at a flow rate of 0.5 L/min. The bubbler was filled with deionized water. The image in Fig. 2.22 shows the characteristic red color of the suspension, indicating nonagglomerated particles. The collection efficiency curve of the bubbler for this size was derived from the intensity of the plasmon absorption peak. The collection efficiency is given by the formula

$$\kappa = 1 - e^{-\gamma V_{\text{liq}}}, \tag{2.25}$$

with γ = 0.06 mL^{-1}, equivalent to a capturing efficiency of κ = 95% in a liquid volume V_{liq} of 50 mL. Spectroscopy also revealed that coagulation in the liquid was so slow that a stabilizing agent was not necessary for this measurement.

Figure 2.22 Suspension of 17 nm Au particles transferred into deionized water by bubbling.

2.4.2 Deposition onto Solid Substrates

In most nanoparticle applications, where particles below 20 nm are needed, such as in the fields of electronics, sensors, or catalysts, for example, for production of catalyst arrays for screening experiments, our nanoparticles have to be deposited onto surfaces. An important advantage of synthesizing nanoparticles in the gas phase is that there are a number of deposition principles with which these building blocks can be deposited in a more or less controlled way, allowing additive manufacturing, production of porous layers, or dense arrangements of individual particles. Inertial impaction, electrostatic deposition, and filtration are three state-of-the-art techniques of collecting the nanoparticles to choose from (cf. Fig. 2.10), depending on the application. For our size range below 20 nm, diffusion is the main transport mechanism in filtration. Inertial impaction can be used in connection with aerodynamic focusing, thereby producing a narrow nanoparticle beam before deposition, which is advantageous for applications such as printing nanoparticles onto electronic boards. Figure 1.2, Chapter 1, shows an example in the form of conducting paths printed onto a contact lens. This application makes use of the liquid-like state of very small metal particles as well as the impaction energy, locally heating the deposit but transferring only a small quantity of heat to the substrate, which enables deposition even on heat-sensitive materials. Impaction of particles in the size range of a few nanometers basically requires a nozzle, behind which the aerosol is released into a volume of a pressure reduced to a few mbar surrounding the substrate.

Electrostatic deposition provides a more gentle way of depositing nanoparticles onto substrates and does not require any pressure drop. Among others, it can be used as a way of surface decorating existing nanomaterials or for the more sophisticated techniques of nanopatterning of surfaces described in Chapter 9.

Finally, by collecting particles by filters, surface filtration can be used for powder collection, where nanopowders are required. Volume filtration by fiberous filters can be applied to create very high-surface-area porous material. The fibers are either decorated by the particles, or the nanoparticulate layer is built up to a thickness exceeding the fiber thickness by far. This is a process that "happens by itself" in a very efficient way within our size range smaller than 20 nm due to the high diffusion coefficient of the particles.

The following Sections 2.4.2.1 to 2.4.2.4 go into more detail regarding deposition by impaction, electrostatic forces, and diffusional deposition. For applications, where equally sized particles are required, the combination of an SDG, an electrostatic classifier, and a deposition module is a solution (e.g., VSP-G1 combined with VSP-S1 and VSP-A).

An approach not treated here but to which Chapter 8 is dedicated forms a focused nanoparticle or atomic cluster beam in high vacuum, which writes patterns on practically any substrate with a high spatial resolution.

2.4.2.1 Electrostatic deposition

Electrostatic precipitators and electrostatic samplers [53] have already been applied in the early days of aerosol technology, mainly for removing particles from the air in cleaning devices and as samplers for analysis of the atmospheric aerosol. Aerosol particles must possess an electric charge in order to enable their electrostatic deposition. As for particles smaller than 20 nm the charging efficiency of conventional aerosol chargers based on ion attachment is typically below 5%, electrostatic deposition is usually not the method of choice for this size range. However, SA delivers a very high fraction of charged particles in the atomic cluster size range. According to Ref. [54] (see also Section 7.2.2.1, Chapter 7), 80% of the clusters can be charged, 40% of each polarity. Around the size of 5 nm there is a lack of data on the charged fraction, which is not easy to determine. In any case, it was possible to obtain a concentration of 10^7 cm^{-3} in the monodisperse charged fraction behind a mobility classifier [35]. SA thus circumvents the necessity of charging, and it is applicable to basic studies on well-defined nanoparticles of a very wide variety of compositions.

Electrostatic deposition of particles in arbitrary patterns has been achieved by writing charged patterns on an insulating substrate, which then attract nanoparticles of the opposite charge for them to selectively deposit on the desired spots. An early example (Zonnevylle et al., 2009) [55] used an electron beam–charged substrate. The SiO_2 substrate was precharged by ions and the charge was reversed along a straight line by an electron beam. Particles that have the opposite charge of the line are electrostatically focused onto this line. As shown in Fig. 2.23, this enabled deposition of 5 nm particles with a precision of ca. 100 nm. The limitation here was that the electron beam ionized the surface on an area much larger than its cross section due to secondary electron generation.

Chapter 9 reviews a more advanced nanoscale patterning technology called electric-field-assisted aerosol lithography (EAAL), which also uses electrostatic focusing to increase the spatial resolution [85]. SA is the most suitable nanoparticle source for this application because besides its versatility a large portion of the particles is charged.

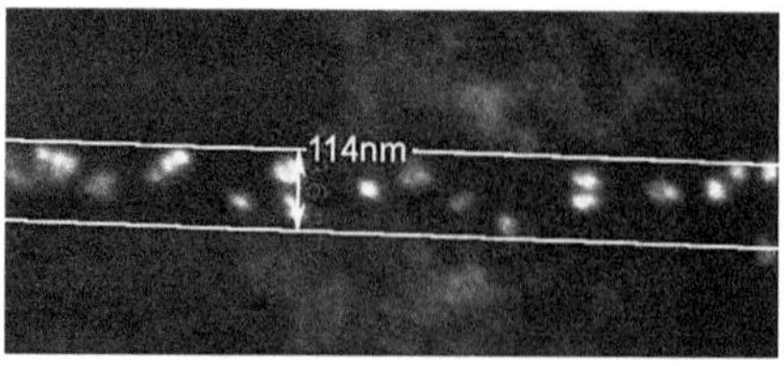

Figure 2.23 HRSEM picture of a line of 5 nm Pd particles electrostatically deposited onto a Si wafer having a thin SiO_2 surface layer (native oxide layer) locally charged to the opposite polarity by an electron beam [55].

2.4.2.2 Deposition by impaction

Figure 2.24a shows the principle of a subsonic low-pressure impactor, similar to the ones introduced by J. Fernandez de la Mora et al. [56, 57]. The aerosol, having been produced at 1 bar, enters the device through a critical orifice, where the flow is expanded into a lower pressure volume. This impactor is equipped with an aerodynamic focusing stage [58] consisting of a series of "lenses" (Fig. 2.24b) in the form of discs that have a hole in the center. In the example of focused deposition in Fig. 2.25 as performed by Peineke [59], their diameters are 19 mm with 3 mm holes and 38 mm spacings between them. The distance between the impactor nozzle and the substrate was adjusted to approximately 3 mm. Figure 2.24c illustrates the effect of aerodynamic focusing, where the particle trajectory deviates from the flow trajectories due to the inertia that the particles possess. To focus and impact 5 nm particles, a pressure in the range of a few mbar is necessary downstream of the critical orifice. The sample holder is attached to an XYZ manipulator, which enables 3D printing of nanoparticulate structures.

Figure 2.26 shows the size of the spot obtained after a couple of seconds of bombardment with 5 nm Pd particles using five lenses. The slight asymmetry of the spot indicates some misalignment of the aerodynamic lenses, and improvement of the resolution would have been possible by optimizing the geometry with respect to the particle size to be focused.

A scanning electron microscopy (SEM) micrograph of a simple 3D pattern of Pd particles is seen in Fig. 2.25. The particles are produced by SA and printed onto a substrate by the impactor in Fig. 2.24 (left). The cracking, visible in higher magnification in Fig. 2.25

(right), is a consequence of reconstruction driven by surface energy reduction shortly after deposition. Successive printing of thin layers avoids cracks. If the substrate is not moved, a cone structure with a sharp tip can be produced, as shown in Fig. 2.27 (left). Sputtering with gold under SEM may lead to destruction of the cone, making fiber-like structures visible (Fig. 2.27, right). This indicates stronger adhesion in the vertical direction than in the lateral direction. Fiber or flake-like structures are bent outward to form a flower-like shape. Possibly, electrical charging effects under SEM promote this effect.

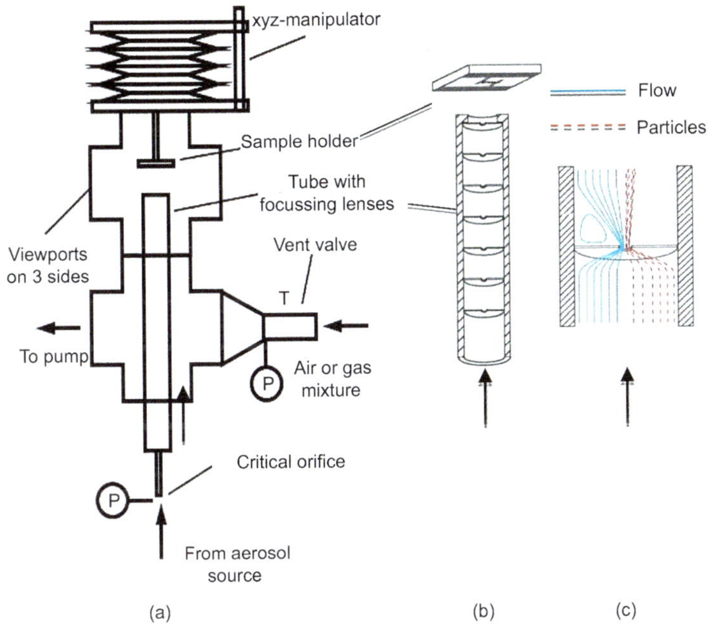

Figure 2.24 Focusing impactor with movable sampling stage for 3D printing of nanoparticulate structures (a). The series of "lenses" for aerodynamic focusing (b) and the mechanism of aerodynamic focusing (c) are illustrated.

The transmission electron microscopy (TEM) micrograph in Fig. 2.28 clearly reveals sintered structures along a certain direction. The explanation for vertical sintering is the impact energy, which locally leads to melting at the primary particle surfaces. Krypton adsorption measurements on such porous structures and layers determined a specific surface area ≥ 100 m^2/g, which is very close to the sum of the surface areas of 5 nm diameter primary particles, if the contact areas are neglected.

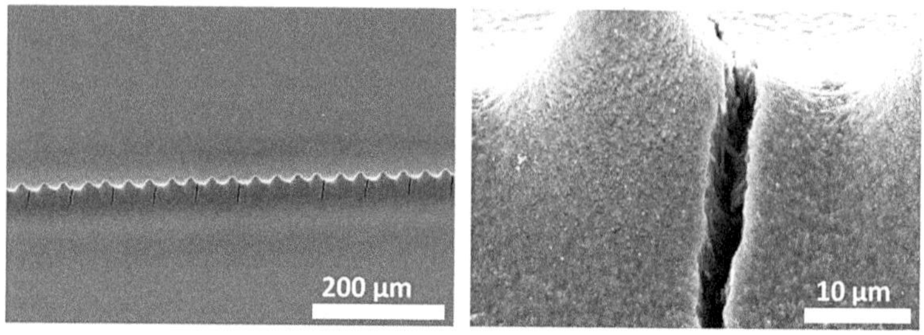

Figure 2.25 3D structure printed onto a substrate with the impactor shown in Fig. 2.24 [59].

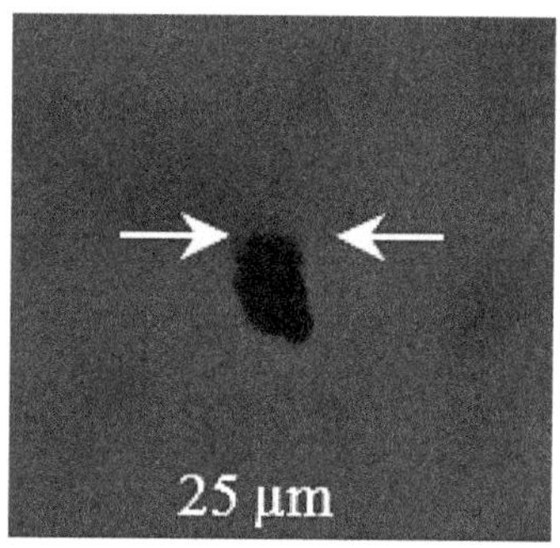

Figure 2.26 Optical micrograph Pd spot on sample holder when the XYZ stage is not moved.

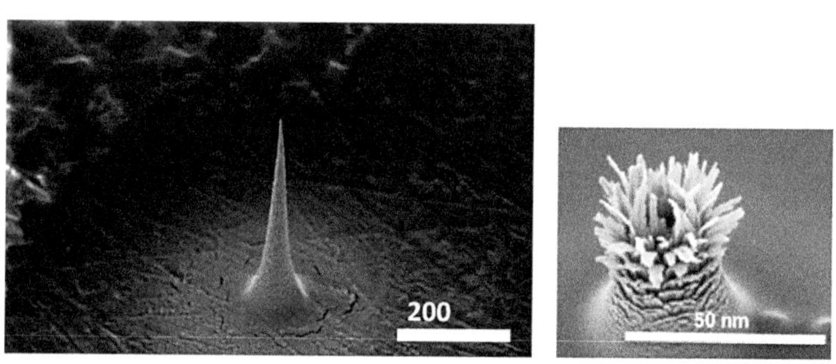

Figure 2.27 Tapered structure produced by depositing 5 nm particles on a single spot by a focusing impactor (left) and such a structure that has been destroyed by gold sputtering under SEM, revealing stronger adhesion in the vertical direction [59].

The vertical sintering indicates that apart from particle size and material, the impaction speed must have a strong influence on the structure of the deposit. In one extreme case, complete fusion is reached, as indicated by the "golden" color of the gold pattern impacted onto the contact lens in Fig. 1.2. In the case of isolated particles the color would be red (cf. Sections 2.4.1 and 2.4.2.3), and in the case of contacting agglomerated particles or powder layers, the multiplicity of plasmon absorption modes leads to a black appearance of the metal. The porous metal structures under the conditions, under which the micrographs in Figs. 2.25–2.28 were deposited, appeared black, indicating limited fusion.

When going down with the impaction velocity, there is a point where diffusion becomes more important than the vertical speed, and the deposits become isotropic, as those shown in Section 2.4.2.3. A publication from the group of Hogan (2014) [61] presents a useful regression formula, with which particulate film porosities can be predicted for the case of singlet particle deposition as a function of the relevant parameters. As an alternative to singlet particle deposition, agglomerates or aggregates (hard agglomerates) can be deposited if pre-agglomeration in the gas phase is allowed. Obvious applications (cf. Chapter 1) of the porous structures shown are catalysts (cf. Chapter 10), chemical sensors (cf. Chapter 11), and deposits for microelectronics (Section 1.4.5, Chapter 1).

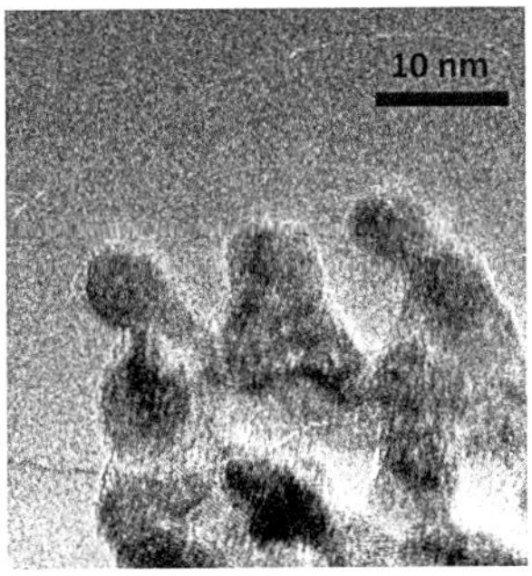

Figure 2.28 TEM close-up in a fibrous deposit, as shown in Fig. 2.27, confirming impact sintering in one direction [8]. Reprinted from Ref. [8], Copyright (2014), with permission from Elsevier.

2.4.2.3 Diffusional deposition onto surfaces

In aerosol technology, pure diffusional deposition is seldom made use of, because it is inefficient in most cases. However, in the size range below 20 nm, it is a very efficient mechanism of bringing particles to surfaces. The model of the Hogan group [61] can also be applied to predict porosity in this special case. For production of porous high-surface-area materials with a maximum of flexibility regarding the composition at the atomic scale as well as the nanoscale, SA can be combined with deposition on filter-like materials.

Polyvinyl alcohol (PVA) nanofibers were produced by electrospinning of the polymer solution and deposited on a coarse grid, and subsequently an aerosol stream carrying particles typically 5 nm in size was passed through this mesh, as described by van Vugt [62] (see Fig. 2.29). To enable fast transport of the particles to the fibers to avoid growth beyond 5 nm, the hollow electrode configuration (Fig. 2.2, right) is chosen. A small inert gas flow through the spark gap guarantees flushing of the gap and particle transport into the main flow. Due to diffusional motion, there is a high probability for each particle to collide with a fiber, where each collision normally leads to sticking. In Fig. 2.30 we see the porous nanoparticulate buildup on the fibers, which can grow to a thickness exceeding the fiber diameter by far. The dominance of diffusional transport with respect to convective transport is evident through the observation that the fiber coverage is almost as thick on the downstream side as on the upstream side. The high efficiency of diffusional deposition is visible by the fact that the fibers that are behind those facing the flow have a thinner layer deposited on them.

The particles coming from SA may consist of any (homogeneous) mixture of materials, and in addition, different aerosols can be passed through the mesh simultaneously, leading to mixtures on a nanoscale. For example, Au particles can be co-deposited with larger particles consisting of TiO_2, a typical carrier material for Au-based (photo) catalysts, as illustrated in Fig. 2.7d using two SDGs. The oxidation step can be done in the gas phase according to Fig. 2.6. This brings a virtually unlimited flexibility for producing catalysts. An example of elemental mapping by energy-dispersive X-ray spectroscopy (EDX) under TEM for a mixture of Au-TiO_2 is shown in Fig. 2.31, where the even distribution of Au and Ti is demonstrated [63]. Annealing leads to coalescence and stabilization of the structure. If the particles are oxides, the organic fibers can then be oxidized away by heating to about 500°C.

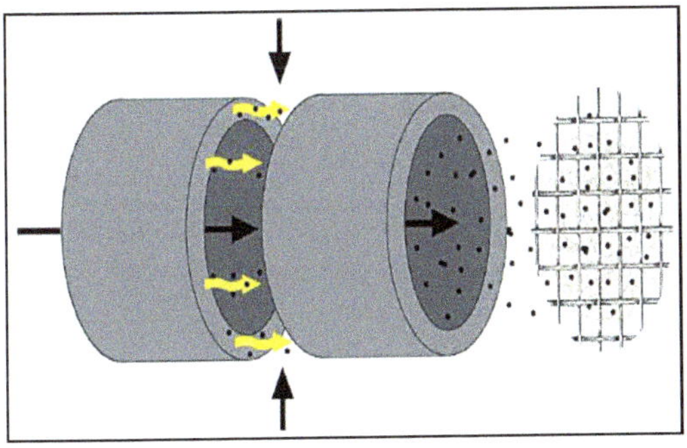

Figure 2.29 Production of Au nanoparticles in the hollow-electrode configuration of the VSPARTICLE G1 spark discharge generator with diffusional deposition on nanofibers. A radial flow through the spark gap guarantees efficient flushing of the space between the electrodes, and the flow through the electrodes enables fast particle transport to the point of deposition in order to limit particle growth [62].

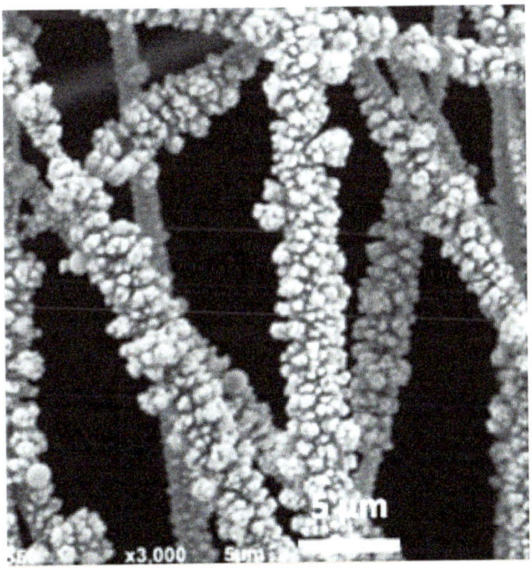

Figure 2.30 SEM picture of 100 nm polymer fibers that have been exposed to diffusional deposition of 5 nm gold particles [62].

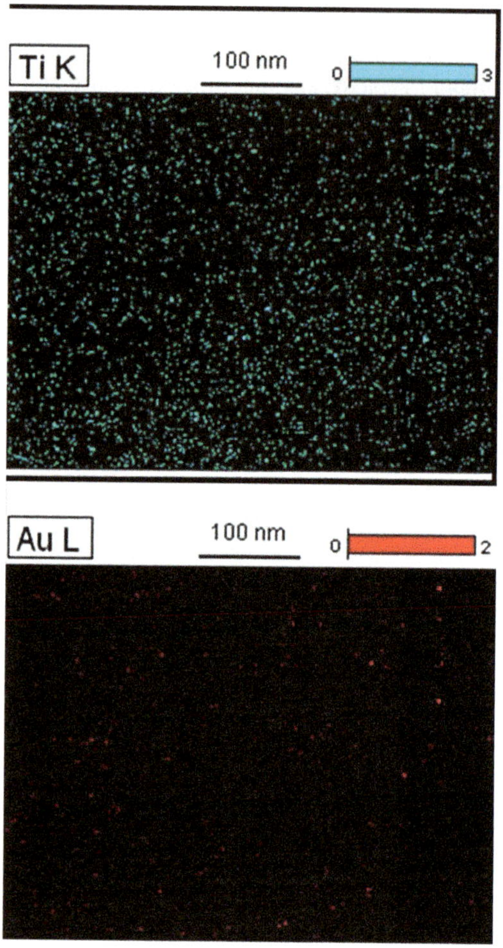

Figure 2.31 EDX elemental map showing distribution of TiO_2 particles (cyan dots, above) and Au particles (red dots, below) in the porous diffusionally deposited layer surrounding the fibers in Fig. 2.30 [63].

If Al particles and Au particles are simultaneously deposited without oxidizing the former, it can be oxidized after deposition. However, this does not form a structure that would be interesting for catalysis with the Au particles decorating the Al_2O_3 "carrier" particles. When exposed to ambient air, a run-away reaction occurs, in which enough heat is released to form a continuous Al_2O_3 phase with the gold particles distributed inside it. A TEM micrograph of this Au-Al_2O_3 nanocomposite is shown in Fig. 2.32 (left), where the

Au nanoparticles imbedded in the Al_2O_3 matrix have maintained the size they had at deposition. The red color of the material in Fig. 2.32 (right) is a consequence of the plasmon resonance and confirms that the singlet gold particles are kept apart by the oxide matrix.

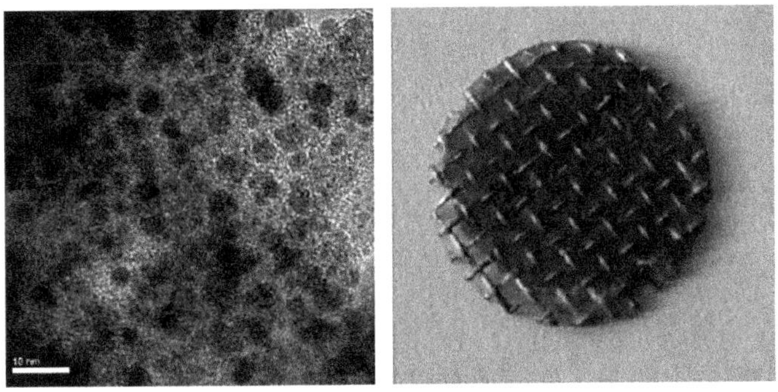

Figure 2.32 TEM micrograph (left) shows Au particles embedded in an Al_2O_3 matrix (white bar: 10 nm). The red color of the deposit on the photo (right) is due to the plasmon mode of noncontacting particles of that size [62].

2.4.2.4 Self-organization of very small particles on substrates

Atomic clusters or singlet particles from SA that are significantly smaller than the primary particle size that particles of the same material would obtain in an agglomerate are liquid-like at ambient temperature. They can be produced and deposited in a setup according to Fig. 2.33, where the mass production rate and the tube length are small enough and the flow rate is large enough to sufficiently limit particle growth. Figure 2.33 shows a flexible solution for deposition on flat surfaces making use of commercial components. In addition to the flexibility concerning material, it has a variable particle growth zone defining the final particle size followed by a deposition module (VSPARTICLE P1), which enables diffusional, electrostatic, or impaction deposition for sample preparation serving basic research. Diffusionally deposited on a smooth flat substrate, particles are usually quite mobile and perform a 2D diffusional motion. While they grow larger by collision and coalescence, their mobility on the substrate is reduced, and eventually their adhesion energy to the substrate becomes significantly larger than kT, so they stay at their positions. Figure 2.34 shows the resulting size distribution, as determined by high-resolution SEM (HRSEM) together with

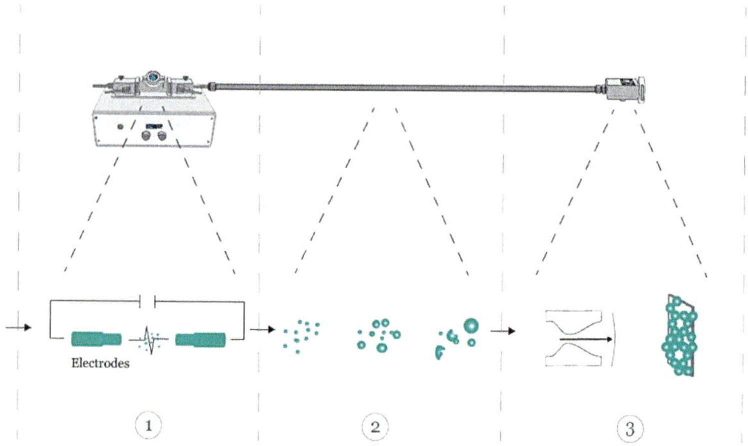

Figure 2.33 Setup for flexible production and deposition of nanoparticles comprising the VSP G1 and the VSP A3 deposition module. The growth zone (2) has a flexible length.

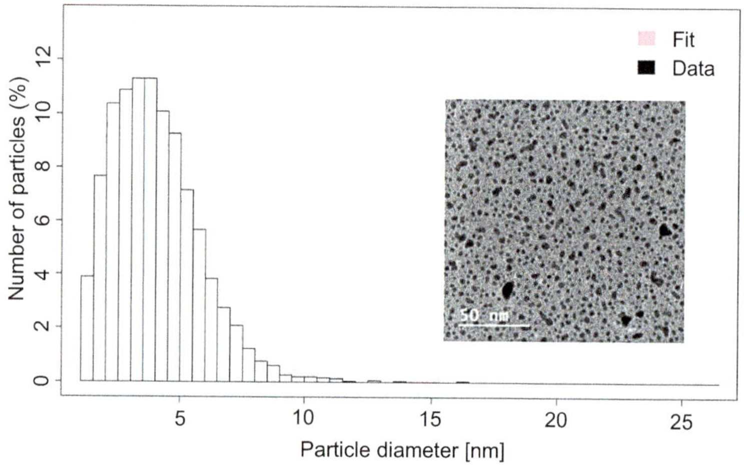

Figure 2.34 Dense layer of individual Au particles obtained by depositing particles presumably smaller than 1 nm in diameter on a Si surface.

the micrograph. The distribution is lognormal, as indicated by the fit. This is the kind of distribution expected from a cluster–cluster growth process, analogous to aerosol coagulation. If the particles are to grow above the critical size (primary particle size) at which they are liquid-like at room temperature, heating of the substrate induces coalescence to obtain "round" particles. This can also be induced by the electron beam under the HRSEM instrument. In our example, the average particle size is below that limit, but at the large end of the distribution, noncircular cross sections are seen. In summary, diffusional deposition of very small liquid-like particles on a flat substrate produces a high coverage of individual (singlet) particles without any dendritic agglomerates. In cases where particle growth fully takes place in the aerosol state, without coalescence on the substrate, fractal-like agglomerates are formed.

2.5 Milestones in Nanoparticle Production by Spark Ablation

The review above is far from complete. In Table 2.2, a brief but more complete overview of technological milestones in SA is given along with the respective references.

Table 2.2 Milestones in spark ablation (SA) and downstream technology

Year	Authors	Ref.	Milestone
1982	Burtscher, H., Schmidt-Ott, A.	2	First report of SA for gas-phase nanoparticle production
1986	H. Burtscher, A. Schmidt-Ott	64	SA combined with electrostatic size selection: equally-sized C particles
1986	H. Burtscher, A. Schmidt-Ott	64	Production of core-shell particles by controlled deposition of Perylene
1988	Schwyn, S., Garwin, E., Schmidt-Ott, A.	1	First detailed description of the SA method; production of 1.4 nm Au particles and 5 nm C particles
1991	Schleicher, B., Jung, T., Burtscher, H.	65	Deposition of equally sized metal particles from SA on highly oriented graphitic carbon (HOPG)

(*Continued*)

Table 2.2 (*Continued*)

Year	Authors	Ref.	Milestone
1991	Siekmann, H. R., Lüder, C., Faehrmann, J., Lutz, H. O., Meiwes-Broer, K.H.	66	Production of atomic cluster beam in vacuum by SA
1993	Helsper, C., Mölter, W., Löffler, F., Wadenpohl, C., Kaufmann, S.	30	First commercial spark discharge generator
1999	Weber, A. P., Seipenbusch, M., Thanner, C., Kasper, G.	67	Introduction of aerosol catalysis using SA
2000	Brown, J. S., Kim, C. S., Reist, P. C., Zeman, K. L., Bennett, W. D.	68	Production of radiolabeled particles
2004	Roth, C., Ferron, G.A., Karg, E., Lentner, B., Schumann, G., Takenaka, S. and Heyder, J.	12	Combining SA with chemical conversion in the aerosol state: synthesis of oxide particles
2007	Simonin, L., Lafont, U., Tabrizi, N., Schmidt-Ott, A. and Kelder, E	69	Oxidation of metal particles is avoided by closed-loop gas supply
2008	Byeon, J. H., Park, J. H., Yoon, K. Y., Hwang, J.	70	Patterning of SA particles on surface using thermophoresis
2009	Tabrizi, N.S., Ullmann, M., Vons, V.A., Lafont, U., Schmidt-Ott, A.	33	Material dependence of mass production rate modeled by Llewellyn-Jones approach
2009	Tabrizi, N.S., Xu, Q., Van Der Pers, N.M., Lafont, U., Schmidt-Ott, A.	17	Spark mixing: producing alloyed nanoparticles by using different metals for anode and cathode
2009	Byeon, J. H., Park, J. H., Yoon, K. Y. and Hwang, J.	71	Production of carbon-encapsulated metal particles using a metal and a carbon electrode
2009	Lafont, U., Simonin, L., Tabrizi, N.S., Schmidt-Ott, A., Kelder, E.M.	18	SA synthesis of alloyed particles by using electrodes of sintered mixed powder

Year	Authors	Ref.	Milestone
2009	Messing, M. E., Dick, K. A., Wallenberg, L. R. and Deppert, K.	72	Synthesis of nanowires by epitaxial growth from catalyst particles from SA
2010	Messing, M. E., Westerstrm, R., Meuller, B. O., Blomberg, S., Gustafson, J., et al.	73	Deposition of non-aggregated equally sized SA particles using mobility classification: model catalysts
2010	Tabrizi, N. S., Xu, Q., van der Pers, N. M., Schmidt-Ott, A.	19	SA synthesis of crystalline mixed nanoparticles from metals immiscible in the bulk
2011	Vons, V. A., de Smet, L. C. P. M., Munao, D., Evirgen, A., Kelder,et al.	74	SA synthesis of semiconductor particles
2011	Vons, V.A., Anastasopol, A., Legerstee, W.J., Mulder, F.M., Eijt, S.W.H. et al.	31	SA synthesis of pure metallic particles from highly oxidizable metal
2012	Saleh, E., Praeger, M., Vaughan, A. S., Stewart, W. and Loh, W. H.	75	Electrostatic focusing and writing with SA particles
2012	Han, K., Kim, W., Yu, J., Lee, J., Lee, H., et al.	76	The effect of modified electrode geometries on the nanoparticle output in SA
2012	Sengar, S. K., Mehta, B. R., Govind	77	Thermal restructuring of SA-produced agglomerates to reach spherical particle shape up to tens of nanometers
2013	Byeon, J. H. and Kim, Y.-W.	78	Combination of SA with spray-drying to produce photocatalytic TiO_2@ graphene nanoflakes
2013	Joe, Y. H., Ju, W., Park, J. H., Yoon, Y. H. and Hwang, J.	79	Antibacterial nanoparticles from SA immobilized in air filters

(Continued)

Table 2.2 (*Continued*)

Year	Authors	Ref.	Milestone
2014	Byeon, J. H. and Kim, Y.-W.	80	Synthesis of hollow Si-Fe particles
2014	Pfeiffer, T.V., Feng, J., Schmidt-Ott, A.	8	Strongly increased mass production rate of SA by applying electronic switching
2015	Palomares, J. M., Kohut, A., Galbács, G., Engeln, R., Geretovszky, Z.	81	Spectroscopic imaging of spark evolution
2015	Choi, H., Kang, S., Jung, W., ho Jung, Y., Park, S. J. et al.	82	Advanced patterning of nanoparticles from SA on surfaces by electrostatic focusing
2015	Mustonen, K., Laiho, P., Kaskela, A., Zhu, Z., Reynaud, O., et al.	83	Aerosol production of single-walled carbon nanotubes (SWCNTs) grown from SA particles
2015	Maisser, A., Barmpounis, K., Attoui, M. B., Biskos, G., Schmidt-Ott, A.	51	Atomic cluster production by SA and electrostatic size classification at atmospheric pressure
2015	Pfeiffer, T.V., Kedia, P., Messing, M.E., Valvo, M., Schmidt-Ott, A.	23	Introduction of "spark coating": particles to be coated passed by the vicinity of a spark, where vapor and atomic clusters deposit on them
2015	Feng, J., Biskos, G., Schmidt-Ott, A.	9	Analytical model predicting particle size from SA
2016	Noh, S. R., Kim, D. S., Park, S. J. and Choi, M.	84	Increased yield of charged particles from SA process by applying electric field
2016	Muntean, A., Wagner, M., Meyer, J., Seipenbusch, M.	20	Analysis of individual alloy nanoparticles confirms: composition identical to electrode composition

Year	Authors	Ref.	Milestone
2017	Kohut, A., Ludvigsson, L., Meuller, B.O., Deppert, K., Messing, et al.	38	Spectroscopic determination of the temperature and quenching rate of the spark plasma
2017	Lee, K., Choi, H., Kim, D. S., Jang, M. S., Choi, M.	85	Nanopatterning by ion- or electric-field-assisted aerosol lithography (EAAL)
2017	Valenti, M., Venugopal, A., Tordera, D., Jonsson, M. P., Biskos, G. et al.	86	Synthesis of alloyed plasmonic nanoparticles: enhancement of photoelectrochemical processes
2017	Sabzehparvar, M., Kiani, F. and Tabrizi, N. S.	87	Production of superparamagnetic particles
2017	Stein, M., Kruis, F.E.	4	Scale-up of discharge-based particle production by parallelization
2018	Domaschke, M., Schmidt, M., Peukert, W.	34	Influence of electrode diameter on mass production rate
2018	Livan, P. and Ozturk, T.	88	Systematic study on carbon encapsulation of metal particles
2018	Feng, J., Ramlawi, N., Biskos, G., Schmidt-Ott, A	36	A model for spark mixing
2018	Hallberg, R. T., Ludvigsson, L., Preger, C., Meuller, B. O., Dick, K. A. et al.	89	Prevention of oxidation in SA by adding H_2
2018	Schlag, L., Isaac, N., Nahrstedt, H., Johannes, T.R., Pezoldt, J. et al.		Self-aligning growth of SA-nanoparticle-based interconnects

2.6 Conclusion

SA is a precursor-less, scalable, versatile method of producing nanoparticles, mainly for the size domain below 20 nm. If suitable measures are taken, high purity can be achieved, which becomes more important the smaller the particles are. A simple electric circuit principle, where the spark itself acts as a switch, fulfills the main requirements for an SDG for research purposes, with a mass production rate up to 6 mg/h for Au [33] (corresponding to 40 mg/h for C; Table 2.1). For higher mass production rates, amounting to 0.5 g/h per electrode pair and/or for better controllability and reproducibility, a switching circuit [8] can be applied. The extreme quenching of the vapor cloud produced by each spark enables easy production of mixed particles that have the composition of the electrodes. Mixing on a nanoscale and coating is possible by combining two or more SA generators. The extreme quenching feature enables simplifying assumptions concerning particle growth, and models that exclusively describe a coagulation process successfully predict the particle size at the SDG's outlet. A simple formula providing an estimate of the particle size as a function of the generator parameters in the range between 2 and 20 nm is given. For applications, where a relative standard deviation of the size distribution of 1.35 is too broad, electrostatic mobility classification can be applied to obtain uniformly sized particles. As part of the particles coming from SA carry a single elementary charge, this can be done without applying any aerosol charging, another advantage with respect to purity. Porous or nonporous layers for sensors or catalysts easily form in diffusional or electrostatic particle deposition, and nanostructured patterns can be deposited/printed using electrostatic or aerodynamic focusing. The versatility in the composition and chemical nanostructure of such assembled nanostructures is unprecedented.

References

1. Schwyn, S., Garwin, E. and Schmidt-Ott, A. (1988). Aerosol generation by spark discharge, *J. Aerosol Sci.*, **19**, pp. 639–642.
2. Burtscher, H. and Schmidt-Ott, A. (1982). Enormous enhancement of van der Waals forces between small silver particles, *Phys. Rev. Lett.*, **48**, pp. 1734–1737.
3. Horvath, H. and Gangl, M. (2003). A low-voltage spark generator for production of carbon particles, *J. Aerosol Sci.*, **34**, pp. 1581–1588.

4. Stein, M. and Kruis, F. E (2018). Scaling-up metal nanoparticle production by transferred arc discharge, *Adv. Powder Technol.*, **29**, pp. 3138–3144.

5. Meuller, B. O., Messing, M. E., Engberg, D. L. J., Jansson, A. M., Johansson, L. I. M., Norlén, S. M., Tureson, N. and Deppert, K. (2012). Review of spark discharge generators for production of nanoparticle aerosols, *Aerosol Sci. Technol.*, **46**, pp. 1256–1270.

6. Ahmad, S., Laiho, P., Zhang, Q., Jiang, H., Hussain, A., Liao, Y., Ding, E. X., Wei, N. and Kauppinen, E. I. (2018). Gas phase synthesis of metallic and bimetallic catalyst nanoparticles by rod-to-tube type spark discharge generator, *J. Aerosol Sci.*, **123**, pp. 208–218.

7. Noh, S., Ryul, Lee, D., Park, S. J., Kim, D. S. and Choi, M. (2017). High throughput nanoparticle generation utilizing high-frequency spark discharges via rapid spark plasma removal, *Aerosol Sci. Technol.*, **51**, pp. 116–122.

8. Pfeiffer, T. V., Feng, J. and Schmidt-Ott, A. (2014). New developments in spark production of nanoparticles, *Adv. Powder Technol.*, **25**, pp. 56–70.

9. Feng, J., Biskos, G. and Schmidt-Ott, A. (2015). Toward industrial scale synthesis of ultrapure singlet nanoparticles with controllable sizes in a continuous gas-phase process, *Sci. Rep.*, **5**, art. no. 15788.

10. Schmidt-Ott, A. (1988). In situ measurement of the fractal dimensionality of ultrafine aerosol particles, *Appl. Phys. Lett.*, **52**(12), pp. 954–956.

11. Kruis, F. E, Fissan, H. and Rellinghaus, B. (2000). Sintering and evaporation characteristics of gas-phase synthesis of size-selected PbS nanoparticles, *Mater. Sci. Eng., B*, **69–70**, pp. 329–334.

12. Roth, C., Ferron, G. A., Karg, E., Lentner, B., Schumann, G., Takenaka, S. and Heyder, J. (2004). Generation of ultrafine particles by spark discharging, *Aerosol Sci. Technol.*, **38**, pp. 228–235.

13. Ghaemi, S., Schmidt-Ott, A. and Scarano, F. (2010). Nanostructured tracers for laser-based diagnostics in high-speed flows, *Meas. Sci. Technol.*, **21**, art. no. 105403.

14. Harra, J., Juuti, P., Haapanen, J., Sorvali, M., Roumeli, E., Honkanen, M., Vippola, M., Yli-Ojanperä, J. and Mäkelä, J. M. (2015). Coating of silica and titania aerosol nanoparticles by silver vapor condensation, *Aerosol Sci. Technol.*, **49**, pp. 767–776.

15. Vollath, D., Szabo, D. V. and Fuchs, J. (1999). Synthesis and properties of ceramic – polymer composites, *NanoStructured Mater.*, **12**, pp. 433–438.

16. Byeon, J. H. Park, J. H. and Hwang, J. (2008). Spark generation of monometallic and bimetallic aerosol nanoparticles, *J. Aerosol Sci.*, **39**, pp. 888–896.

17. Tabrizi, N. S., Xu, Q., van der Pers, N. M., Lafont, U. and Schmidt-Ott, A. 2009). Synthesis of mixed metallic nanoparticles by spark discharge, *J. Nanopart. Res.*, **11**(5), p. 1209.

18. Lafont, U., Simonin, L., Tabrizi, N. S., Schmidt-Ott, A. and Kelder, E. M. (2009). Synthesis of nanoparticles of Cu, Sb, Sn, Sb and Cu $_2$Sb by densification and atomization process, *J. Nanosci. Nanotechnol.*, **9**, pp. 2546–2552.

19. Tabrizi, N. S., Xu, Q., van der Pers, N. M. and Schmidt-Ott, A. (2010). Generation of mixed metallic nanoparticles from immiscible metals by spark discharge, *J. Nanopart. Res.*, **12**, pp. 247–259.

20. Muntean, A., Wagner, M., Meyer, J. and Seipenbusch, M. (2016). Generation of copper, nickel, and CuNi alloy nanoparticles by spark discharge, *J. Nanopart. Res.*, **18**, art. no. 229.

21. Feng, J., Geutjens, R., Thang, N. V., Li, J., Guo, X., Kéri, A., Basak, S., Galbács, G., Biskos, G., Nirschl, H., Zandbergen, H. W., Brück, E. and Schmidt-Ott, A. (2018). Magnetic phase transition in spark-produced ternary LaFeSi nanoalloys, *ACS Appl. Mater. Interfaces*, **10**, pp. 6073–6078.

22. Anastasopol, A., Pfeiffer, T. V., Schmidt-Ott, A., Mulder, F. M. and Eijt, S. W. H. (2011). Fractal disperse hydrogen sorption kinetics in spark discharge generated Mg/NbO_x and Mg/Pd nanocomposites, *Appl. Phys. Lett.*, **99**, art. no. 194103.

23. Pfeiffer, T. V., Kedia, P., Messing, M. E., Valvo, M. and Schmidt-Ott, A. (2015). Precursor-less coating of nanoparticles in the gas phase, *Materials*, **8**, pp. 1027–1042.

24. Knutson, E. O. and Whitby, K. T. (1975). Aerosol classification by electric mobility: apparatus, theory, and applications, *J. Aerosol Sci.*, **6**, pp. 443–451.

25. Fernández de la Mora, J. and Kozlowski, J. (2013). Hand-held differential mobility analyzers of high resolution for 1–30 nm particles: design and fabrication considerations, *J. Aerosol Sci.*, **57**, pp. 45–53.

26. Ehara, K., Hagwood, C. and Coakley, K. J. (1996). Novel method to classify aerosol particles according to their mass-to-charge ratio—aerosol particle mass analyser, *J. Aerosol Sci.*, **27**, pp. 217–234.

27. Tajima, N., Fukushima, N., Ehara, K. and Sakurai, H. (2011). Mass range and optimized operation of the aerosol particle mass analyzer, *Aerosol Sci. Technol.*, **45**, pp. 196–214.

28. Broßell, D., Valenti, M., Bezantakos, S., Schmidt-Ott, A. and Biskos, G. (2015). The nano-particle mass classifier (nano-PMC): development, characterization, and application for determining the mass, apparent density, and shape of particles with masses down to the zeptogram range, *Aerosol Sci. Technol.*, **49**, pp. 495–507.

29. Wentzel, M., Gorzawski, H., Naumann, K.-H., Saatho, H. and Weinbruch S. (2003). Transmission electron microscopical and aerosol dynamical characterization of soot aerosols, *J. Aerosol Sci.*, **34**, pp. 1347–1370.

30. Helsper, C., Mölter, W., Löffler, F., Wadenpohl, C. and Kaufmann, S. (1993). Investigations of a new aerosol generator for the production of carbon aggregate particles, *Atmos. Environ.*, **27A**, pp. 1271–1275.

31. Vons, V. A., Anastasopol, A., Legerstee, W. J., Mulder, F. M., Eijt, S. W. H. and Schmidt-Ott, A. (2011). Low-temperature hydrogen desorption and the structural properties of spark discharge generated Mg nanoparticles, *Acta Mater.*, **59**, pp. 3070–3080.

32. Llewellyn-Jones, F. (1950). Electrode erosion by spark discharges, *Br. J. Appl. Phys.*, **1**, 60–65.

33. Tabrizi, N. S., Ullmann, M., Vons, V. A., Lafont, U. and Schmidt-Ott, A. (2009). Generation of nanoparticles by spark discharge, *J. Nanopart. Res.*, **11**, pp. 315–332.

34. Domaschke, M., Schmidt, M. and Peukert, W. (2018). A model for the particle mass yield in the aerosol synthesis of ultrafine monometallic nanoparticles by spark ablation, *J. Aerosol Sci.*, **126**, pp. 133–142.

35. Feng, J., Biskos, G. and Schmidt-Ott, A. (2015). www.nature.com/articles/srep15788#supplementary-information.

36. Feng, J., Ramlawi, N., Biskos, G. and Schmidt-Ott, A. (2018). Internally mixed nanoparticles from oscillatory spark ablation between electrodes of different materials, *Aerosol Sci. Technol.*, **52**, pp. 505–514.

37. Stein, M. and Kruis, F. E. (2016). Optimization of a transferred arc reactor for metal nanoparticle synthesis, *J. Nanopart. Res.*, **18**, art. no. 258.

38. Kohut, A., Ludvigsson, L., Meuller, B. O., Deppert, K., Messing, M. E., Galbács, G. and Geretovszky, Z. (2017). From plasma to nanoparticles: optical and particle emission of a spark discharge generator, *Nanotechnology*, **28**, art. no. 475603.

39. Kohut, A., Wagner, M., Seipenbusch, M., Geretovszky, Z. and Galbács, G. (2018). Surface features and energy considerations related to the erosion processes of Cu and Ni electrodes in a spark discharge nanoparticle generator, *J. Aerosol Sci.*, **119**, pp. 51–61.

40. Feng, J. (2016). PhD thesis, Delft University of Technology, https://repository.tudelft.nl/islandora/object/uuid:fb6c0122-587b-471d-8009-b52ef9b69b07?collection=research.

41. Kodas, T. and Hampden-Smith, M. (1999). *Aerosol Processing of Materials* (Wiley, USA).

42. Dahneke, B. (1983). Simple kinetic theory of Brownian diffusion in vapors and aerosols, in Meyer, R. E. (ed.) *Theory of Dispersed Multiphase Flow,* Academic Press, New York, pp. 97–133.

43. Otto, E., Fissan, H., Park, S. and Lee, K. (1999). The log-normal size distribution theory of brownian aerosol coagulation for the entire particle size range: Part II. Analytical solution using dahneke's coagulation kernel, *J. Aerosol Sci.*, **30**, pp. 17–34.

44. Park, S., Lee, K., Otto, E. and Fissan, H. (1999). The log-normal size distribution theory of brownian aerosol coagulation for the entire particle size range: Part I. Analytical solution using the harmonic mean coagulation kernel, *J. Aerosol Sci.*, **30**, pp. 3–16.

45. Ouyang, H., Gopalakrishnan, R. and Hogan, J. C. (2012). Nanoparticle collisions in the gas phase in the presence of singular contact potentials, *J. Chem. Phys.*, **137**, art. no. 064316.

46. http://dx.doi.org/10.1063/1.4742064.

47. Gotsmann, B. and Fuchs, H. (2001). Dynamic force spectroscopy of conservative and dissipative forces in an Al-Au(111) tip-sample system, *Phys. Rev. Lett.*, **86**, pp. 2597–2600.

48. Yu, M. and Yueyan, L. (2016). Methods of moments for resolving aerosol dynamics, in Volkov, K. (ed.) *Aerosols: Science and Case Studies*, InTech, Rijeka, Chapter 2.

49. Rogak, S. N. and Flagan, R. C. (1992). Coagulation of aerosol agglomerates in the transition regime, *J. Colloid Interface Sci.*, **151**, pp. 203–224.

50. Lee, K. W., Chen, J. and Gieseke, J. A. (1984). Log-normally preserving size distribution for Brownian coagulation in the free-molecule regime, *Aerosol Sci. Technol.* **3**, p. 53.

51. Maisser, A., Barmpounis, K., Attoui, M. B., Biskos, G. and Schmidt-Ott, A. (2015). Atomic cluster generation with an atmospheric pressure spark discharge generator, *Aerosol Sci. Technol.*, **49**, pp. 886–894.

52. Schmidt-Ott, A., Baltensperger, U., Gäggeler, H. W. and Jost, D. T. (1990). Scaling behaviour of physical parameters describing agglomerates, *J. Aerosol Sci.*, **21**(6), pp. 711–717.

53. Liu, B. Y. H., Whitby, K. T. and Yu, H. H. S. (1967). Electrostatic aerosol sampler for light and electron microscopy, *Rev. Sci. Instrum.*, **38**, pp. 100–102.

54. Methling, R.-P., Senz, V., Klinkenberg, E.-D., Diederich, T., Tiggesbäumker, J., Holzhüter, G., Bansmann, J. and Meiwes-Broer, K. (2001). Magnetic studies on mass-selected iron particles, *Eur. Phys. J. D*, **16**, pp. 173–176.

55. Zonnevylle, A. C., Hagen, C. W., Kruit, P. and Schmidt-Ott, A. (2009). Directed assembly of nano-particles with the help of charge patterns created with scanning electron microscope, *Microelectron. Eng.*, **86**, pp. 803–805.

56. Fernandez de la Mora, J. (1996). Drastic improvements on the resolution of aerosol size spectrometers via aerodynamic focusing: the case of variable-pressure impactors, *Chem. Eng. Commun.*, **151**, pp. 101–124.

57. Fernandez de la Mora, J., Liedtke, L. and Schmidt-Ott, A. (2003). Mass and size determination of nanometer particles by means of mobility analysis and focussed impaction, *J. Aerosol Sci.*, **34**, pp. 79–98.

58. Wang, X., Kruis, F. E. and McMurry, P. H. (2005). Aerodynamic focusing of nanoparticles: I. Guidelines for designing aerodynamic lenses for nanoparticles, *Aerosol Sci. Technol.*, **39**, p. 611.

59. Peineke, C. (2008). Production and deposition of well defined aerosol nanoparticles for studies on basic properties, PhD thesis, Delft University of Technology, Delft.

60. Peineke, C. and Schmidt-Ott, A. (2007). Highly porous nanostructured materials from impacted nanoparticles, *Proc. PARTEC*, Nuremberg.

61. Lindquist, G. J., Pui, D. Y. H. and Hogan, C. J. (2014). Porous particulate film deposition in the transition regime, *J. Aerosol Sci.*, **74**, pp. 42–51.

62. van Vugt, A. W. (2014). A clean and flexible method of catalyst synthesis by efficient deposition of nanoparticles on electrospun nanofibres (Unpublished master's thesis), Delft University of Technology, Delft, Netherlands.

63. Braams, P. (2015). A clean and flexible catalyst synthesis method (Unpublished master's thesis), Delft University of Technology, Delft, Netherlands.

64. Burtscher, H. and Schmidt-Ott, A. (1986). In situ measurement of adsorption and condensation of a polyaromatic hydrocarbon on ultrafine C particles by means of photoemission, *J. Aerosol Sci.*, **17**(3), p. 456.

65. Schleicher, B., Jung, T., Hug, H. and Burtscher, H. (1991). Ultrafine particles adsorbed on HOPG measured by STM, *Z. Phys. D: At. Mol. Clusters*, **19**(4), pp. 327–331.

66. Siekmann, H. R., Lüder, C., Faehrmann, J., Lutz, H. O. and MeiwesBroer, K. H. (1991). The pulsed arc cluster ion source (PACIS), *Z. Phys. D: At. Mol. Clusters*, **20**(1), pp. 417–420.

67. Weber, A. P., Seipenbusch, M., Thanner, C. and Kasper, G. (1999). Aerosol catalysis on nickel nanoparticles, *J. Nanopart. Res.*, **1**(2), pp. 253–265.

68. Brown, J. S., Kim, C. S., Reist, P. C., Zeman, K. L. and Bennett, W. D. (2000). Generation of radiolabeled "soot-like" ultrafine aerosols suitable for use in human inhalation studies, *Aerosol Sci. Technol.*, **32**(4), pp. 325–337.

69. Simonin, L., Lafont, U., Tabrizi, N., Schmidt-Ott, A. and Kelder, E. (2007). Sb/o nano-composites produced via spark discharge generation for Li-ion battery anodes, *J. Power Sources*, **174**(2), pp. 805–809.

70. Byeon, J. H., Park, J. H., Yoon, K. Y. and Hwang, J. (2008b). Site-selective catalytic surface activation via aerosol nanoparticles, for use in metal micropatterning, *Langmuir*, **24**(11), pp. 5949–5954.

71. Byeon, J. H., Park, J. H., Yoon, K. Y. and Hwang, J. (2009). Ambient spark generation to synthesize carbon-encapsulated metal nanoparticles in continuous aerosol manner, *Nanoscale*, **1**, pp. 339–343.

72. Messing, M. E., Dick, K. A., Wallenberg, L. R. and Deppert, K. (2009). Generation of size-selected gold nanoparticles by spark discharge—for growth of epitaxial nanowires, *Gold Bull.*, **42**(1), pp. 20–26.

73. Messing, M. E., Westerstrm, R., Meuller, B. O., Blomberg, S., Gustafson, J., Andersen, J. N., Lundgren, E., van Rijn, R., Balmes, O., Bluhm, H. and Deppert, K. (2010). Generation of Pd model catalyst nanoparticles by spark discharge, *J. Phys. Chem. C*, **114**(20), pp. 9257–9263.

74. Vons, V. A., de Smet, L. C. P. M., Munao, D., Evirgen, A., Kelder, E. M. and Schmidt-Ott, A. (2011b). Silicon nanoparticles produced by spark discharge, *J. Nanopart. Res.*, **13**(10), p. 4867.

75. Saleh, E., Praeger, M., Vaughan, A. S., Stewart, W. and Loh, W. H. (2012). The direct writing and focusing of nanoparticles generated by an electrical discharge, *J. Nanopart. Res.*, **14**(11), p. 1220.

76. Han, K., Kim, W., Yu, J., Lee, J., Lee, H., Woo, C. G. and Choi, M. (2012). A study of pin-to-plate type spark discharge generator for producing unagglomerated nanoaerosols, *J. Aerosol Sci.*, **52**, pp. 80–88.

77. Sengar, S. K., Mehta, B. R. and Govind (2012). Size and alloying induced changes in lattice constant, core, and valance band binding energy in Pd-Ag, Pd, and Ag nanoparticles: effect of in-flight sintering temperature, *J. Appl. Phys.*, **112**(1), p. 014307.

78. Byeon, J. H. and Kim, Y.-W. (2013). Gas-phase self-assembly of highly ordered titania@graphene nanoflakes for enhancement in photocatalytic activity, *ACS Appl. Mater. Interfaces*, **5**(9), pp. 3959–3966.

79. Joe, Y. H., Ju, W., Park, J. H., Yoon, Y. H. and Hwang, J. (2013). Correlation between the antibacterial ability of silver nanoparticle coated air filters and the dust loading, *Aerosol Air Qual. Res.*, **13**(3), pp. 1009–1018.

80. Byeon, J. H. and Kim, Y.-W. (2014). Ambient plasma synthesis of Si-Fe hollow nanoparticles and their biocompatibility and lithium storage capacity, *Adv. Mater. Interfaces*, **1**(5), p. 1300134.

81. Palomares, J. M., Kohut, A., Galbcs, G., Engeln, R. and Geretovszky, Z. (2015). A time-resolved imaging and electrical study on a high current atmospheric pressure spark discharge, *J. Appl. Phys.*, **118**(23), p. 233305.

82. Choi, H., Kang, S., Jung, W., ho Jung, Y., Park, S. J., Kim, D. S. and Choi, M. (2015). Controlled electrostatic focusing of charged aerosol nanoparticles via an electrified mask, *J. Aerosol Sci.*, **88**, pp. 90–97.

83. Mustonen, K., Laiho, P., Kaskela, A., Zhu, Z., Reynaud, O., Houbenov, N., Tian, Y., Susi, T., Jiang, H., Nasibulin, A. G. and Kauppinen, E. I. (2015). Gas phase synthesis of non-bundled, small diameter single-walled carbon nanotubes with near-armchair chiralities, *Appl. Phys. Lett.*, **107**(1), p. 013106.

84. Noh, S. R., Kim, D. S., Park, S. J. and Choi, M. (2016). Enhanced yield of positively charged particles from a spark discharge generator via in situ corona discharges, *J. Aerosol Sci.*, **101**, pp. 188–195.

85. Lee, K., Choi, H., Kim, D. S., Jang, M. S. and Choi, M. (2017). Vertical stacking of three-dimensional nanostructures via an aerosol lithography for advanced optical applications, *Nanotechnology*, **28**(47), p. 475302.

86. Valenti, M., Venugopal, A., Tordera, D., Jonsson, M. P., Biskos, G., Schmidt-Ott, A. and Smith, W. A. (2017). Hot carrier generation and extraction of plasmonic alloy nanoparticles, *ACS Photonics*, **4**(5), pp. 1146–1152.

87. Sabzehparvar, M., Kiani, F. and Tabrizi, N. S. (2018). Spark discharge generation of superparamagnetic nickel oxide nanoparticles, *Mater. Today*, **5**(7), Part 3, pp. 15821–15827.

88. Livan, P. and Ozturk, T. (2018). Carbon encapsulation of elemental nanoparticles by spark discharge, *J. Mater. Sci.*, **53**(20), pp. 14350–14360.

89. Hallberg, R. T., Ludvigsson, L., Preger, C., Meuller, B. O., Dick, K. A. and Messing, M. E. (2018). Hydrogen-assisted spark discharge generated metal nanoparticles to prevent oxide formation, *Aerosol Sci. Technol.*, **52**(3), pp. 347–358.

90. Schlag, L., Isaac, N., Nahrstedt, H., Johannes, T. R., Pezoldt, J. and Jacobs, H. (2018). Self-aligning growth of nanoparticle-based interconnects, *IEEE 13th Nanotechnology Materials and Devices Conference (NMDC)*, pp. 1–4.

Chapter 3

Spark Plasma Diagnostics

Attila Kohut,[a,b] Gábor Galbács,[b,c] and Zsolt Geretovszky[a,b]
[a]*Department of Optics and Quantum Electronics, University of Szeged, Dóm sq. 9, Szeged, H-6720, Hungary*
[b]*Department of Materials Science, Interdisciplinary Excellence Centre, University of Szeged, Dugonics sq. 13, Szeged, H-6720, Hungary*
[c]*Department of Inorganic and Analytical Chemistry, University of Szeged, Dóm sq. 7, Szeged, H-6720, Hungary*
galbx@chem.u-szeged.hu; gero@physx.u-szeged.hu

The present chapter discusses the basic methodology of optical plasma diagnostics and the information that can be gained by its application to nanoparticle spark discharge generators (SDGs). To obtain a good understanding of the information obtained by these methods, we explain basic features of the spark discharge used for nanoparticle production here, some of which are treated in more detail in Chapter 4.

3.1 Introduction

Spark discharge is a rapid, transient process that occurs when the voltage between two electrodes reaches the breakdown voltage of

Spark Ablation: Building Blocks for Nanotechnology
Edited by Andreas Schmidt-Ott

ISBN 978-981-4800-82-2 (Hardcover), 978-0-367-81709-1 (eBook)
www.jennystanford.com

the electrode gap. The voltage required for the breakdown is quite high, usually in the kV to tens of kV range, and depends mainly on the quality and pressure (p) of the gas, as well as the interelectrode gap (d). When the breakdown voltage is reached, a conducting channel (spark plasma channel) will bridge the two electrodes. In small gaps and/or at low pressures (up to ca. $p \cdot d < 1000$ Torr·cm) the breakdown mechanism can be described by the multiplication of avalanches theory, also known as the Townsend mechanism [1] (see Chapter 4). The formation of the spark is accompanied by a cracking sound, generated by the shock wave, which is a steep rise in pressure caused by the intense release of Joule heat when high current (in the hundreds to thousands ampere range) passes through the spark channel. Concomitantly, the spark channel also emits strong electromagnetic radiation.

The characteristics of a spark discharge can be very different, depending on the properties of the discharge circuit. An important distinction between different types of sparks can be made via the presence or absence of local thermodynamic equilibrium (LTE) in the discharge plasma. Spark plasmas initiated under atmospheric pressure in a discharge circuit of ~nF capacitance, small resistance (in the order of ohms) and small gaps (in the order of millimeters)—conditions typical to spark discharge generators (SDGs)—are considered to be in LTE, and hence follow the laws of equilibrium thermodynamics and can be characterized by a single temperature at each point of space (i.e., the electron and heavy particle temperatures are locally equal) [2] (see also Chapter 4).

The morphological, electrical, and optical characteristics of spark plasmas can vary greatly. For example, analytical spark plasmas used in atomic spectroscopy are constructed to repetitively atomize and thermally excite a relatively small amount of electrode material (sample) in a very reproducible way and therefore are optimized for high temperature, strong emission, good precision. This is often achieved by using unipolar discharge circuits [3, 4]. Conversely, plasmas in SDGs are typically optimized for efficient electrode material removal, which is usually achieved by continuous operation of free running, bipolar, oscillatory discharge circuits [5].

In the fundamental research of plasmas, it is customary to use a set of experimental tools to obtain diagnostic data. These tools include electric or magnetic probes (e.g., Lagmuir probes or ballpen probes), active or passive spectroscopy (based on absorption, emission, or fluorescence measurement modes), and imaging. In the case of spark discharge plasmas, measurement of the current in the discharge loop or the voltage in the electrode gap also provides valuable information. Due to the highly transient nature of spark plasmas (in both space and time), these measurements also have to be performed with spatial and temporal resolution, typically at submillimeters and submicroseconds, respectively. The tools have to be chosen so that (i) they exert minimum disturbance to the plasma under study and (ii) they are robust, so SDGs cause little, if any, interference in their operation. These requirements are most easily met by using optical tools of diagnostics, that is, via spectrometers and cameras.

In the following sections, we will discuss some of the experimental considerations in connection with spark plasma diagnostics by optical techniques and introduce the major plasma diagnostic methods, based on noninvasive passive spectroscopy and imaging, and applicable to spark discharge plasmas. Last, but not least, the main properties of spark plasmas used for nanoparticle (NP) generation and their evolution in time and space will also be discussed.

3.2 Primary Stages in the Time Evolution of Spark Plasmas

The formation of a spark discharge plasma starts by the so-called breakdown of the gas environment. This breakdown process is facilitated by charge carriers randomly generated by cosmic rays or natural radioactivity. If an electric field is present, due to voltage applied to the electrodes, these electrons and ions will be attracted to the electrodes of opposing polarity and a weak electric current will flow. If the voltage is increased above a certain threshold level, the ionization of the gas and hence the current will increase drastically, accompanied by an intense light emission. The specifics of the electrical (discharge) circuit and the ambient gas determine how the discharge proceeds after the breakdown. The main discharge types that can form are glow, arc, spark, and corona discharges [1].

The temporal evolution of high-current, microsecond-long sparks is generally characterized by four stages: (i) prebreakdown, (ii) breakdown, (iii) arc, and (iv) afterglow [6, 7]. These four stages have distinctly different physical characteristics and, as a consequence, different contributions to the ablation of the electrode material and hence to NP production.

3.2.1 Prebreakdown and Breakdown Stages

Electrons, emitted from the cathode or formed in the gas due to ionizing radiation, gain sufficient energy to initiate collisional ionization. If the electric field of the space charge of the primary electron avalanche is sufficiently high, a thin conducting channel is formed, which bridges the electrodes [1]. In SDG setups, where a few-millimeter-long electrode gap is used at atmospheric pressure, the Townsend mechanism is likely to occur [1]. It is hard to make a sharp distinction between the prebreakdown and the following breakdown stages. Both of these stages are characterized by the optical emission of the ambient gas, but when, for example, sparks are initiated in nitrogen, the emission spectrum during the prebreakdown stage is dominated by molecular bands of N_2, whereas in the breakdown stage, the main emitters are ions of the gas accompanied by an intense continuous radiation [6]. The breakdown stage completes when a conducting channel bridges the electrodes, allowing high current to flow through the gap.

The temporal length of the breakdown and prebreakdown stages depends on the electrical parameters of the discharge circuit—most notably on its net inductance. It has been shown that the smaller the inductance, the shorter the duration of the breakdown [7] (see also Chapters 2 and 5). SDGs usually keep the inductance to a minimum (e.g., only parasitic inductance of the cabling is present) [5, 8], which suggests that the breakdown of the electrode gap should complete very quickly, typically on the nanosecond timescale. The breakdown was proven to complete in less than 100 ns both by time-resolved emission spectroscopy and by electric measurements [9, 10].

3.2.2 Arc Stage

The discharge enters the arc stage when a large electrical current (in excess of hundreds of amperes) builds up between the electrodes, thereby also causing a large drop in the gap voltage. The large current locally heats up the electrodes at the entry points of the current filaments, which leads to the erosion (evaporation, sputtering, ablation, etc.) of the electrode material [11]. Hence, the arc stage is optically characterized by a concomitant light emission of the species of the ambient gas and the electrode material [6]. The end of the arc stage is defined by the moment when the electrical current ceases, that is, up to a few tens of microseconds after the start of the discharge, which is oscillatory, except for the case when a damping resistor is added to the circuit, as described in Chapter 5.

The arc stage took its name from the stationary arc discharge due to their similarities. Despite its transient nature, under the conditions typical for SDGs (most importantly at high currents and atmospheric pressure) spark plasma reaches LTE in typically less than a microsecond after the onset of the breakdown [6, 12, 13].

3.2.3 Afterglow Stage

Light emission from the spark gap does not cease when the electrical current terminates, but emission can still be observed for a relatively long time afterward. This late emission may originate from metastable species of the ambient gas and/or excited atomic species (primarily of the electrode material) lingering in and around the spark gap. The emission in this stage is facilitated by the lower gas temperature and the decreased pressure behind the spark generated shock wave, which jointly reduce the collision rate and hence the rate of de-excitation. The afterglow stage, by definition, lasts until the emission of the spark fully ceases, which occurs typically somewhere between tens of microseconds and a couple of milliseconds [6, 9, 10]. Since the spark plasma recombines during the afterglow stage, the fulfilment of LTE conditions is no longer granted. Nevertheless, LTE may hold while the electron concentration is sufficiently high [10].

3.3 Experimental Considerations Related to Spark Plasma Monitoring

Popular SDG designs usually operate in a "free running" discharge mode, that is, without any means of electronic regulation that would control when and at what energy the breakdown will set off [5]. This allows for keeping the total resistance and inductance of the discharge circuit minimal, which is beneficial from an energetic point of view [14]; however, it also leads to less control over the initiation of individual spark events. This is usually not an issue during NP production but poses a serious challenge when plasma diagnostics is attempted. Since sparking is a transient process occurring at the nanosecond–microsecond timescale, this means that nanosecond temporal resolution is needed for plasma monitoring investigations. Several experimental techniques are available that possess the required high temporal resolution, such as laser interferometry [15], Schlieren shadowgraphy [16], and rotating slit optical emission spectroscopy [17, 18].

In recent years, spark plasmas have been mostly investigated by fast imaging and temporally resolved emission spectroscopy [9, 10, 19]. The central instrument in these measurements is an intensified charge-coupled device (ICCD) camera, either as a stand-alone imaging device or as a detector of a spectrometer arrangement. The high time resolution of the experiments requires fast data collection electronics and precise synchronization between the experimental equipment and the sparking process. Due to the above-mentioned unregulated fashion of conventional sparking, the triggering of the devices can only be done passively by using a properly chosen electrical or optical signal. This trigger signal is usually formed either from the sharp voltage drop or from the high current peak associated with the initiation of the discharge (i.e., formation of the conducting channel) [9, 10]. Alternatively, some spectroscopy signal (e.g., abrupt increase in emission intensity or absorption at a particular wavelength characteristic of the atoms or ions of the ambient gas) may also be used.

As an illustration, Fig. 3.1 schematically shows how synchronization of an ICCD camera with respect to, for example, the voltage signal can be arranged. Please also note that ICCD cameras have a low readout rate. Even though the exposure time can be as short as a few nanoseconds, the readout of images acquired by the camera can only occur a few times in every second. Thus, the time evolution of a spark plasma can only be monitored in a sequential way when each ICCD frame is taken with a progressively increasing gate delay, but of consecutive spark events. This necessitates highly reproducible sparking.

It has to be borne in mind that the use of electrical signals for triggering will not allow the study of the (pre)breakdown stage of the spark plasma, which precedes the birth of these electrical signals. In addition to this, optical signals propagate (e.g., through a fiber optic cable) significantly faster than electrical signals in a conductor. The overall time delay of electronic triggering, in typical coaxial cables, is in the range of several tens of nanoseconds, which is a value similar to that of the wake-up time (insertion delay) of electronic instruments.

A word of caution is also due here with respect to the electromagnetic interference (EMI) caused by an oscillatory spark plasma. The strong, high-frequency oscillating current that flows in the discharge circuit practically makes the electrical gap (and the discharge circuit) an antenna, which emits strong electromagnetic impulses in the radiofrequency range [20]. This can well influence the operation (or even endanger the integrity) of any fine electronic instrumentation (cameras, data collection devices, oscilloscopes, computers, etc.) placed near the SDG. In addition to this, the strong magnetic field generates induction currents in electrical signal cables, leading to spurious signal spikes. In our own experience, EMI effects are present in a few meters' (up to 5–10 m) distance from the SDG. Extensive cable shielding (earthing), encasing the SDG in a Faraday cage, keeping a safe distance, etc., can help to reduce EMI, but complications should be expected, as its frequency range is high. Magnetic shielding is also very complicated and inefficient under these conditions, so the best solution in plasma monitoring is to use optical signals guided by fiber optic cables. This introduces some time delay in data acquisition (e.g., a 10 m fiber optic cable gives rise to about a 35–50 ns delay), but luckily this, as well as the insertion delay of (opto)electronic instrumentation, can be compensated for by storing the optical signal in an optical delay line.

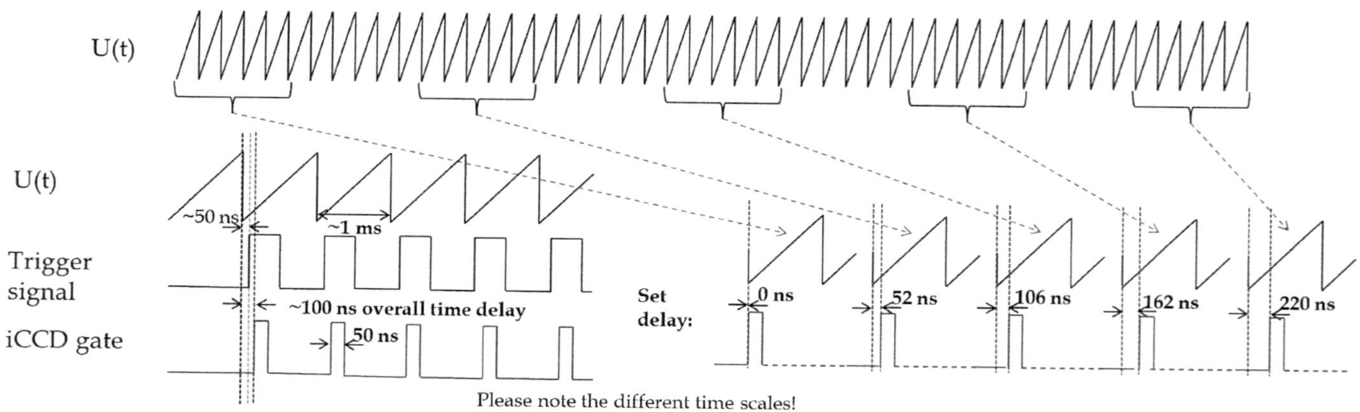

Figure 3.1 Illustrative timing diagram for an ICCD camera with respect to the voltage signal of the sparking ($U(t)$) in the particular case when a 50 ns gate width is used with varying delay.

3.4 The Methodology of Spark Plasma Diagnostics

The overall objective of plasma diagnostics is to obtain information about key physical plasma parameters and the concentration of species, as well as their spatial profiles and dynamics—or plasma properties for short. Consequently, plasma modeling and diagnostic calculations work with a range of parameters, distributions, and profiles, such as temperature, pressure, magnetic and electric field strengths, particle fluxes, concentration of particles, etc. Technically speaking, of course there is no universal technique that could provide all this information; therefore a pool of methods is employed, and their output information is combined. The choice of the employed methods is made also with regard to the type of plasma and the invasiveness of the method [21–24].

Within the frame of the present section, we will give a brief overview of the most basic plasma diagnostic techniques based on noninvasive passive spectroscopy and imaging and applicable to spark discharge plasmas. The organization of the subsections is according to the plasma parameter obtainable. We also present an elementary description of the equilibria assumed and calculations needed for obtaining the intended information. An important simplification, generally attempted in the related literature, is the assumption of LTE in the spark plasma. This assumption, as it will be illustrated by our results in a later section, is valid for practically the whole arc and some part of the afterglow stages of the discharge, which play the most important role in NP generation. It should also be emphasized that most diagnostic techniques require optically thin plasma, one that is nearly homogeneous along the line of sight and remains steady state for the duration of observation. If this condition cannot be fulfilled, spatial and temporal resolution of the plasma is a must or some inversion technique (e.g., Abel inversion) also needs to be applied.

3.4.1 Morphology

Imaging is a relatively simple technique, which is essentially based on the recording of the spatial and/or temporal distribution of plasma emission, with or without wavelength selection. If used without wavelength selection, imaging can provide information about the morphological changes of the spark discharge occurring over time. In addition to the shape and symmetry of the discharge plasma, imaging data also allows the determination of the spatial position of the plasma filament(s), expansion of the plasma channel, the duration of the discharge, etc. In case it is used with wavelength selection, imaging can contribute to the selective determination of the distribution of chemical species (atoms, ions or molecules, or heavy particles in general), as well as their concentration and velocity. Please note that to obtain such data, it is usually necessary to perform point-by-point (or pixel-by-pixel) evaluation of the imaging data using calculations described in Sections 3.4.2–3.4.4. For the purposes of this methodological chapter, we refer to this above large pool of information as "morphology."

As an illustration to the above said things, snapshots are shown in Fig. 3.2 that illustrate the spark morphology at the beginning of the arc stage. All images were taken under identical experimental conditions on different spark events. In this particular case the reproducibility of the shape of the emitting channel was monitored by imaging and shown to vary quite a bit. In a similar manner, fast imaging can also be used to follow the variation of plasma morphology in space, time, or both.

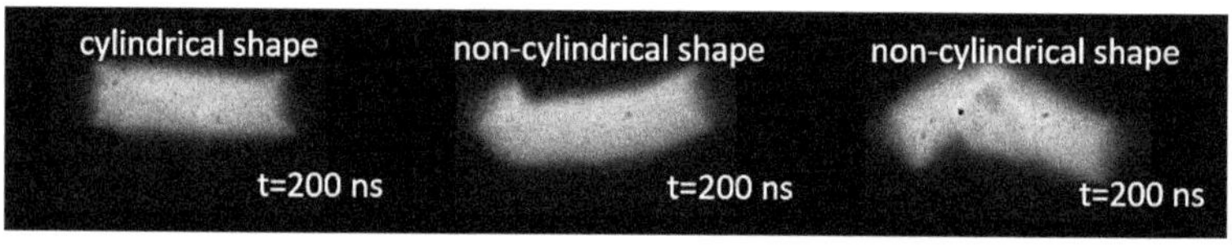

Figure 3.2 Images of spark plasma filaments acquired with a 50 ns gate width, 200 ns after the onset of the breakdown (Cu electrodes in N_2 gas, 2.0 mm electrode gap, 5 L/min gas flow, 400 mJ capacitor energy).

3.4.2 Plasma Temperature

Since plasmas are often transient, energetic, reactive, and nonequilibrium phenomena, they typically cannot be characterized by a single "plasma" temperature, as is conventional in low-temperature physical or chemical systems. Due to the nonexistence of an overall thermal equilibrium, the states of different species present can only be described by several temperature values. Tentatively, the temperature of electrons (T_e), which are the most mobile charge carriers, is always highest in electrical discharges. The temperature of other "heavy particles" (also called, for example, T_g or gas temperature) is usually lower than T_e, especially in low-pressure plasmas, where collisions are rare [25]. Only when the electric field strength decreases and/or the pressure is raised will the temperature of all species converge to a common value. In this case the plasma reaches LTE and all species can be described by a single temperature in a given spatial position.

Depending on the conditions, there are multiple emission spectroscopy-based possibilities to obtain temperature estimations for a plasma. These methods usually assume that the population of energy levels of certain species (atom, ion, or molecule) follows the Boltzmann distribution (which is a consequence of the LTE condition). This assumption facilitates temperature measurement via the assessment of the ratio of the population of levels by measuring the emission intensity ratio of spectral features. In low-temperature molecular plasmas (ca. $<$10,000 K, in, for example, O_2 or N_2), where the degree of thermal dissociation and ionization is low, the intensity ratios of rotational or vibrational bands of molecules can be measured, thereby obtaining the T_{rot} or T_{vib} temperatures. At the same time, in higher-temperature plasmas (approx. between 10,000 and 100,000 K), where the ionization degree is high and as a consequence the line emission of atoms and ions is strong, the ratio of spectral line intensities can be used as a basis of calculation. This latter approach results in the so-called excitation (T_{exc}) or ionization (T_i) temperatures. Other temperature measurement methods use the line-to-continuum ratio or the Doppler broadening of spectral lines [23, 24]. Due to their importance in spark plasma diagnostics, we hereby briefly outline the execution of the two most popular temperature measurement methods, namely the Boltzmann and the Saha–Boltzmann method.

The Boltzmann plot method starts with the postulation that if the excitation of the emitting species obeys the Boltzmann distribution, then

$$\ln\left(\frac{I_{ij}\lambda_{ij}}{cg_i A_{ij}}\right) - \ln\left(\frac{hn}{Z}\right) - \frac{E_i}{k_B T}, \tag{3.1}$$

where A_{ij} is the transition probability; Z is the partition function; g_i and E_i are the statistical weight and the energy of the excited level (denoted by subscript i), respectively; k_B, h, and c are the Boltzmann constant, Planck's constant, and speed of light, respectively; T is the excitation temperature[a]; λ_{ij} is the wavelength of the transition; and n is the number concentration of the species under investigation [26]. Thus, if, based on an experimentally recorded intensity value, $\ln\left(\frac{I_{ij}\lambda_{ij}}{cg_iA_{ij}}\right)$ is plotted as a function of E_i for different spectral lines of a selected species, the data points will scatter around a straight line with a slope of $-\frac{1}{k_\mathrm{B}T}$. This experimentally determined slope can then be used to calculate T. Needless to say, all constants for the involved transitions in Eq. 3.1 have to be known for this method to be used; consequently its applicability and accuracy are limited by the availability and accuracy of these constants. Obviously, the slope of the line is better defined, and hence the temperature determination is more accurate, if the spread of energy levels involved is broader, which requires an intensity-calibrated spectrometer with a broad spectral coverage accompanied by sufficient resolution to resolve the line-rich spectra of spark plasmas.

In this form, the Boltzmann plot method can only be used for a single species in a given ionization level. However, Eq. 3.1 can be generalized to a given species in different ionization levels to obtain the Saha–Boltzmann method (see, for example, Refs. [13, 26]). This necessitates considering the ionization (Saha) equilibrium, too:

$$n_+ = \frac{1}{n_\mathrm{e}} 2\frac{Z_+}{Z}\left(\frac{2\pi m_\mathrm{e}k_\mathrm{B}T}{h^2}\right)^{\frac{3}{2}} \exp\left(-\frac{W_+ - \Delta W}{k_\mathrm{B}T}\right) n \equiv Sn, \qquad (3.2)$$

where n_e is the number concentration of electrons, W_+ is the ionization energy of the studied species, and ΔW is the lowering of the ionization energy [27]. The + subscript refers to ions. By combining Eqs. 3.1 and 3.2 the following generalized equation can be written for a given species at different ionization levels:

$$\ln\left(\frac{I_{ij}\lambda_{ij}Z}{hcg_iA_{ij}S}\right) = \ln(n) - \frac{E_i}{k_\mathrm{B}T}, \qquad (3.3)$$

---

[a]Due to the LTE condition the excitation temperature is assumed to describe the temperature of the plasma; therefore it will be denoted by T.

where S can be calculated from Eq. 3.2 for each ionization level as a function of temperature and electron concentration. The temperature can then be determined by reading the slope of the straight line fitted to the data points. Although the applicability and general accuracy of the Saha–Boltzmann method is similar to that of the Boltzmann method, it is more practical to use because it incorporates data points from ions, which obviously widens the energy spread and hence increases the reliability of the temperature determination. The validity of the LTE assumption can also be inferred from the linearity of the plot. Please note though that Eq. 3.3 requires the a priori knowledge of the electron concentration and that both Z and S also depend on temperature; hence solving the Saha equation for the temperature is an iterative process. Figure 3.3 exemplifies a typical Saha-Boltzmann plot derived from temporally resolved emission spectra acquired in an SDG.

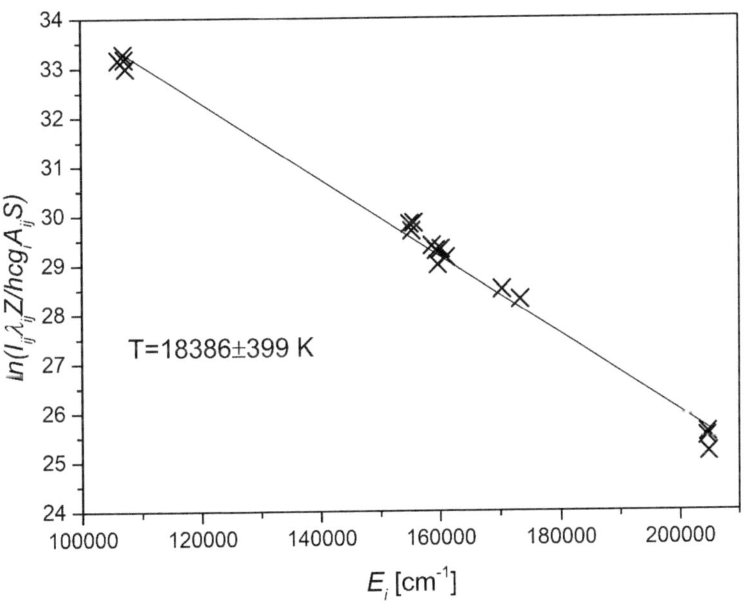

Figure 3.3 A typical Saha–Boltzmann plot obtained in the spark gap about 1000 ns after the onset of the breakdown (Cu electrodes in N_2 gas, 2.0 mm electrode gap, 5 L/min gas flow, 25 mJ capacitor energy).

3.4.3 Electron Concentration

Another important plasma diagnostic parameter is the number concentration of electrons (N_e). It is commonly used as an indicator of plasma behavior and reproducibility [28], and it also defines a necessary condition for the presence of LTE in a plasma known as the Griem [29] or the McWhirter [30] criterion.

A widely used optical method for deriving the electron concentration in a plasma is based on the Stark effect, which relies on the splitting of energy levels in an external electrical field, which dominantly depends on electron impacts, while the electric microfields of ions have a minor contribution [31]. Since the splitting of energy levels results in a strong overlap, in most practical cases the Stark effect manifests itself as the broadening of spectral lines (hence also known as "Stark broadening"). The method for calculating the electron concentration based on Stark broadening is predominantly used for the atomic lines of hydrogen or argon [21–24], for which the Stark broadening of spectral lines is a largely linear function of N_e. For example, in argon gas, according to Konjevic [31]

$$W(N_e,T_e) = w_e(T_e) \cdot N_e \cdot 10^{-17}\text{cm}^3, \tag{3.4}$$

where W is the full width at half maximum (FWHM) of the Lorentzian line profile associated to Stark broadening, N_e is the electron concentration, w_e is the electron impact FWHM at $N_e = 10^{17}$ cm^{-3}, and T_e is the electron temperature. As seen in Eq. 3.4, the electron concentration can be derived from the Stark width of a spectral line if both T_e and $w_e(T_e)$ are known. It should be noted that prior to applying Eq. 3.4 the measured spectral line profile should be corrected for the effects of other broadening mechanisms, such as Doppler, collisional, and instrumental broadening. In most practical cases the Stark width of the spectral line can be obtained by the deconvolution of the measured line profile [10]. When the Stark component of the width of a measured spectral line is deduced, the electron concentration can be estimated by using the appropriate electron impact width. Tabulated electron impact width values for Ar lines can be found, for example, in Refs. [32, 33]. Unfortunately, this data is scarce for some gases (e.g., N_2) relevant to SDGs.

The accuracy of electron number density determination (perhaps 10%–20%) is comparable to the accuracy in the measurement of the FWHM of the spectral line, which, in turn, depends on the resolution of the spectrometer. However, if the whole line profile is compared to a theoretically calculated line profile, an improved accuracy (ca. 10%) can be achieved [21–23].

3.4.4 Concentration of Plasma Species

It was pointed out earlier that the intense optical emission of the spark plasma is mainly attributed to the spontaneous emission of excited atoms and ions that exist in the spark gap. The emission intensity of a spectral line alone cannot provide direct information on the concentration of the emitting species (for that further information such as the temperature is inevitably needed), but it can be used to identify species that exist in the spark gap and to provide a rough idea about their spatial distribution and dynamics. Such information is essential for particle formation models and can particularly be useful when dissimilar elemental electrodes, or composite electrodes, are used.

Emission spectra can be used to estimate the concentration of the species in the observed plasma volume, most importantly that of the electrode material in an SDG. This, for example, requires the solution of plasma equations derived under LTE conditions [34, 35]. In an optically thin equilibrium (LTE) spark plasma, the following assumptions can be utilized: (i) atmospheric pressure prevails in the spark gap, (ii) the discharge is only composed of electrons and the neutral and singly charged ionic species of the ambient gas, as well as that of the electrode material (here denoted by M), and (iii) the net charge of the discharge plasma is zero. These, for the case of Ar ambient, can be expressed via the following equations [10]:

$$n_{\mathrm{M}^+} + n_{\mathrm{Ar}^+} = n_{\mathrm{e}} \tag{3.5}$$

$$p = (n_{\mathrm{M}^0} + n_{\mathrm{Ar}^0} + n_{\mathrm{M}^+} + n_{\mathrm{Ar}^+} + n_{\mathrm{e}})k_{\mathrm{B}}T \tag{3.6}$$

$$I_{\mathrm{M}^0} = h\nu_{\mathrm{M}^0} A_{\mathrm{M}^0} \frac{1}{Z_{\mathrm{M}^0}} g_{\mathrm{M}^0} \exp\left(\frac{-E_{\mathrm{M}^0}}{k_{\mathrm{B}}T}\right) n_{\mathrm{M}^0} \tag{3.7}$$

$$I_{\mathrm{Ar}^0} = h\nu_{\mathrm{Ar}^0} A_{\mathrm{Ar}^0} \frac{1}{Z_{\mathrm{Ar}^0}} g_{\mathrm{Ar}^0} \exp\left(\frac{-E_{\mathrm{Ar}^0}}{k_{\mathrm{B}}T}\right) n_{\mathrm{Ar}^0} \tag{3.8}$$

$$\frac{n_{\mathrm{e}} n_{\mathrm{M}^+}}{n_{\mathrm{M}^0}} = 2\frac{Z_{\mathrm{M}^+}}{Z_{\mathrm{M}^0}} \left(\frac{2\pi m_{\mathrm{e}} k_{\mathrm{B}} T}{h^2}\right)^{\frac{3}{2}} \exp\left(-\frac{W_{\mathrm{M}^+} - \Delta W}{k_{\mathrm{B}}T}\right) \tag{3.9}$$

$$\frac{n_e n_{Ar^+}}{n_{Ar^0}} = 2\frac{Z_{Ar^+}}{Z_{Ar^0}}\left(\frac{2\pi m_e k_B T}{h^2}\right)^{\frac{3}{2}} \exp\left(-\frac{W_{Ar^+} - \Delta W}{k_B T}\right) \tag{3.10}$$

where n_{M^0}, n_{Ar^0}, n_{M^+}, n_{Ar^+}, and n_e are the concentration of M (i.e., electrode metal) and Ar atoms and singly charged ions, and electrons, respectively. p denotes the pressure in the gap; T is the equilibrium plasma temperature; I, ν, g, and E are the intensity, frequency, statistical weight, and energy of the upper state of a specific M I and Ar I spectral line, respectively. Z is the partition function, h is the Planck's constant, k_B is the Boltzmann constant, m_e is the mass of the electron, and W is the first ionization energy; ΔW is the decrease in the ionization energy according to Griem [27]. Equation 3.5 expresses the electronic neutrality criterion; Eq. 3.6 is Dalton's law; Eqs. 3.7 and 3.8 describe the emission intensity of an M I and Ar I line, respectively; while Eqs. 3.9 and 3.10 correspond to a pair of Saha equations describing the ratio of the concentration of ions to atoms.

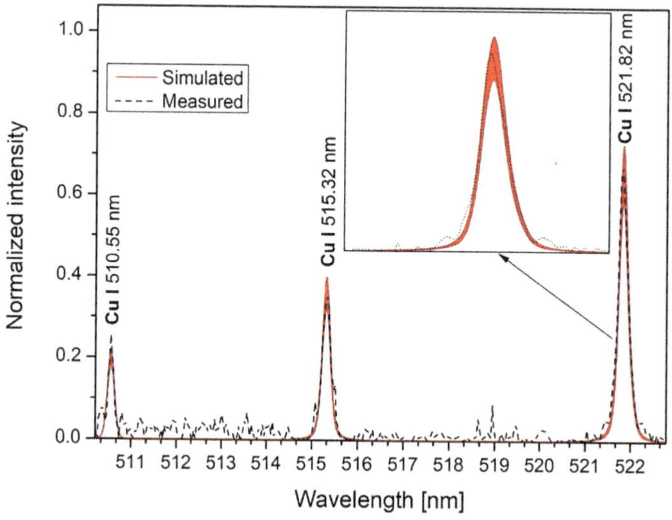

Figure 3.4 Comparison of the emission spectrum measured during the afterglow in an SDG producing Cu NPs in an Ar atmosphere with the spectrum simulated under the same conditions (2 mm gap, 100 Hz spark repetition rate, 5 L/min gas flow, 25 mJ capacitor energy). From Ref. [10].

One way for solving Eqs. 3.5–3.10 is to calculate the intensity ratios of spectral lines of species of the gas and electrode material in a given temperature and electron concentration range and compare them to experimentally measured intensity ratios [10]. The results of such calculations are exemplified by Fig. 3.4,[b] where a measured spectrum is compared with a simulated one; the latter is based on the parameters derived from Eqs. 3.5–3.10.

3.5 Plasma Properties of Spark Discharge Generators

The following sections illustrate the application of plasma diagnostic methodology to SDGs. The results shown here refer to the arc and afterglow stages, the ones most relevant from the point of view of NP generation. Data shown here were obtained during the generation of copper NPs in different carrier gases (either argon or nitrogen), but similar results were obtained with different electrode materials, such as Ni, Ag, Au, etc. [36].

3.5.1 SDG Plasma Morphology

After the formation of a conducting channel between the two electrodes, that is, at the beginning of the arc stage, the plasma is a thin, intense, approximately cylindrical object with a diameter in the submillimeter range [19]. This is illustrated in Fig. 3.5, which shows the image of the spark acquired 50 ns after the voltage drop in the gap.

Due to the transient nature of the spark plasma, its morphology varies in time. Snapshots shown in Fig. 3.6 illustrate the evolution of spark morphology over time. Please note that every image shown in Fig. 3.6 was autoscaled in intensity in order to make the shape of the discharge discernible at every delay, since the total emitted intensity varies considerably. The images prove that as time passes and more

[b]Please note that spectroscopic notation of spectral lines is used throughout this chapter in parallel with the standard notation of neutral and charged species. Therefore, spectral lines arising from neutral species (e.g., Cu^0 and N^0) are denoted by the Roman numeral I (e.g., Cu I and N I), while the lines emitted by singly charged ions (e.g., N^+) are denoted by the numeral II (e.g., N II), and so on.

energy is dissipated in the electrode gap, the diameter of the initially thin and nearly cylindrical channel increases. At about 500 ns, the channel starts to partially lose its cylindrical shape. After about 10 μs, when the current oscillations and consequently the energy input cease (i.e., the discharge reaches the end of the arc stage), the discharge channel spreads greatly and loses its cylindrical shape almost completely. After this point, that is, in the afterglow stage, the decaying plasma expands in 3D and gradually fades away [19].

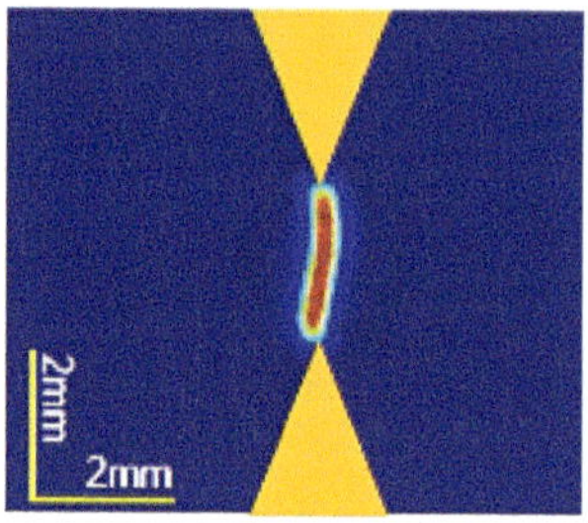

Figure 3.5 Image acquired at a delay of 50 ns for a spark discharge between tipped-end Cu electrodes placed at 2 mm distance in a 10 L/min of co-axial N_2 flow, with a capacitor energy of about 275 mJ. The electrodes are schematically shown by yellow triangles. The upper electrode was initially the anode. The coloration of the image follows a decreasing intensity in the order of red-yellow-green-blue.

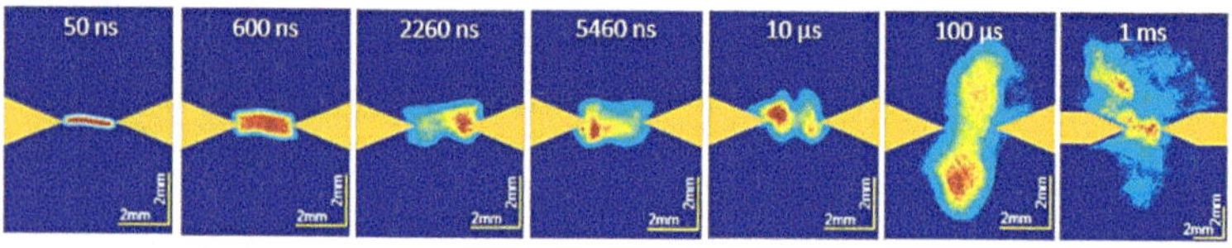

Figure 3.6 Images of spark discharges initiated between tipped-end Cu electrodes placed at 2 mm distance in a 10 L/min of co-axial N_2 flow, with a capacitor energy of about 275 mJ, acquired at the gate delays indicated. The electrodes are schematically shown by yellow triangles. The left electrode was initially the anode. The autoscaled intensity decreases in the red-yellow-green-blue order on every snapshot. Reprinted from Ref. [19], with the permission of AIP Publishing.

Besides the above given qualitative information on overall plasma morphology, the processing of temporally resolved images also allows for deriving the channel width quantitatively. Figure 3.7

depicts the temporal evolution of the averaged channel width for up to 15 μs. Please note that the large scatter of data points at later times (at delays beyond about 3.5 μs) is due to the gradual loss of the well-defined cylindrical shape of the channel. Still, the tendencies are clear and conclusive. The discharge sets off as a narrow channel with a width below 500 μm, and then it grows monotonically as a function of time. At around 3.5 μs the channel width stabilizes around a value of (1.5 ± 0.2) mm. From the slope of the curve the expansion velocity of the plasma front can be derived. It was shown that up until about 500 ns after the breakdown, the expansion velocity of the plasma front is nearly constant and is in the range of about 0.36–0.48 mm/μs [19].

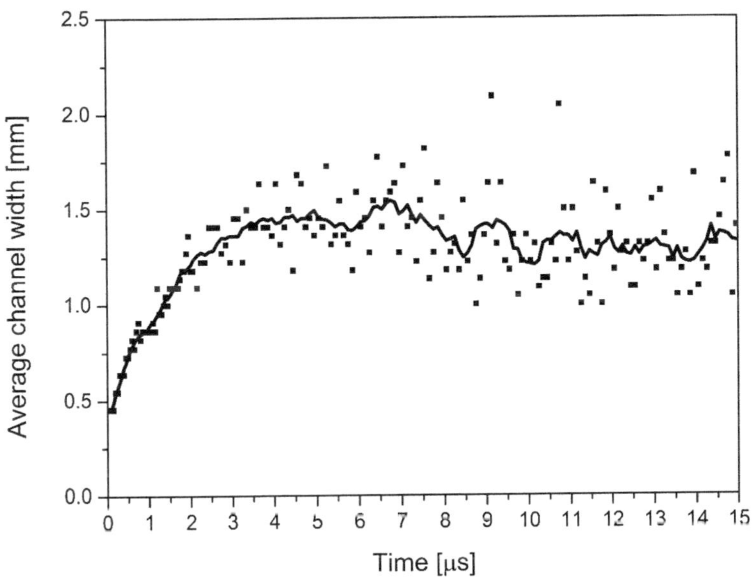

Figure 3.7 Average channel width for spark discharges ignited between tipped-end Cu electrodes placed at a 2 mm distance in a 10 L/min of co-axial N_2 flow, with a capacitor energy of about 275 mJ. The dots show the measured values of the channel width, while the solid line is obtained by arithmetic averaging using a 9-point-wide moving window.

This is a supersonic speed in N_2 at room temperature [37]. Such fast growth of the plasma channel, together with the inherent change in temperature and gas pressure, induces a shock wave [38], which is confirmed also by the audible, snapping sound during sparking. The values obtained in the SDG agree fairly well with the typical values published for the radial velocity of shock waves measured in other spark plasma sources [15].

Another peculiarity of the snapshots of the spark plasma shown in Fig. 3.6 is the presence of a "moving" axial region that exhibits distinctly higher emission intensity than that of the surrounding area. The axial position of these hot spots varies over time in correlation with the instantaneous current, in a way that they are always situated near the momentary cathode. The presence of the hot spot is likely to be the result of a region with nonzero charge, known as the "cathode fall," which forms in the gap that effectively shields the electric field. Beyond the cathode fall region, that is, a bit further toward the anode, the plasma tends to be more intense in emission due to the dissipation of a substantial amount of energy gained by the electrons in the fall region. When the polarity changes due to oscillations, the existing hot spot quickly collapses and another one builds up around the opposite electrode [19].

3.5.2 SDG Plasma Temperature

As briefly mentioned in Section 3.4.2, the determination of the plasma temperature is complicated by the conditions existing in spark discharges, but the situation is greatly simplified when LTE can be assumed. In this case, all species of the plasma can be described locally by a single temperature. This temperature of high-current, microsecond-long, atmospheric pressure sparks has been extensively investigated in the literature by means of optical emission spectroscopy. Peak values as high as 30,000–40,000 K have been reported [39, 40], but reports on temperatures around 20,000 K are typical [13, 15].

The graph shown in Fig. 3.8 presents the time evolution of the temperature derived in an SDG in the center of the gap. The values shown in Fig. 3.8 were obtained by using the Saha–Boltzmann plot method. The temperature in the 0.5–2.5 μs temporal range—that is, during the arc stage—was found to have a peak value of ~(20,000 ± 480) K, and its decreasing trend followed the decrease of the amplitude of the current oscillations [10]. The Saha–Boltzmann plot method can be used for plasma temperature determination in the arc stage, but this method is not applicable after the current ceased (when the emission from ionized species largely disappear from the spectrum); therefore at later times only the Boltzmann plot method can be employed. The temperature can also be deduced from the simple plasma composition model described in Section 3.4.4. Both methods result in similar temperatures, with values around 10,000 K in the afterglow stage, independently of the carrier gas used [9, 10].

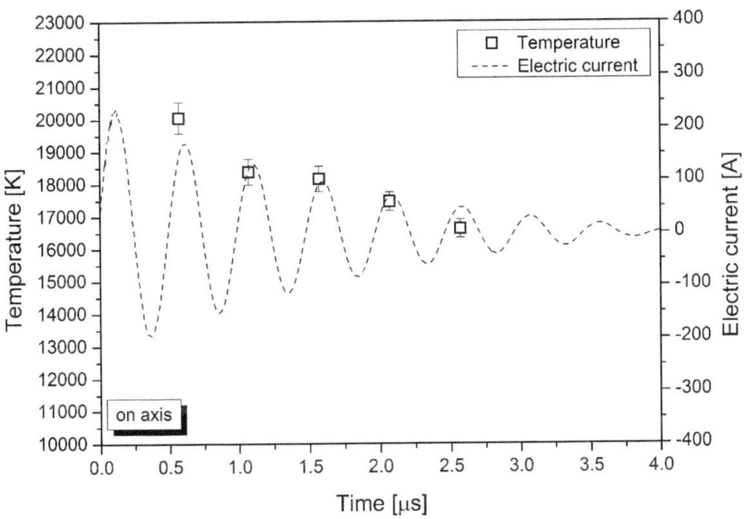

Figure 3.8 Variation of the plasma temperature in the arc stage derived from emission spectra of a spark plasma produced between Cu electrodes in Ar gas (2 mm gap, 100 Hz spark repetition rate, 5 L/min gas flow, 25 mJ capacitor energy). From Ref. [10]. © IOP Publishing. Reproduced with permission. All rights reserved.

Figure 3.9 shows the temporal variation of the instantaneous temperature derived by the Boltzmann plot method in the afterglow. The peak temperature was found to be about 12,000 K at the beginning of the afterglow stage, which decreased linearly down to ca. 9000 K in the temporal window investigated. From the data shown in Fig. 3.9, the cooling (sometimes also called quenching) rate of the plasma was estimated to be $(5.5 \pm 0.6) \times 10^8$ K s^{-1} [9].

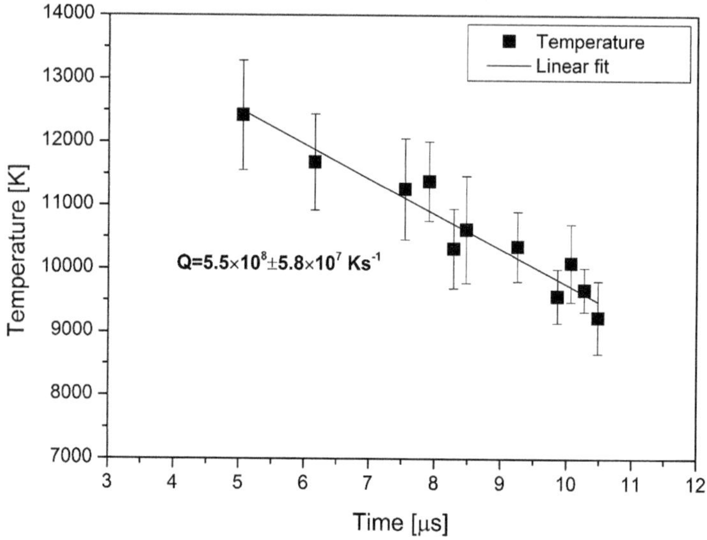

Figure 3.9 Variation of the plasma temperature in the afterglow stage derived from emission spectra of a spark plasma produced between Cu electrodes in N_2 gas (2 mm gap, 180 Hz spark repetition rate, 1.68 L/min gas flow, 180 mJ capacitor energy).

3.5.3 Electron Concentration in a Spark Plasma

The electron concentrations in various types of spark discharge plasmas have been studied and reported in the literature. In high-current, atmospheric pressure unipolar sparks ignited between Cu and W electrodes in Ar, a peak value of $2.5 \pm 0.3 \times 10^{17}$ cm^{-3} has been reported [13]. Similar values can be found in the earlier literature for oscillatory spark discharges [40]. The electron concentration in spark plasmas used for NP generation has only been recently investigated by means of optical emission spectroscopy. The method based on the Stark broadening of spectral lines—described in Section 3.4.3—was applied in the case of an SDG producing copper NPs in an argon atmosphere [10].

The temporal evolution of the electron concentration in the spark channel (reported here for three points around the center of the spark gap) during the arc stage is shown in Fig. 3.10. As seen, the concentration decreases from $\sim 7 \times 10^{17}$ cm^{-3} to $\sim 2 \times 10^{17}$ cm^{-3} along with the decay of the amplitude of the oscillating current. The overlapping character of the three datasets plotted in Fig. 3.10 also proves that the electron concentration is fairly uniform across the plasma channel [10].

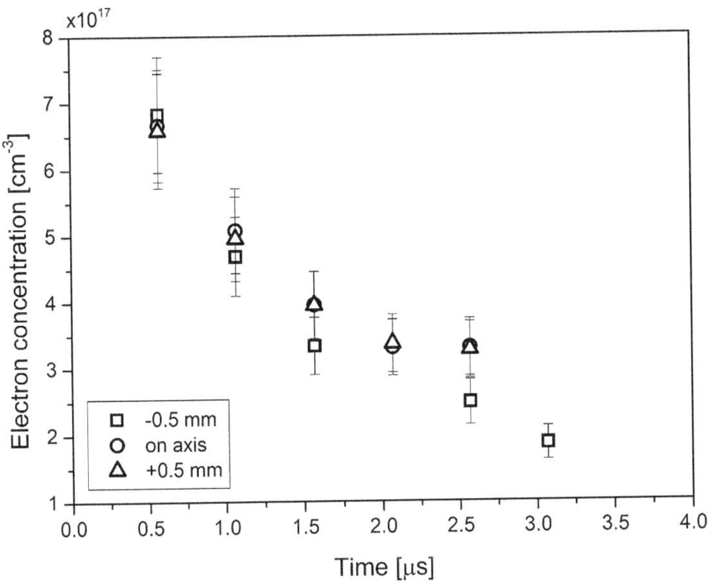

Figure 3.10 Electron concentration of the spark plasma during the arc stage at three different, axially central spatial locations (on axis and two radially symmetric positions, Cu electrodes in Ar gas, 2 mm gap, 100 Hz spark repetition rate, 5 L/min gas flow, 25 mJ capacitor energy). From Ref. [10].

The electron concentration has a fairly high value in the arc stage, which drops when the electric current ceases, and the discharge enters the cooler afterglow stage. As a consequence, the broadening of spectral lines will also be less pronounced and hence harder to detect. This hinders the determination of the electron concentration by the Stark method in the afterglow stage. With the application of equilibrium plasma models—like the one previously described in Section 3.4.4—the electron concentration was found to be around 3.5×10^{16} cm^{-3} midtime in the afterglow stage (at around 20 μs), so it is about 1 order of magnitude lower than it was at the end of the arc stage (c.f. Fig. 3.10) [10]. From the temporally resolved data obtained in the afterglow stage, the rate of change of the electron concentration was also estimated to be $\frac{dn_e}{dt} \approx -1.4 \times 10^{22} \frac{\text{cm}^{-3}}{\text{s}}$. Applying the Griem criterion [13], this value can be used to give a rough estimate on the width of the temporal window, in which LTE potentially holds in the spark plasma. According to this estimation, it was found that LTE does hold until around 24 μs after the breakdown in the studied system, which is somewhat longer than half of the total duration of the afterglow stage [10].

3.5.4 Composition of the Spark Plasma

Optical emission spectroscopy also allows for the composition of the plasma to be deduced. Figure 3.11 represents the typical emission spectrum of the spark acquired at the beginning of the arc stage. As seen, emission is dominated by the spectral lines of the excited atomic and ionic species of the ambient gas (here N_2 in Fig. 3.11A and Ar in Fig. 3.11B). During the arc stage the composition of the plasma also evolves. The temporal behavior of the intensity of the emission from several, mostly ionic species[c] measured in the center point of the spark gap is illustrated in Fig. 3.12. The emission from Ar^+ and Ar^{2+} follows the current oscillations in the discharge loop and is only detectable as long as electric current flows between the electrodes.[d] This is due to the high excitation and ionization energies of these species and thus can only be sustained by the discharge when and where the temperature is highest in the gap [6]. This implies that, for example, argon ions are present and emit in the close proximity of the axis of the current-conducting channel bridging the electrodes. As a consequence, light emission of these ions can be used to monitor the properties (e.g., position, spread, temperature) of the current-carrying channel. Similarly to the emission of the ionic species, atomic argon emission is also detectable from the very beginning of the arc stage (see

[c]Several spectral lines of each species have been investigated. It was confirmed that the intensity of spectral lines of the same species has similar temporal and spatial evolution. Therefore one line, detectable with a good signal-to-noise ratio, was selected from each species as the representative spectral line of the given species.

[d]As a first approximation, a typical, unregulated, or free-running SDG can be electronically modeled as a series RLC circuit operating in the underdamped regime. This means that both current and voltage oscillate with exponentially decaying amplitude with a certain phase shift with respect to each other.

Fig. 3.12). It means that argon atoms and ions coexist in the conducting spark channel. However, they have different spreading rates, and argon atoms spread farther apart from the electrodes' axis than ions as time passes [10].

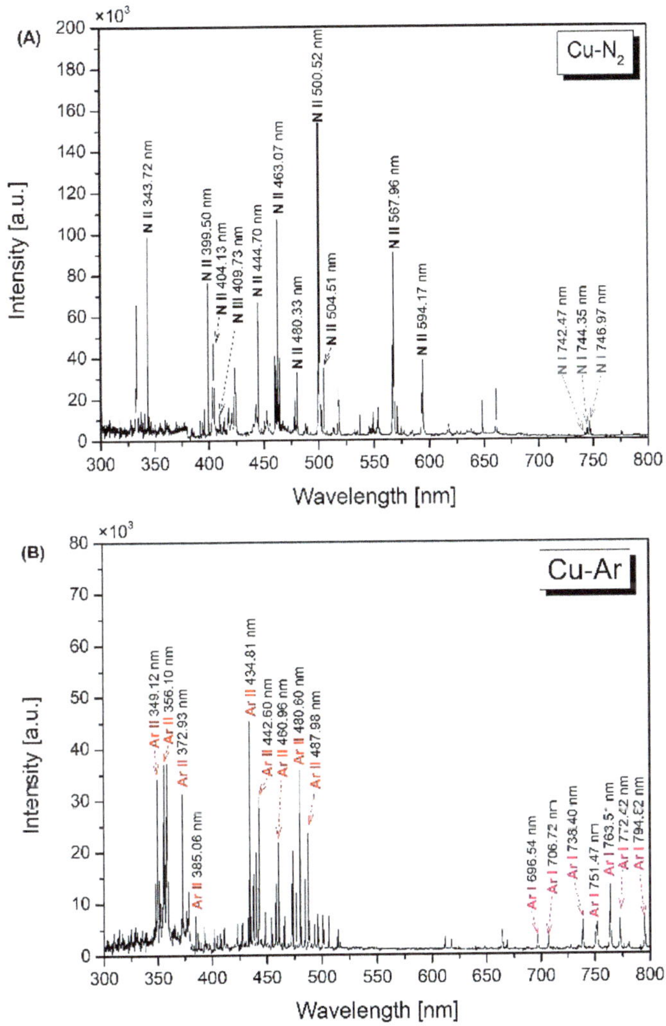

Figure 3.11 Emission spectrum of the spark acquired at the beginning of the arc stage of a spark discharge at 500 ns, used for the generation of Cu NPs in a nitrogen (A) and an argon (B) atmosphere (2 mm gap, 5 L/min gas flow rate, 100 Hz spark repetition rate). The capacitor energy was 110 mJ and 25 mJ in N_2 and Ar, respectively. The spectra were taken in the middle of the gap.

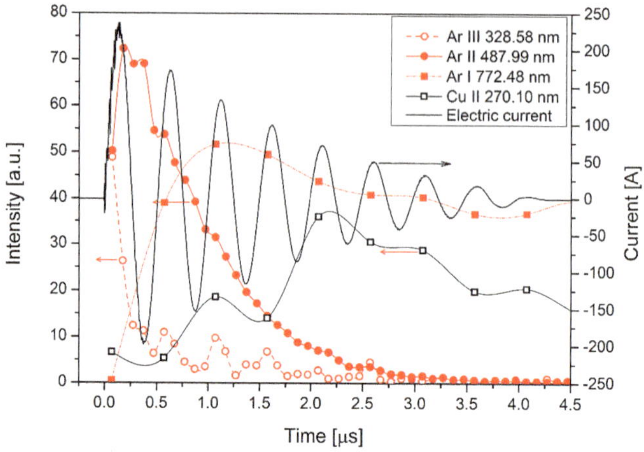

Figure 3.12 Time evolution of the intensity of some selected spectral lines measured at the center of the gap during the arc stage. Spectroscopic data points are connected to guide the eyes only (2 mm gap, 5 L/min gas flow rate, 100 Hz spark repetition rate, 25 mJ capacitor energy).

Ions of the electrode material (here Cu) can also be detected in the spark gap during the arc stage; however, their emission is low in intensity compared to those of the ambient gas (here Ar). The time at which the light emission from Cu^+ peaks (approx. 2.2 μs in the particular case shown in Fig. 3.12) differs significantly from that of various argon species (approx. 0.3 μs and 1.1 μs for Ar^+ and neutral argon, respectively). It appears that atomic copper is even less abundant in the gap during the arc stage than Cu^+ ions. Its amount only starts to ramp up in the first 5 μs and peaks in the afterglow stage (see Fig. 3.13B), thereby indicating that atomic Cu, that is, the source of the generated NPs, may be, at least partially, produced by the recombination of Cu ions with electrons. The emission spectra also prove that copper species are much less abundant in the gap at early times of the discharge.

Figure 3.13A shows an emission spectrum recorded 20 μs after the onset of the breakdown, that is, during the afterglow, in an SDG producing Cu NPs in a N_2 atmosphere. It is apparent that the spectrum consists of atomic spectral lines of both the electrode material and the carrier gas, that is, the sustained radiation, experienced after the arc stage, is produced by excited atomic species. The evolution of spectral intensity of Ar I and Cu I lines is shown in Fig. 3.13B. The Ar I emission peaks within the arc stage of the discharge (more clearly resolved at around 1.1 μs in Fig. 3.13B), and it is sustained until approx. 50 μs. The emission of copper atoms ceases approximately at the same time as argon emission but reaches its peak intensity considerably later, at around 20 μs after the breakdown [10].

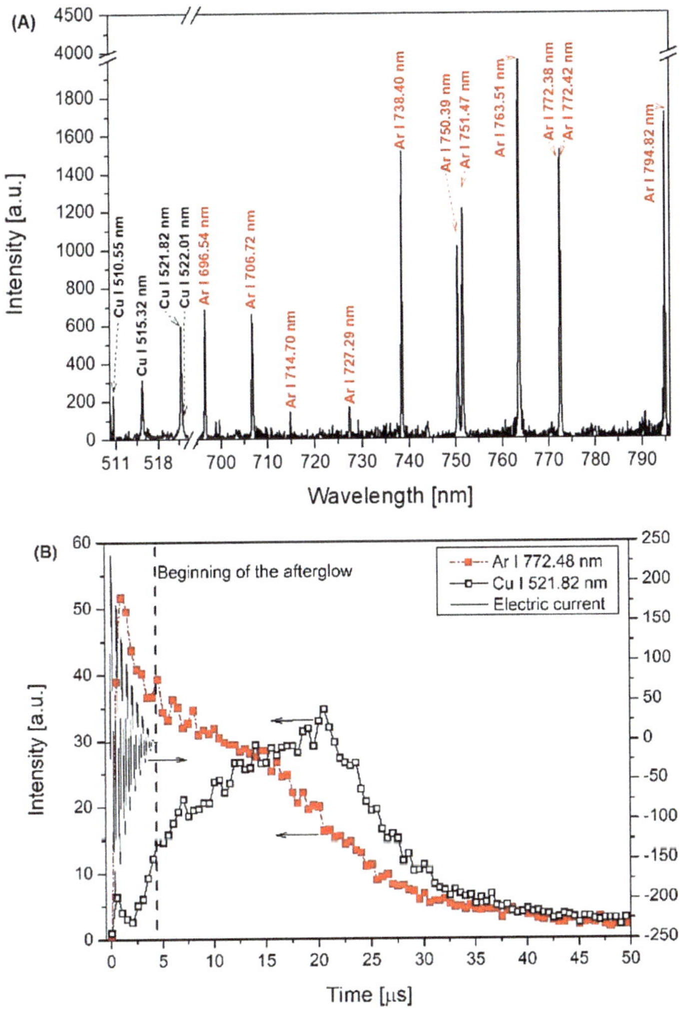

Figure 3.13 Typical emission spectrum measured in the spark gap during the afterglow stage (A). Time evolution of the intensity of some selected atomic spectral lines measured at the center of the gap during the afterglow stage (B) (2 mm gap, 5 L/min Ar gas flow rate, 100 Hz spark repetition rate, 25 mJ capacitor energy). (B) From Ref. [10].

The temporal evolution of the emission intensity of excited species was shown here for the case of copper electrodes, but similar temporal behavior was seen for other electrodes, too. Figure 3.14 illustrates that the intensity of atomic spectral lines of four electrode materials, namely copper, nickel, gold, and silver, evolves very similarly over time [36].

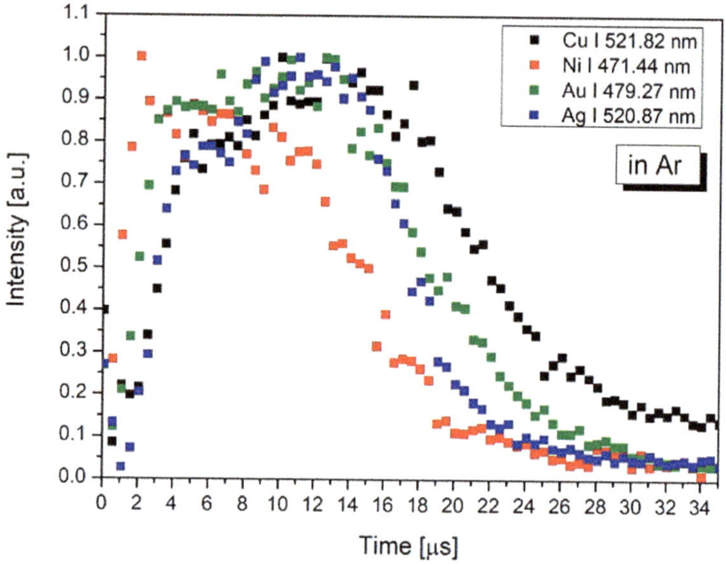

Figure 3.14 Temporal evolution of the intensity of selected atomic spectral lines of Cu, Ni, Au, and Ag emitted from the center of the gap in the case of an argon atmosphere (2 mm gap, 5 L/min N_2 flow, 100 Hz spark repetition rate, capacitor energy in the range of 55–85 mJ).

An estimation for the concentration of the electrode material in a given plasma volume can be given by using the equations of the equilibrium plasma model described in Section 3.4.4. For instance, it was found that the total Cu concentration (i.e., the sum of atomic and ionic forms) is about 10^{15} cm^{-3} [10]. It should also be noted that the number concentration of Ar species was found to be more than 2 orders of magnitude higher than that of Cu, that is, the electrode material is highly diluted in the gap.

By acquiring the spectra at different spatial positions, the distribution of the species emitting in the spark plasma can also be investigated. The distribution of the Cu species estimated at different

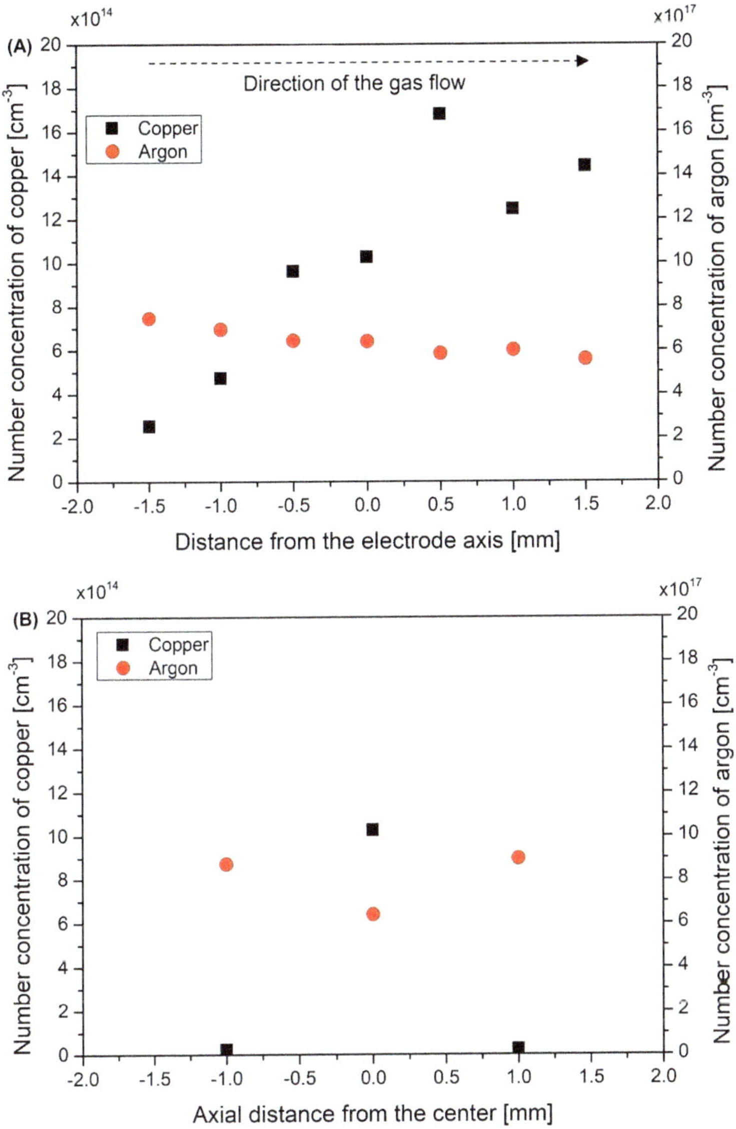

Figure 3.15 Spatial distribution of the total number concentration of copper and argon species (i.e., the sum of the atomic and ionic species) perpendicular to (A) and along (B) the direction of the electrodes' common axis, 20.5 μs after the breakdown (Cu electrodes in Ar gas, 2 mm gap, 5 L/min flow, 100 Hz spark repetition rate, 25 mJ capacitor energy). From Ref. [10]. © IOP Publishing. Reproduced with permission. All rights reserved.

spatial positions in the electrode gap is shown in Fig. 3.15. As seen in Fig. 3.15A the number concentration of copper increases, while the argon concentration—which is about 2–3 orders of magnitude higher than that of the copper, as mentioned earlier—decreases in the direction of the gas flow, which is probably due to the fact that the majority of copper is in ionic form whose concentration increases with increasing temperature, which was also found to increase along the same direction [10]. At the same time, the argon concentration exhibits slight and symmetrical changes in the direction coaxial with the electrodes, while the total concentration of copper species varies significantly in the −1.0 mm to +1.0 mm axial range. It is, by a factor of about 50, smaller close to the electrode tips compared to the value obtained in the middle of the gap. The results are summarized in Fig. 3.15B. The two distributions shown in Fig. 3.15 suggest that at 20.5 μs, the copper plume is concentrated in the middle of the gap and exhibits increasing concentration along the direction of the carrier gas flow [10].

3.6 Conclusions

We have illustrated above that useful information promoting a deeper understanding of processes occurring in the spark plasma of SDGs can be obtained via optical plasma diagnostic methods. Temporally and spatially resolved optical emission spectra of the spark facilitate the reconstruction of the distribution of electron concentration, temperature, and the concentration of the electrode material and gas species. Optical emission spectroscopy, complemented with imaging, consequently allows for the in situ investigation of the spark plasma.

The production of particles in an SDG begins with the erosion of the electrode material, which predominantly occurs during the arc stage, when the energy input to the plasma is highest. The high current during this period results in a relatively high temperature and electron concentration in the order of 15,000–20,000 K and 10^{17}–10^{18} cm^{-1}, respectively. Although it is rarely stated explicitly, the majority of studies assume the dominance of thermal processes, for example, when they quantitatively describe electrode erosion by Joule heating [41–45]. Even if Joule heating is not the dominant mechanism (cf. Chapter 4, which suggests a strong role of cation bombardment), it is macroscopically the resistance [14], whereas on the microscopic level, the temperature and electron concentration [2] that play an important role in the material erosion process. This duality calls for a detailed spark plasma ablation model. The creation of such a model has been recently attempted, and promising predictions were shown for the quantity and size of NPs formed by sparking [46]. Optical plasma diagnostic methods, briefly outlined in the present chapter, provide a powerful set of tools that can generate data for the validation of spark ablation models.

At the end of the arc stage, the vapor of the electrode material produced in the gap starts to cool down, hence approaching the conditions where particle formation sets off via the nucleation process. Simply speaking, the hottest part of this cooling phase is the afterglow stage of the plasma. The afterglow ends when the conditions are no longer sufficient for maintaining the spontaneous emission of plasma species. Since the timescale of the afterglow stage is much shorter than that of particle formation [47], plasma emission cannot provide direct information on the NP formation process; nonetheless, important data can be obtained via plasma diagnostics about the conditions preceding nucleation, for example, rapid quenching of the eroded material vapor, which causes extremely high supersaturation and hence pushes the critical nucleus size down to the atomic range [47, 48]. The variation of the temperature during the afterglow stage—derived from temporally resolved emission spectra—can be used to estimate the rate of cooling of the atomic vapor, which is equivalent to the quenching rate [9]. Another important quantity affecting the generated NPs is the mass of the initial vapor produced by a single spark, which is usually estimated from electrode mass loss measurements (gravimetry) [47, 48]. Temporally resolved emission spectroscopy together with plasma diagnostic calculation methods provide an independent way for obtaining such value, hence further strengthening the reliability of any conclusion relying on this quantity.

In addition to the above examples, the correlation between the emission of the spark plasma and the NP generation process was also demonstrated by showing that the alteration of key control parameters, which are typically used to optimize NP generation (e.g., gap size, carrier gas flow rate, charging current, RLC parameters of the discharge circuit), clearly affect the emission spectra, obviously via the key properties of the spark plasma [9, 10, 19]. This may also contribute to our fundamental understanding of the NP formation process on the one hand and the particle output of an SDG on the other.

Acknowledgments

The authors kindly acknowledge the financial support of the Ministry of Human Capacities of Hungary through project 20391-3/2018/FEKUTSTRAT.

References

1. Raizer, Y. P. (1991). *Gas Discharge Physics* (Springer-Verlag, Berlin Heidelberg).
2. Fridman, A. A. and Kennedy, L. A. (2011). *Plasma Physics and Engineering* (CRC Press, Boca Raton).
3. Scheeline, A. and Coleman, D. M. (1987). *Anal. Chem.*, **59**, p. 1185A.
4. Walters, J. P. (1968). *Anal. Chem.*, **40**, p. 1540.
5. Meuller, B. O., Messing, M. E., Engberg, D. L. J., Jansson, A. M., Johansson, L. I. M., Norlén, S. M., Tureson, N. and Deppert, K. (2013). *Aerosol Sci. Technol.*, **46**, p. 1256.
6. Boumans, P. W. J. M. (1972). Excitation of spectra, in Grove, E. L. (ed.) *Analytical Emission Spectroscopy* (Marcel Dekker, New York).
7. Walters, J. P. and Malmstadt, H. V. (1965). *Anal. Chem.*, **37**, p. 1484.
8. Pfeiffer, T. V., Feng, J. and Schmidt-Ott, A. (2014). *Adv. Powder Technol.*, **25**, p. 56.
9. Kohut, A., Ludvigsson, L., Meuller, B. O., Deppert, K., Messing, M. E., Galbács, G. and Geretovszky, Zs. (2017). *Nanotechnology*, **28**, p. 475603.
10. Kohut, A., Galbács, G., Márton, Z. and Geretovszky, Zs. (2017). *Plasma Sources Sci. Technol.*, **26**, p. 045001.
11. Scheeline, A., Norris, J. A., Travis, J. C., Devoe, J. R. and Walters, J. P. (1981). *Spectrochim. Acta, Part B*, **36**, p. 373.
12. Mandelstam, S. (1959). *Spectrochim. Acta*, **297**.
13. Bye, C. A. and Scheeline, A. (1993). *Appl. Spectrosc.*, **47**, p. 2022.

14. Kohut, A., Villy, L. P., Ajtai, T., Geretovszky, Zs. and Galbács, G. (2018). *J. Aerosol Sci.*, **118**, p. 59.

15. Reinmann, R. and Akram, M. (1997). *J. Phys. D: Appl. Phys.*, **30**, p. 1125.

16. Mathews, S. E., Niemczyk, T. M. and Walters, J. P. (1980). *Appl. Spectrosc.*, **34**, p. 200.

17. Bardócz, Á. and Varsanyi, F. (1956). *Nature*, **177**, p. 222.

18. Walters, J. P. and Goldstein, S. A. (1984). *Spectrochim. Acta, Part B*, **39**, p. 693.

19. Palomares, J. M., Kohut, A., Galbács, G., Engeln, R. and Geretovszky, Zs. (2015). *J. Appl. Phys.*, **118**, p. 233305.

20. Nigam, S., Aneesh, K., Navathe, C. P. and Gupta, P. D. (2011). *Rev. Sci. Instrum.*, **82**, p. 024702.

21. Hutchinson, I. H. (2002). *Principles of Plasma Diagnostics* (Cambridge University Press, Cambridge).

22. Ovsyannikov, A. A. and Zhukov, M. F. (2000). *Plasma Diagnostics* (Cambridge International Science).

23. Wiese, W. L. (1991). *Spectrochim. Acta, Part B*, **46**, p. 831.

24. Robinson, D. and Lenn, P. D. (1967). *Appl. Opt.*, **6**, p. 983.

25. Kunze, H.-J. (2009). *Introduction to Plasma Spectroscopy* (Springer, Berlin, Heidelberg.

26. Aragón, C. and Aguilera, J. A. (2008). *Spectrochim. Acta, Part B*, **63**, p. 893.

27. Griem, H. R. (1962). *Phys. Rev.*, **128**, p. 997.

28. Bye, A. and Scheeline, A. (1995). *J. Quant. Spectrosc. Radiat. Transfer*, **53**, p. 75.

29. Griem, H. R. (1963). *Phys. Rev.*, **131**, p. 1170.

30. McWhirter, R. W. P. (1965). Spectral intensities, in Huddlestone, R. H. and Leonard, S. L. (eds.) *Plasma Diagnostic Techniques*, Academic Press, New York, pp. 201–264.

31. Konjević, N. (1999). *Phys. Rep.*, **316**, p. 339.

32. Aparicio, J. A., Gigosos, M. A., Gonzalez, V. R., Perez, C., de la Rosa, M. I., Mar, S. and Departamento de Optica, U. de V. V. S. (1998). *J. Phys. B: At. Mol. Opt. Phys.*, **31**, p. 1029.

33. Sansonetti, J. E. and Martin, W. C. (2005). *J. Phys. Chem. Ref. Data*, **34**, p. 559.

34. Bachmann, B., Kozakov, R., Gött, G., Ekkert, K., Bachmann, J.-P., Marques, J.-L., Schöpp, H., Uhrlandt, D. and Schein, J. (2013). *J. Phys. D: Appl. Phys.*, **46**, p. 125203.
35. Charfi, B. (2013). *J. Spectrosc.*, **1**, p. 6.
36. Kohut, A. (2018). On the plasma and electrode erosion processes in spark discharge nanoparticle generators (PhD Dissertation), University of Szeged.
37. Meier, K. and Kabelac, S. (2016). *J. Chem. Eng. Data*, **61**, p. 3941.
38. Braginski, S. I. (1958). *J. Exp. Theor. Phys.*, **34**, p. 1548.
39. Bardócz, Á. and Vörös, T. (1966). *J. Quant. Spectrosc. Radiat. Transfer*, **6**, p. 35.
40. Diermeier, R. and Krempl, H. (1970). *Spectrochim. Acta, Part B*, **25**, p. 69.
41. Scheeline, A. (1990). *Mikrochim. Acta*, **100**, p. 247.
42. Jones, L. (1950). *Br. J. Appl. Phys.*, **60**.
43. Feng, J., Ramlawi, N., Biskos, G. and Schmidt-Ott, A. (2018). *Aerosol Sci. Technol.*, **52**, p. 505.
44. Muntean, A., Wagner, M., Meyer, J. and Seipenbusch, M. (2016). *J. Nanopart. Res.*, **18**, p. 229.
45. Kohut, A., Wagner, M., Seipenbusch, M., Geretovszky, Zs. and Galbács, G. (2018). *J. Aerosol Sci.*, **119**, p. 51.
46. Voloshko, A. (2016). Formation of nanoparticles by means of spark discharge at atmospheric pressure (PhD Dissertation), University Jean Monnet.
47. Feng, J., Huang, L., Ludvigsson, L., Messing, M. E., Maisser, A., Biskos, G. and Schmidt-Ott, A. (2016). *J. Phys. Chem. C*, **120**, p. 621.
48. Feng, J., Biskos, G. and Schmidt-Ott, A. (2015). *Sci. Rep.*, **5**, p. 15788.

Chapter 4

The Physics and Equivalent Circuit Modeling of Spark Ablation

Emiliano Trodini,[a] **Ian M. Richardson,**[a] **and Andreas Schmidt-Ott**[b,c]

[a]*Faculty of Mechanical Maritime and Materials Engineering, Delft University of Technology, NL-2628 CD Delft, The Netherlands*

[b]*Faculty of Applied Sciences, Delft University of Technology, NL-2629 HZ Delft, The Netherlands*

[c]*Energy Environment and Water Research Center, The Cyprus Institute, Nicosia 2121, Cyprus*

e.trodini@tudelft.nl

Spark ablation is governed by an interplay of rather complex physical processes. To achieve a clear yet succinct treatment, we describe the physics of ablation via its electrical behavior in this chapter. First, some physical considerations are established (Section 4.1), discussing the properties that are most relevant for developing a spark equivalent circuit model (ECM) that directly mirrors the physical structure and properties of the plasma (Section 4.2). The introduction of two simplified versions of such a model allows the derivation of equivalent circuit parameters from the measured spark voltage and current signals. Thus, the physically based theoretical circuit can be related to the real spark electrical

Spark Ablation: Building Blocks for Nanotechnology
Edited by Andreas Schmidt-Ott

ISBN 978-981-4800-82-2 (Hardcover), 978-0-367-81709-1 (eBook)
www.jennystanford.com

behavior. This provides important indications to the design of power supply units for spark discharge generators (SDGs), which crucially determine particle properties [1]. Finally, we discuss the physics of the electrode ablation process, identifying the three main contributing mechanisms and their significance (Section 4.3).

4.1 Spark Physical Characteristics

In this section we provide a brief discussion of the characteristics that are relevant for developing a spark ECM, namely partial local thermodynamic equilibrium (pLTE, Section 4.1.1), gas breakdown (Section 4.1.2). and quasi-neutrality (Section 4.1.3).

4.1.1 Partial Local Thermodynamic Equilibrium

A spark differs from an arc by being a markedly unstable and irreversible discharge [2]. Conversely, an arc is a process that does not deviate significantly from the steady state and reversibility. This contrast is particularly evident at the onset of a discharge, in the so-called breakdown stage. Breakdown lasts for a few tens of nanoseconds in SDGs and is characterized by rapid changes of ionization degree, gas pressure, gap voltage, and electron temperature. Electrode ablation begins at the end of this stage, when the spark converges toward an arc-like behavior. More precisely, the plasma approaches pLTE; that is, ions and neutrals have approximately the same temperature, which is a function of position and may differ from the electron temperature [3].

In noble gasses, the concentrations of electrons and gas atoms (including both ions and neutrals) measured in an SDG gap after current quenching have approximately the same magnitude, implying a high degree of ionization [4]. This confirms the tendency of the spark toward arc-like quasi-equilibrium conditions during the final discharge stage. This consideration justifies the application of arc models for constructing the physically based spark ECM[a] (see Section 4.2.1).

[a]It should be kept in mind that arc equivalent circuit models have reduced validity during the spark breakdown stage (see Section 4.1.2). Moreover, pLTE is just an asymptotic behavior that is approached but rarely fully reached by the sparks in SDGs. However, the absence of literature on spark ECMs reflecting their relatively complex physics makes it convenient to employ arc models.

4.1.2 Breakdown Process

In Section 4.1.1, it was shown that the application of arc models to sparks is particularly critical for the breakdown stage. To assess the level of approximation involved, it is necessary to analyze the breakdown process in more depth. During the so-called prebreakdown stage, electrons are extracted from their parent atoms and accelerated toward the anode. Meanwhile, the freshly generated ions do not move significantly due to their lower mobility [2]. Consequently, the center of mass of the electrons is displaced from that of the ions, creating an electric dipole. The dipole electric field sums up with the external one, enhancing ionization in an avalanche fashion, except in the region in between the electron and ion clouds, in which the external field is attenuated. If the dipole field is sufficiently large, it cancels out the longitudinal component of the external field. This causes the formation of an equipotential channel between the electron cloud and the cathode; that is, a so-called streamer. Conversely, if the gap distance or applied voltage is not large enough, the electron cloud enters the anode before forming a streamer, leaving an unbalanced positive space charge between the electrodes [5]. This latter process is known as Townsend or avalanche breakdown.

The transition from avalanche to streamer occurs when the following empirical relation is fulfilled [2]:

$$\alpha(E_0)d_{\text{gap}} \geq \ln\left(\frac{9\pi\varepsilon_0 E_0}{2q\alpha(E_0)^2}\right), \tag{4.1}$$

where d_{gap} is the gap distance, $E_0 = V/d_{\text{gap}}$ is the spatial average of the externally applied electric field, V is the electrode voltage, ε_0 is the dielectric constant, q is the elementary charge, and $\alpha(E_0)$ is the first Townsend ionization coefficient. Typical magnitudes in SDGs are $d_{\text{gap}} \approx 1$ mm, $E_0 \approx 2 \times 10^4 V/\text{cm}$, $\alpha_{\text{N}}(E_0) \approx 23\ \text{cm}^{-1}$, and $\alpha_{\text{Ar}}(E_0) \approx 100\ \text{cm}^{-1}$ for nitrogen and argon, respectively [5, 6, 7]. Adopting these values, Eq. 4.1 becomes $\alpha_{\text{N}}(E_0)d_{\text{gap}} \approx 2.3 \ngeq 19.5$ for nitrogen and $\alpha_{\text{Ar}}(E_0)d_{\text{gap}} \approx 10 \ngeq 25$ for argon. Consequently, streamers do not develop significantly in SDGs. Instead, the breakdown stage follows Townsend's description. However, the Townsend discharge current is proportional to the electrode voltage, while both theory and experimental observation indicate that sparks in SDGs have descending current–voltage characteristics (CVC), as in the case of arcs [2, 8]. It follows that the Townsend behavior lasts only for the initial stage of discharge. Namely, the unbalanced space charge channel converges toward arc-like equilibrium conditions within a few tens of nanoseconds after a current begins to flow, while the total discharge duration is of the order of a few microseconds. Spark conditions just after a Townsend breakdown deviate from arc-like quasi-equilibrium more than in the case of streamer formation. This implies a limited validity of the arc models that will be used to assemble the physically based spark equivalent circuit (Section 4.2.1) during the first ~50 ns of the discharge [5].

4.1.3 Quasi-Neutrality and Sheath Formation

Free charges in a plasma naturally tend to assume a configuration where the mutually exerted electrostatic forces are balanced. This tendency translates into establishment of the quasi-neutrality condition: the plasma becomes neutral in each subregion with dimensions above the Debye length λ_D (see below); that is, $\langle n_e \rangle = \langle n_i \rangle$, where $\langle n_e \rangle$ and $\langle n_i \rangle$ are the electron and ion concentration, respectively, averaged over the sub-region volume [9]. Quasi-neutrality is a fundamental prerequisite for the applicability of arc ECMs to spark discharges.

Quasi-neutrality does not hold at distances smaller than λ_D from the electrode walls, in the so-called sheath space charge layers. Electrons are depleted from those regions, owing to their higher mobility relative to ions [9]. As a consequence of this imbalance, a voltage drop arises across the sheath layers, accelerating the charges that pass through the sheath. The radical difference in electrical characteristics between the quasi-neutral plasma and the sheath layers will be reflected in the physically based ECM. This division permits discrimination between phenomena occurring close to electrode walls from those located in the inner plasma, which may be advantageous for investigating ablation.

To reach quasi-neutrality and form the sheath layers, a plasma should extend over length scales larger than the Debye length

$$\lambda_D = \left(\frac{q^2 n_e}{\varepsilon_0 T_e} + \frac{q^2 n_i}{\varepsilon_0 T_i} \right)^{-1/2}. \tag{4.2}$$

Adopting values typical of SDG experiments in argon, the electron temperature is $T_e \approx 2 \times 10^4$ K and the electron concentration is $n_e \approx 5 \times 10^{17}$ cm^{-3} [4, 9]. Neglecting for simplicity the corresponding ionic quantities, denoted by the subscript *i*, allows to find an upper bound for λ_D of 14 nm. This length is much smaller than any spark characteristic dimension, since both gap spacing d_{gap} and effective radius r are in the millimeter range [5, 10]. Therefore, the spatial requirement for the establishment of quasi-neutrality and space charge sheaths (which typically extend for several Debye lengths) is easily fulfilled. Considering that the charge concentration in the space charge layer is lower than in the bulk plasma, the screening capability of the sheaths is expected to be less than that calculated above. Thus, the sheath thickness d_s of sparks in nanoparticle (NP) generators can be assumed to be in the order of some hundreds of nanometers.

A further condition for achieving quasi-neutrality is having enough discharge time for the charge carriers to assume the required configuration. The corresponding time constant can be expressed as [9]

$$\tau_{pe} = \frac{1}{\omega_{pe}} = \left(\frac{\varepsilon_0 m_e}{n_e q^2}\right)^{1/2}, \tag{4.3}$$

where w_{pe} is the electron plasma frequency and m_e is the electron mass. Ideally, the electron concentration n_e in the above formula should refer to the beginning of the breakdown stage; that is, when the electrode gap is bridged by a conductive channel. Owing to the practical difficulties of such a measurement, the electron concentration $n_e = 5 \times 10^{17}$ cm^{-3} evaluated 0.5 μs after breakdown is used as an indication, giving $\tau_{pe} \sim 2.5 \times 10^{-14}$s [4]. This time is negligible compared to a spark duration of some microseconds typical of SDGs, implying the establishment of quasi-neutrality and sheath layers.

The characteristic time for ions to respond to the applied field is given by the inverse ion plasma frequency $\tau_{pi} = 1/\omega_{pi} \approx 6.76$ ps, which is obtained by inserting the argon atomic mass m_i and the ion concentration n_i instead of the corresponding electron quantities into Eq. 4.3 and assuming quasi-neutrality; that is, $n_i = n_e = 5 \times 10^{17}$ cm^{-3}. The ion response time τ_e is much shorter than the period of the sheath voltage waveform, which is around 1 μs [5]. Therefore, the ions are capable of reacting to the instantaneous electric field across the sheath layers, and their motion has to be represented by the ECM.

4.2 Spark Equivalent Circuit Models

The ECM of an SDG discharge loop is represented in Fig. 4.1: a current source I_{gen} feeds the biasing capacitor C_{bias}, which then discharges through a spark impedance Z_{spark} via the parasitic inductance L_c (mostly due to cabling). The present section describes possible ECMs for the impedance Z_{spark}. On the basis of the physical framework developed in Section 4.1, a physically rooted spark ECM is developed. It turns out to be extremely challenging to implement this model into a fitting algorithm. Therefore, the simplest and most widely applied constant-resistance ECM and the time-varying R-C ECM, which is the most advanced SDG electrical model to date, are also presented.

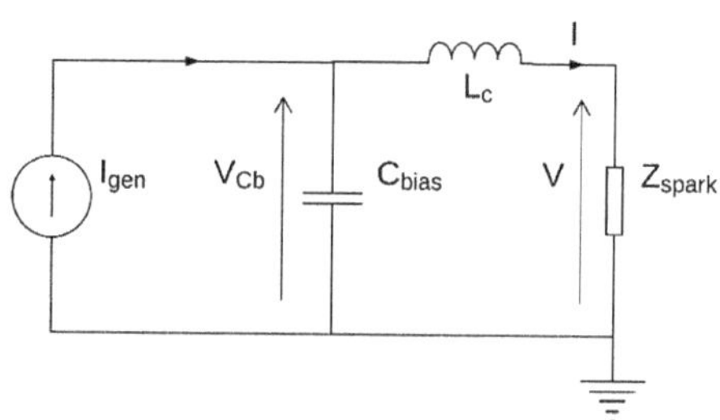

Figure 4.1 Equivalent circuit representation of an SDG discharge loop, made of a current source I_{gen} that charges a bias capacitor C_{bias}, which is connected in parallel with the spark impedance Z_{spark} and a series parasitic inductance L_c.

4.2.1 Physically Based Equivalent Circuit

A time-varying spark equivalent circuit that conforms to the plasma physical conditions described in Section 4.1 is shown schematically in Fig. 4.2. It is divided into three blocks of parallel elements: an inner $(R$-$L)//C$ circuit representing the quasi-neutral plasma is connected to the electrodes via two mirror-symmetric circuits representing the sheath layers [5]. In this section, the characteristics of each single circuit element are considered.

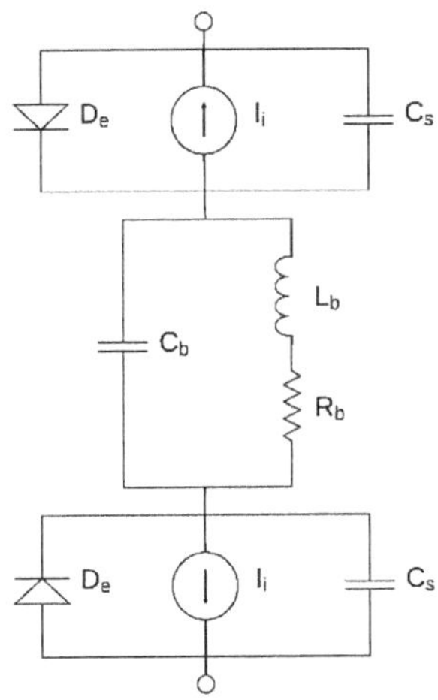

Figure 4.2 Diagram of the physically based spark equivalent circuit: an internal block represents the quasi-neutral plasma, which is connected to the electrodes via two mirror-symmetric blocks representing the sheath layers.

4.2.1.1 Quasi-neutral plasma equivalent circuit

The ECM for the quasi-neutral bulk plasma is made of a capacitance C_b in parallel connection with a resistance R_b and a series inductance L_b (see Fig. 4.2). The empty capacitor C_b depends only on spark geometry [9]:

$$C_b(t) = \frac{\varepsilon_0 A(t)}{d_{gap}}, \tag{4.4}$$

where $A(t)$ is the cross-sectional area of the plasma column. Instead of having the parallel R-L circuit elements included in the model, the bulk plasma material properties can also be incorporated into C_b by multiplying it with the relative dielectric constant $\varepsilon_r(t)$. Despite being more compact, however, this latter representation does not differentiate between the contributions of different physical agents to the bulk plasma electrical behavior as in the three-element ECM of Fig. 4.2.

The inductor L_b models the inertia of the electrons flowing through the plasma and is defined as

$$L_{\mathrm{b}}(t)=\omega_{\mathrm{pe}}^{-2}(t)C_{\mathrm{b}}^{-1}(t)=\frac{d_{\mathrm{gap}}m_{\mathrm{e}}}{q^{2}n_{\mathrm{e}}(t)A(t)}. \tag{4.5}$$

The spark inductance can also be calculated using only plasma geometry as input, employing an expression that is given in Ref. [11].

The resistor R_{b} models electron friction and is defined as [9, 12]

$$R_{\mathrm{b}}(t)=\frac{d_{\mathrm{gap}}}{A(t)}\frac{1}{\sigma_{\mathrm{b,DC}}(t)}=\frac{d_{\mathrm{gap}}^{2}}{\mu(t)Q(t)}, \tag{4.6}$$

where $\mu(t)$ is electron mobility, $Q(t) = d_{\mathrm{gap}}\, qA(t)n_{\mathrm{e}}(t)$ is the total charge in the discharge volume, and $\sigma_{\mathrm{b,DC}}(t)$ is the DC conductivity of the bulk plasma. The latter parameter may be estimated from the well-known Spitzer model, transforming Eq. 4.6 into the following expression [9, 13]:

$$R_{\mathrm{b}}(t)=\frac{d_{\mathrm{gap}}}{A(t)}\frac{\pi q^{2}\sqrt{m_{\mathrm{e}}}\ln\Gamma(t)}{(4\pi\varepsilon_{0})^{2}(k_{\mathrm{B}}T_{\mathrm{e}}(t))^{2}}, \tag{4.7}$$

where k_{B} is the Boltzmann constant and $\Gamma(t) = (k_{\mathrm{B}}T_{\mathrm{e}}(t))^{3/2}/\left(4q^{3}\sqrt{\pi n_{\mathrm{e}}(t)}\right)$.

4.2.1.2 Sheath equivalent circuit

The sheath ECM described in this section implies a set of assumptions that are discussed in Ref. [14]. The complete list is not given here for brevity; instead, only the two most notable assumptions are discussed. First, the time average of the voltage across each sheath $\bar{V}_{\mathrm{s}}$ is assumed to be large, namely $q\bar{V}_{\mathrm{s}} \gg k_{\mathrm{B}}T_{\mathrm{e}}$, implying that the space charge layers are heavily depleted of electrons. Second, inelastic collisions in the sheaths are neglected, owing to a layer thickness $d_{\mathrm{s}} \approx 100$ nm that is shorter than the ion mean free path $\lambda_{\mathrm{i}} = k_{\mathrm{B}}T_{\mathrm{i}}/(\sqrt{2}\sigma_{\mathrm{i,coll}}P) \geq 1\ \mu\mathrm{m}$ [5]. To estimate this lower bound for λ_{i}, an ion temperature $T_{\mathrm{i}} = 1000$ K, a pressure $P = 10^{5}$ Pa, and a collision cross section $\sigma_{\mathrm{i,coll}} = 10^{-19}\ \mathrm{m}^{2}$ have very conservatively been assumed for an arc-stage discharge.

The diode D_{e} in Fig. 4.2 models the rectifying behavior of the electrons that, due to their high mobility, tend to immediately cancel any negative potential V_{s} of the bulk plasma relative to the electrodes [9]. The electronic current can be formulated as [15]

$$I_e(t) = \frac{qA(t)u_e(t)n_p(t)}{4}\exp\left(\frac{qV_s(t)}{k_B T_e(t)}\right), \tag{4.8}$$

where $u_e = \sqrt{8k_B T_e(t)/\pi m_e}$ is the electron mean velocity and n_p is the charge concentration at the sheath edge; that is, the plasma surface closest to a given electrode in which the quasi-neutrality relationship $\langle n_e \rangle = \langle n_i \rangle = n_p$ is still valid. The charge concentration n_p at the sheath edge differs from that in the bulk plasma n_0 because of the presence of an intermediate layer of thickness $\sim\lambda_i$, the so-called presheath. In this region, charge depletion occurs while maintaining quasi-neutrality. The charge concentration at the sheath edge can be expressed as [9]

$$n_p(t) = n_0(t)\exp\left(\frac{q\bar{V}_{ps}(t)}{k_B T_e(t)}\right) \approx n_0(t)e^{-1/2}, \tag{4.9}$$

where the presheath voltage averaged over a discharge period is approximated as $\bar{V}_{ps}(t) \approx k_B T_e(t)/(2q)$.

The current source I_i in Fig. 4.2 models the ion flux across the space charge layer, being straightforwardly defined as [15]

$$I_i(t) = qA(t)\,\mathbf{u}_b(t)\,n_p(t), \tag{4.10}$$

where the Bohm velocity $\mathbf{u}_b(t) = \sqrt{k_B T_e(t)/m_i}$ is assumed to be the velocity of the ions that enter the sheath. An alternative formula is provided by the well-known Child–Langmuir law [9]:

$$I_i(t) = \frac{4}{9}\varepsilon_0 A(t)\left(\frac{2q}{m_i}\right)^{1/2}\frac{V_s^{3/2}(t)}{d_s^2(t)}, \tag{4.11}$$

where this latter expression shows more explicitly the dependence of the ionic flux on the instantaneous value of the sheath voltage, as discussed in Section 4.2.1.2.

The sheath capacitor C_s in Fig. 4.2 represents the displacement current $I_d = C_s dV_s/dt$ (i.e., a movement of charges within the plasma but not across its boundaries) induced by time variations of sheath thickness and is defined simply as [15]

$$C_s(t) = \frac{\varepsilon_0 A(t)}{d_s(t)}, \tag{4.12}$$

where the sheath thickness can be estimated by combining Eqs. 4.10 and 4.11 as

$$d_s(t) \approx \frac{2}{3}\left(\frac{\varepsilon_0}{n_p(t)}\right)^{1/2}\left(\frac{2V_s^3(t)}{qk_B T_e(t)}\right)^{1/4} . \tag{4.13}$$

4.2.2 Constant Spark Resistance

The circuit model presented in Section 4.2.1, in spite of its physical accuracy, is too complex for practical implementation, at least when most of its input variables (i.e., r, n_e, T_e, V_s, and d_s) are unknown. The simplest and most widely applied solution in SDG research is to model the spark as a constant resistance [1, 16–18]. This assumption allows an analytical solution for the second-order differential equation given by the Kirchhoff's voltage law of the discharge loop in Fig. 4.1, yielding

$$V_{Cb}(t) = V_{bd}\left(1+\frac{1}{\tau^2\omega_s^2}\right)^{1/2}\exp\left(-\frac{t}{\tau}\right)\cos(\omega_s t-\varphi) , \tag{4.14}$$

$$I(t) = \omega_s C_{bias} V_{bd}\left(1+\frac{1}{\tau^2\omega_s^2}\right)\exp\left(-\frac{t}{\tau}\right)\sin(\omega_s t) , \tag{4.15}$$

where V_{bd} is the voltage at the beginning of the breakdown stage, τ is the damping coefficient, ω_s is the angular oscillation frequency, and φ is a phase factor. The latter three parameters are given by

$$\tau = \frac{2L_c}{R_{spark}} , \tag{4.16}$$

$$\omega_s = \sqrt{\frac{1}{L_c C_{bias}}\left(\frac{C_{bias}R_{spark}^2}{4L_c}\right)} , \tag{4.17}$$

and

$$\varphi = \arctan\left(\frac{1}{\tau\omega_s}\right) . \tag{4.18}$$

The magnitude of the spark resistance can easily be derived experimentally from the measurement of either the current or the capacitor-voltage waveform, obtaining $Z_{spark} = R_{spark}$ from the damping coefficient τ through Eqs. 4.14–4.16.

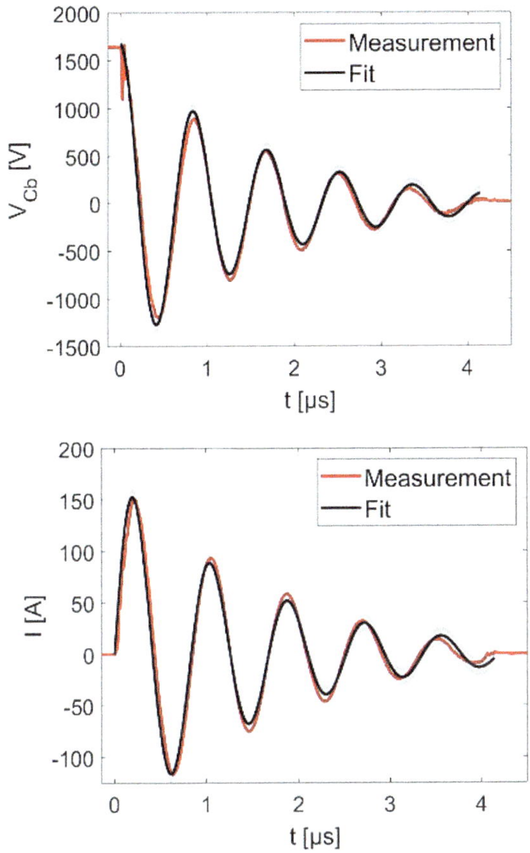

Figure 4.3 Comparison between waveforms resulting from measurement (red) and constant-resistance ECM (black) of the capacitor voltage (left) and the discharge current (right). The uncertainty interval (shaded gray) corresponds to the mean relative error, which is 16.3% for V_{Cb} and 16.2% for I. The measurements were performed on an SDG with Cu electrodes in N5.0 Ar gas, with electrode diameter d_{el} = 4.2 mm, gas flow rate ϕ = 5 slm, electrode spacing d_{gap} = 1.2 mm, biasing capacitor C_{spark} = 14 nF, bias current I_{bias} = 10 mA, parasitic inductance L_c = 1.19 μH, parasitic resistance R_{par} = 296 mΩ, spark resistance R_{spark} = 1.30 Ω, and decay constant τ = 1.55 μs.

Figure 4.3 shows a comparison between the capacitor voltage V_{Cb} and the current I measured in an SDG with the corresponding best fit obtained from the constant-resistance ECM of Eqs. 4.14–4.18 [5]. The two signals are fairly well represented by the model. Therefore, they have little sensitivity to the time variations of the spark resistance, being mostly determined by its DC value $R_{spark} \approx 1\ \Omega$. Moreover, the fit accuracy of Fig. 4.3, obtained in the absence of reactive elements in the spark ECM, manifests the small influence of such elements on the bias voltage and discharge current waveforms [5]. Instead, the behavior of these signals is determined by the reactive elements of the external circuit, that is, the biasing capacitor C_{bias} and the cabling inductance L_c, according to Eqs. 4.14–4.18. Specifically, the frequency ω_s and the attenuation τ of the discharge current I are fixed by the external circuit design and have little dependence from the physical conditions of the spark gap. It is worth noting that L_c depends on the actual wiring of the circuit, and hence, a very accurate design of the discharge loop circuit is needed to control spark electrical behavior. In particular, it is important for the relative ablation of the electrodes, especially in mixing applications (see Chapter 5). L_c and R_{spark} can be varied for modifying the damping of spark oscillations (see Eq. 4.16) in order to tune particle production.

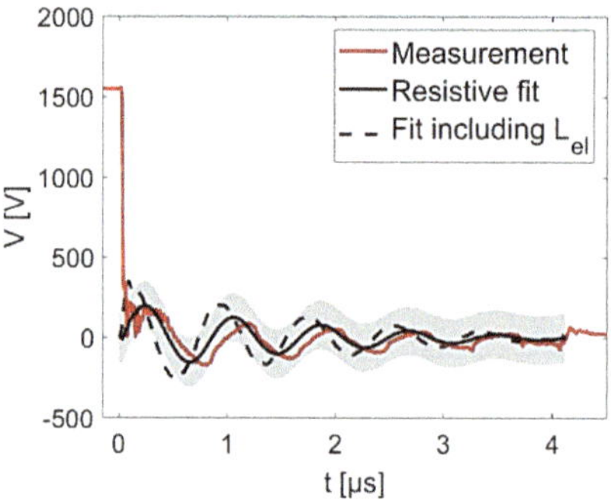

Figure 4.4 Comparison between the electrode voltage waveforms resulting from measurement (red, conditions as in Fig. 4.3), the constant-resistance ECM (black), and adding a constant electrode inductance in series to spark resistance (dashed black). The uncertainty interval (shaded gray) corresponds to the mean relative error of the constant-resistance model, which amounts to 94.4%.

Figure 4.4 shows the spark voltage waveform derived from measurement and from the constant-resistance ECM using $V = R_{spark}I$. The poor agreement between the two signals indicates that a more refined model is needed to accurately describe the spark voltage. The evolving phase shift of the modeled signal with respect to the measured one suggests that a time-dependent reactive contribution should be included in the ECM. Furthermore, a large resistance that rapidly falls upon breakdown is required to represent the voltage decay. These considerations motivate modeling of the spark by a time-varying *R-C* equivalent circuit, which translates into the representation of the discharge loop that is shown in Fig. 4.5.

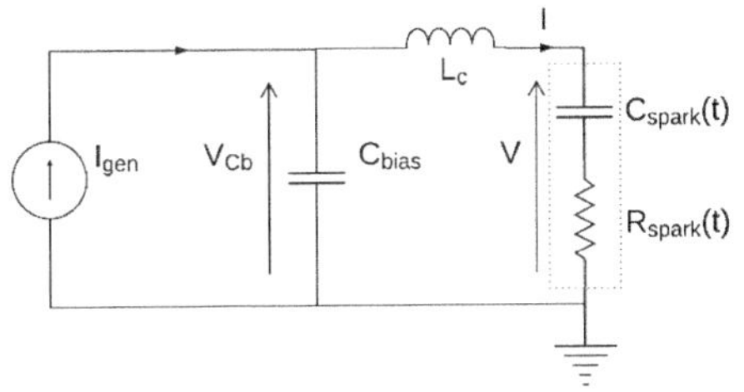

Figure 4.5 Equivalent circuit representation of an SDG discharge loop (cf. Fig. 4.1). The time-varying spark impedance (enclosed in the dotted box) is modeled with a capacitance $C_{\text{spark}}(t)$ and a series resistance $R_{\text{spark}}(t)$.

4.2.3 Time-Varying *R-C* Equivalent Circuit

A time-varying series *R-C* spark equivalent circuit (see Fig. 4.5) can be derived with an iterative graphical interpolation of the V/I curve, which is described in Ref. [5] and not reported here for brevity. This model correctly resembles the behavior of an SDG spark voltage, as demonstrated by Fig. 4.6. The voltage waveform in the figure was modeled from the spark resistance and capacitance time dependences shown in Fig. 4.7. The computation of spark capacitance starts at time t_{0c}, together with spark voltage ringing, as indicated by the green circle in Figs. 4.6 and 4.7; that is, C_{spark} is assumed to be infinite for previous times.

Figure 4.7 enables a physical interpretation of spark evolution based on the equivalent circuit elements. Having the physically based ECM of Fig. 4.2 as a reference, spark resistance R_{spark} is associated with the quasi-neutral plasma, while spark capacitance C_{spark} is mostly due to the cathode sheath [5]. This interpretation implies the following assumptions:

1. The sheath diodes D_e are ideal, which occurs for $V_s >> k_B T_e/q \approx$ 2V (see Eq. 4.8). As a consequence, anode sheath thickness d_s and voltage V_s are negligibly small compared to corresponding cathode parameters.

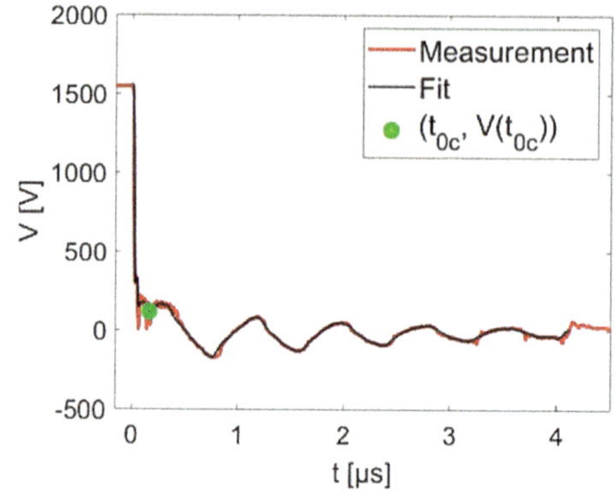

Figure 4.6 Comparison between the electrode voltage waveforms resulting from measurement (red, conditions as in Fig. 4.3) and the *R-C* time-varying ECM (black). The uncertainty interval (shaded gray) corresponds to the mean relative error, which amounts to 11.2%. The time at which C_{spark} becomes finite is indicated by a green circle.

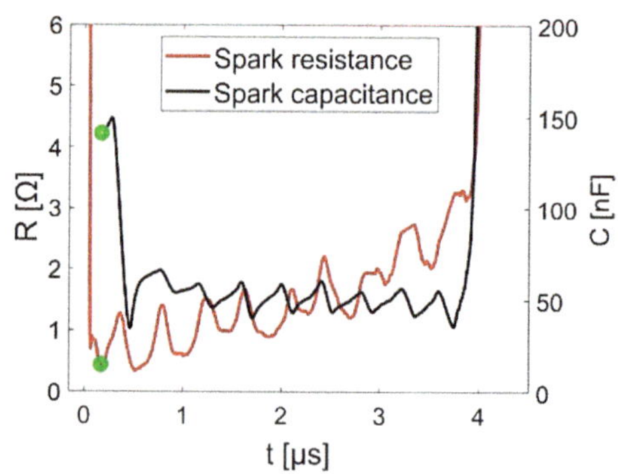

Figure 4.7 Time evolution of the spark resistance (red) and series capacitance (black) corresponding to the modeled voltage waveform of Fig. 4.6. The time at which C_{spark} becomes finite is evidenced with a green circle.

2. The ionic current I_i is negligible with respect to the displacement current represented by C_{spark}. Conservatively assuming $T_e = 17{,}000$ K, $n_e = 2 \times 10^{17}$ cm^{-3}, and $r = 500$ µm, results in a lower bound of $I_i \approx 20$ A (see Eq. 4.10).
3. Bulk plasma inductance L_b produces a negligible voltage, which, taking $r = 500$ µm, occurs for $n_e \geq 10^{12}$ cm^{-3} (see Eq. 4.5).

4. The capacitance of the bulk plasma C_b is negligible compared to that of the cathode sheath, which is true for $d_s << d_{gap} \approx$ 1 mm (see Eqs. 4.4 and 4.12).

The validity of the assumptions listed above seems to be guaranteed for any practical condition in SDG experiments, except for the second one. Indeed, neglecting the ionic current $I_i \approx 20$ A when the total spark current is $I \approx 100$ A is not fully appropriate. This may affect the accuracy of quantitative predictions of spark parameters that employ the definition of R_{spark} and C_{spark} from the physically based ECM discussed in Section 4.2.1 (see Ref. [5] for more details). Nevertheless, from a qualitative perspective, the interpretation $R_b = R_{spark}$ and $C_s = C_{spark}$ remains the most sensible option, given the above approximations and the topology of the time-varying ECM.

This physical interpretation can be employed to understand the time history of the equivalent circuit elements shown in Fig. 4.7. Upon breakdown, the spark resistance rapidly falls, owing to a rise of the plasma radius and temperature, corresponding to the increase in conductivity described by Eqs. 4.6 and 4.7 [19]. At time t_{0c}, when C_{spark} becomes finite, the resistance reaches its first and smallest minimum. At this point, both temperature and conductivity reach their maximum value over the whole discharge, while the plasma column is still performing its first expansion. For the above reasons, t_{0c} is interpreted as the time when the unbalanced space charge left after a Townsend-type breakdown has formed an equipotential channel between the electrodes, that is, the beginning of the arc stage (see Section 4.1). For subsequent times, R_{spark} can be seen as an oscillating function superimposed onto an exponentially increasing one. The oscillations follow the periodicity of the discharge current I, while retaining the time-varying phase features of the spark voltage V. This behavior is mostly due to the fluctuations of spark radius r, which, in turn, follow those of the current [5]. More precisely, to allow a current rise, the plasma cannot accelerate electrons beyond a certain limit, which is given by the collision frequency. Therefore, the plasma has to expand and ionize the surrounding gas to provide additional charge carriers for the current flow and vice versa. The exponentially growing resistance component is due to the progressive decrease of spark temperature and charge density as the discharge progresses. This process is accompanied by a channel expansion (lowering the resistance), but the exponential trend is nevertheless dominant.

The spark capacitance C_{spark} is assumed to become finite at time $t_{0\text{c}}$ because before that point in time spark electrical characteristics could be accurately modelled as being purely resistive. After its formation, C_{spark} keeps a large value for ~100 ns because spark impedance behavior is inductive during that interval (see Fig. 4.4, where the modeled voltage is directly proportional to the current). The following rapid fall of C_{spark} marks the transition to a capacitive spark impedance. Next, C_{spark} oscillates around an almost stable value, having the same frequency as the spark voltage and resistance (cf. Figs. 4.6 and 4.7). Such fluctuations reflect the dynamics of space charge geometry, that is, the time variations of the channel radius r and sheath thickness d_s (see Eq. 4.12). Toward the end of the discharge, the spark capacitance blows up, behaving as a short circuit and bringing the spark again into a purely resistive regime. This may be due to a small charge concentration that reduces the plasma screening capabilities (see Eq. 4.2), resulting in an enlargement of sheath thickness.

4.3 Ablation Mechanisms

Electrode ablation, or more precisely bringing electrode material into the vapor phase, is the desired effect of spark discharges in SDGs. Short discharges allow rapid quenching of the vapor produced, leading to extremely high supersaturation. The result is the production of very small particles, where the lower limit is the single atom (see Chapter 7). As explained in various publications (see Chapter 5), the important capability of SDGs of producing virtually any mixture of materials is also linked to the limited duration of the sparks as opposed to arc discharges. The ablation process can arise from different origins, which are qualitatively discussed in this section. It has to be remarked that none of these origins is associated with ordinary diffusion-based thermal conduction of the gas, transferring thermal energy from the hot spark to the electrode material. This stems from the full ionization that is to be assumed in the sheath zones as well as the high electric fields present here.

Unfortunately, there are no analytic models in spark ablation literature that allow a quantitative description of the different ablation mechanisms. Nevertheless, the spark parameters having a direct influence on each ablation mechanism can be identified. Finally, some indications on which of the mechanisms play a dominant role in determining electrode mass loss are also given.

4.3.1 Shock-Wave Expansion

Spark ignition is accompanied by the emission of a shock wave, which releases the temperature and pressure gradients that build up during gas breakdown. When this wave hits an electrode surface, it may produce NPs through fragmentation (due to momentum transfer) or evaporation (due to heat transfer). The radius of the shock wave r_{sw} coincides with that of the plasma column r from breakdown until the time t_{sep}, when the wave enters a regime of free expansion characterised by $r_{sw} \propto \sqrt{t}$ [20]. The time t_{sep} is approximately twice the time of the first discharge-current peak. Shock-wave velocity and acceleration for $t < t_{sep}$ have no statistical correlation with the ablation on SDG electrodes in various experimental conditions [5]. This assigns shock-wave expansion to a secondary role in determining SDG electrode mass loss.

However, the occasionally observed parasitic phenomenon referred to as "splashing" [21] and already described before any interest in NP production had arisen [22] may very well be triggered by the shock wave. It is a mechanism producing large round particles up to several microns in size that are evidently not formed from vapor but from droplets emitted from molten ponds on the electrodes. This phenomenon has rarely been investigated systematically, although a seemingly successful attempt has been made to suppress it by avoiding multiple occurrence of sparks at the same positions [23].

4.3.2 Electron Condensation

Electrode evaporation in SDGs can be caused by the energy transfer associated with electron flow, in a process known as electron condensation. When an electron is extracted from the cathode, its potential energy rises by an amount that is close to the electrode workfunction Φ. Then, the electron is accelerated through the cathode sheath and the quasi-neutral plasma column, arriving after many collisions at the anode sheath edge in thermal equilibrium with the electron plasma, that is, with kinetic energy equal to $k_B T_e$. Finally, the acceleration through the anode sheath adds a contribution of qV_{s+} to the electron kinetic energy, where V_{s+} is the anode sheath voltage. When the electron is absorbed by the anode, it releases the part of its total energy in excess of the anode thermal energy $k_B T_+$. This energy eventually assumes the form of heat and contributes to evaporating the anode. The power flux due to electron condensation that is received by the anode can be expressed as

$$P_+(t) = \left(\frac{\Phi}{q} + \frac{k_B}{q}(T_e(t) - T_+(t)) + V_{s+}(t) \right) I_{e+}(t) . \tag{4.19}$$

It is worth noting that the conduction current at the anode I_{e+} is not equal to the externally-measurable discharge current I, because the latter includes also the displacement current component I_d, representing charges moving within the plasma and not across the electrode boundary. No analytical expression is available in spark ablation literature to relate the power flux in Eq. 4.19 to the ablation rate. However, among the different mechanisms treated in this section, electron condensation has the highest degree of statistical correlation with observed SDG electrode mass loss rate [5]. Therefore, electron condensation can be regarded as an influential contribution to SDG electrode ablation, especially when anode mass loss forms a significant contribution to the total ablation.

4.3.3 Cation Bombardment

Cations are accelerated toward the cathode by the electric field in the neighboring sheath layer. Recall from Section 4.1.3 that ions respond to the instantaneous electric field, owing to the large magnitude of their plasma frequency ω_{pi}. Cation impact can result in NP production by momentum or heat transfer to the cathode. The power flux received by the cathode can be defined as

$$P_-(t) = V_{s-}(t) I_{i-}(t) - \frac{\Phi}{q} I_{e-}(t) , \tag{4.20}$$

where V_{s-} is the cathode sheath voltage; I_{i-} and I_{e-} are the ionic and electronic current at the cathode, respectively; and the second term on the right-hand side represents cathode cooling due to electron emission (see Section 4.3.2).

Under typical SDG experimental conditions, the most ablated electrode is the one that starts with the negative polarity [5]. This electrode is the cathode during breakdown and for the first half-cycle of the oscillating spark voltage, which is the interval when the spark electrical power is at its maximum. Feng et al. have even shown that the experimental electrode ablation ratio in an oscillatory discharge agrees with a model in which *only* the momentary cathode is ablated (see Chapter 5) [24]. This leads to the conclusion that among the ablation mechanisms herewith considered, only cation bombardment is consistent with these observations, implying the dominance of its contribution to the overall electrode mass loss. At least this is true for conditions under which SDGs are normally operated (take the conditions of Fig. 4.3 as a reference), but this may not be the case if gas, energy per spark, or other important parameters strongly deviate from these.

4.4 Conclusion

There is a rather well-established qualitative physical understanding of the spark discharges as used for NP production. However, quantitative modeling suffers from unknown parameters, difficult to get access to experimentally. A simple RCL circuit model consisting of the bias capacitance C_{bias} (used to store the energy released in each spark), the parasitic inductance of the circuit, L_c, and representation of the spark as a constant resistance provides a good fit to the measured capacitance voltage for the oscillatory discharges usually encountered in spark ablation systems and desired in spark mixing applications (see Chapter 5). However, this model fails to describe the voltage measured across the spark as a function of time, $V(t)$. Modelling the spark impedance by a time-dependent resistance and a time-dependent capacitance in series allows a perfect fit of the measured $V(t)$ and corresponds to a dominant resistive component $R(t)$ in the core of the plasma and a dominant capacitive component $C(t)$ (sheath capacitance) near the electrode surfaces, in agreement with the physical picture. The time dependences $C(t)$ and $R(t)$ can qualitatively be understood by the physical evolution of the spark process.

Shock-wave expansion, electron condensation, and cation bombardment are the physical mechanisms to be considered in spark ablation. Experiments under typical SDG conditions, where cathode ablation is clearly dominant, lead to the conclusion that cation acceleration in the cathode sheath layer is the dominant ablation mechanism, which expresses that within the sheath the driving force is predominantly electric due to the high electric field within the sheath and total ionization. However, the sheath layer is very thin, which means that a model assuming pure thermal transfer from the bulk plasma to the electrode as suggested in Chapter 3 may also be justified for a model aimed at predicting ablation from conveniently measurable input parameters. Electron condensation may also have a significant influence, but shock-wave expansion seems, at most, to play a role in the parasitic effect of splashing.

References

1. Pfeiffer, T. V. (2014). Towards the industrial application of spark ablation for nanostructured functional materials, Technische Universiteit Delft, Delft.
2. Bazelyan, E. M. and Raizer, Y. P. (1998). *Spark Discharge* (CRC Press, USA).
3. Kroesen, G. M. W., Schram, D. C., Timmermans, C. J. and de Haas, J. C. M. (1990). The energy balance of a plasma in partial thermodynamic equilibrium, *IEEE Trans. Plasma Sci.*, **18**, pp. 985–991.
4. Kohut, A., Galbács, G., Marton, Z. and Geretovszky, Z. (2017). Characterization of a copper spark discharge plasma in argon atmosphere used for nanoparticle generation, *Plasma Sources Sci. Technol.*, **26**, p. 045001.
5. Trodini, E. (2017). Modelling the electrical signals in spark discharge generators for electrode ablation control, ETH Zurich, Zurich.
6. Harrison, M. A. (1957). Townsend's first ionization coefficient in nitrogen, *Phys. Rev.*, **105**, p. 366.
7. Kruithof, A. A. (1940). Townsend's ionization coefficients for neon, argon, krypton and xenon, *Physica*, **7**, p. 519.
8. Hontanon, E., Palomares, J. M., Stein, M., Guo, X., Engeln, R., Nirschl, H. and Kruis, F. E. (2013). The transition from spark to arc discharge and

its implications with respect to nanoparticle production, *J. Nanopart. Res.*, **15**, p. 1957.

9. Piel, A. (2010). *Plasma Physics* (Springer, New York).
10. Palomares, J. M., Kohut, A., Galbács, G., Engeln, R. and Geretovszky, Z. (2015). A time-resolved imaging and electrical study on a high current atmospheric pressure spark discharge, *J. Appl. Phys.*, **118**, p. 233305.
11. Persephonis, P., Vlachos, K., Georgiades, C. and Parthenios, J. (1992). The inductance of the discharge in a spark gap, *J. Appl. Phys.*, **71**, p. 4755.
12. Persephonis, P., Giannetas, V., Georgiades, C., Parthenios, J. and Ioannou, A. (1995). The influence of the external circuit on arc-discharge of a spark-gap: its application to a pulsed gas laser, *IEEE J. Quant. Electron.*, **31**, p. 567.
13. Robiscoe, R. T. and Sui, Z. (1985). Circuit model of surface arcing, *J. Appl. Phys.*, **58**, p. 1744.
14. Metze, A., Ernie, D. W. and Oskam, H. J. (1986). Application of the physics of plasma sheaths to the modeling of RF plasma reactors, *J. Appl. Phys.*, **60**, p. 3081.
15. Edelberg, E. A. and Aydil, E. S. (1986). Modelling of the sheath and the energy distributions of ions bombarding RF-biased substrates in high density plasma reactors and comparison to experimental measurements, *J. Appl. Phys.*, **60**, p. 3081.
16. Vons, V. A. (2010). Spark discharge generated nanoparticles for hydrogen storage applications, Technische Universiteit Delft, Delft.
17. Meuller, B. O., Messing, M. E., Engberg, D. L. J., Jansson, A. M., Johansson, L. I. M., Norlen, S. M., Tureson, N. and Deppert, K. (2012). Review of spark discharge generators for production of nanoparticle aerosols, *Aerosol Sci. Technol.*, **46**, p. 1256.
18. Tabrizi, N. S., Ullmann, M., Vons, V. A., Lafont, U. and Schmidt-Ott, A. (2009). Generation of nanoparticles by spark discharge, *J. Nanopart. Res.*, **11**, p. 315.
19. Kushner, M. J., Kimura, W. D. and Byron, S. R. (1985). Arc resistance of laser-triggered spark gaps, *J. Appl. Phys.*, **58**, p. 1744.
20. Akiyama, H., Kristiansen, M., Krompholz, H. and Maas, B. (1988). Current-voltage characteristics of a high-current pulsed discharge in air, *IEEE Trans. Plasma Sci.*, **16**, p. 312.
21. Pfeiffer, T. V., Feng, J. and Schmidt-Ott, A. (2014). New developments in spark production of nanoparticles, *Adv. Powder Technol.*, **25**, p. 56.

22. Petr, R. A. and Burkes, T. R. (1980). Acoustic phenomena in erosion of spark-gap electrodes, *Appl. Phys. Lett.*, **36**, p. 536.

23. Feng, J. (2016). Scalable gas-phase synthesis of nanoparticles: from fundamentals to applications, Technische Universiteit Delft, Delft.

24. Feng, J., Ramlawi, N., Biskos, G. and Schmidt-Ott, A. (2018). Internal nanoparticle mixing by oscillatory spark ablation between electrodes of different materials, *Aerosol Sci. Technol.*, **52**, p. 505.

25. Mitchell, E. W. and Mitchell, J. W. (1951). The work functions of copper, silver and aluminium, *Proc. Roy. Soc. London*, **210**, p. 70.

Chapter 5

Generation of Mixed Nanoparticles by Spark Ablation of Alloys and Spark Mixing

Alex Muntean,[a] Gábor Galbács,[b,c] Andreas Schmidt-Ott,[d,e] and Martin Seipenbusch[a,f]

[a]*Institute of Chemical Process Engineering, University of Stuttgart, Böblinger Straße 78, 70199 Stuttgart, Germany*
[b]*Department of Inorganic and Analytical Chemistry, University of Szeged, Dóm sq. 7, 6720 Szeged, Hungary*
[c]*Department of Materials Science, Interdisciplinary Excellence Centre, University of Szeged, 6720 Szeged, Dugonics sq. 13, Hungary*
[d]*Delft University of Technology, Faculty of Applied Sciences, 2629 HZ Delft, The Netherlands*
[e]*Energy Environment and Water Research Center, The Cyprus Institute, Nicosia 2121, Cyprus*
[f]*ParteQ GmbH, Brunnenstraße 12, 76361 Malsch, Germany*
seipenbusch@parteq.net

One of the great advantages of particle production by means of spark ablation is the rather unique option of generating alloyed or mixed nanoparticles. The present chapter focusses mainly on the dependence of the feedstock composition on production rate, actual

Spark Ablation: Building Blocks for Nanotechnology
Edited by Andreas Schmidt-Ott

ISBN 978-981-4800-82-2 (Hardcover), 978-0-367-81709-1 (eBook)
www.jennystanford.com

particle composition, and the usability for certain applications. Alloyed identical feedstock electrodes with specific compositions are treated as well as spark mixing using different materials for the cathode and the anode.

5.1 Introduction

Mixed nanoparticles are of great interest simply because the number of elements is limited, but mixing gives access to myriads of combinations, each of them having different properties, thereby increasing enormously the range of possibilities. As already pointed out in Chapters 1 and 2, the capability of easily producing mixed particles is perhaps the strongest feature of spark ablation, distinguishing it from other particle production methods. In some cases, it has already been shown that mixed nanoparticles possess outstanding characteristics compared to any monometallic ones. The ease of forming mixed phases by spark ablation is due to the extremely fast evaporation process and rapid quenching of the vapor in comparison to furnace generators, glowing-wire aerosol generators, or flame synthesis [1]. This nanoscale mixing can be achieved with metals that are miscible [2] or immiscible [3] in the bulk. In the latter case, they very often form alloys in the nanoparticulate phase [3]. The feasibility of mixing on the nanoscale increases the variability of spark ablation enormously. Thus, spark ablation can be used to tailor nanostructured materials for specific applications, for example, as advanced catalysts or plasmonic particles with a tuned plasmon resonance frequency (see Chapter 12).

Three options of producing bimetallic nanoparticles have been reported: Different elemental or two already alloyed identical feedstock electrodes can be used. For materials that are immiscible in the bulk, powders of micron-sized particles can be mixed and sintered to form the electrodes [11]. In the latter case and in the case of two different electrodes it is the spark process that aids the mixing of materials on a nano- or atomic scale, and the term "spark mixing" has been introduced for these methods [13]. Byeon et al. showed for the first time that one can produce binary mixtures with two different metal electrodes [4], although it was not clear at that point whether an alloy had been formed.

5.2 Mixed Particle Production Using Mixed Feedstocks

In the present chapter, copper-nickel as well as platinum-palladium as bimetallic nanoparticle systems are investigated. Copper-nickel alloys are widely used in the chemical or maritime industry due to their chemical, anticorrosion, and catalytic properties [5]. As these properties are very sensitive to the actual particle composition, it is of special interest to investigate the composition–property relations in the nanoparticles.

Platinum-palladium alloys are deployed in the automobile industry as exhaust gas catalysts, for example. Here, the great endeavor is the maximum conversion rate of harmful gases like CO or NO. Therefore, it is necessary to control the catalyst composition to achieve an optimal exhaust gas treatment.

5.2.1 Production Rate and Primary Particle Size

Principles can be applied here that have already been applied with some success for pure metals. At the same time, alloys provide a means of further testing these principles, because physical constants of the electrode materials can be varied in a quasi-continuous manner through varying the mixing ratio.

5.2.1.1 The bimetallic copper-nickel system

To compare the effects of the particle composition on the production rate, the composition of the feedstock electrodes is first determined, for example, by inductively coupled plasma mass spectrometry (ICP-MS). In Table 5.1, the average copper-to-nickel ratios of three measurements are listed and compared to the specified ratios by the producer. It can be seen that the mean compositions match the feedstock very well.

The composition of the alloy electrode has a significant influence on the particle formation and the production rate. In order to investigate the impact on the production rate, the following experiment was carried out.

Table 5.1 Comparison of the calculated and measured Cu/Ni ratios

Sample	Feedstock Cu/Ni ratio, specified (mass ratio)	Measured nanoparticle composition, Cu/Ni mass ratio
Cu/Ni 80/20	4.33	4.83
Cu/Ni 60/40	1.624	1.684
Cu/Ni 40/60	0.722	0.744
Cu/Ni 20/80	0.27	0.28

Nanoparticles were generated for four hours under constant conditions: as a carrier gas, nitrogen was used at a flow rate of 2 lpm and a purity of 99.999%, and the frequency amounted to 60 Hz, whereby the spark energy varied between 80 and 100 mJ with a spark duration of presumably about 1 μs [6]. The erosion rate was assessed gravimetrically from the electrode weight loss. Tailored feedstock electrodes with desired compositions, as well as commercially available alloys like Monel 400 (Ni:Cu = 65:33) and Cupronickel (Ni:Cu = 31:68) are used. Figure 5.1 summarizes the erosion rate results.

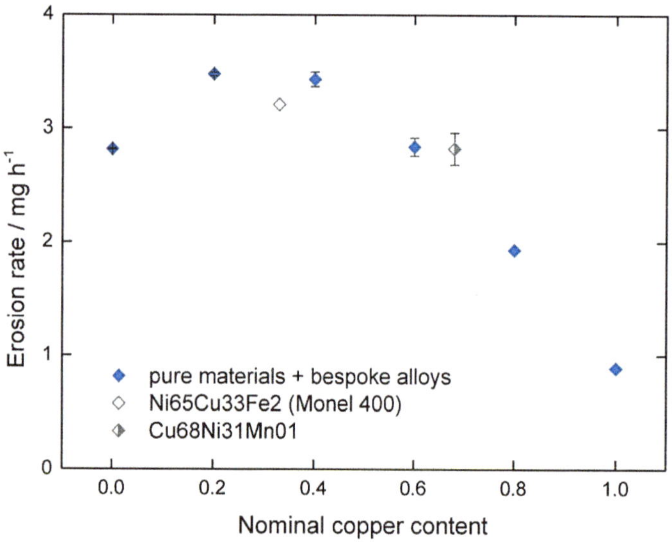

Figure 5.1 Erosion rates of copper, nickel, and different copper-nickel alloyed electrodes.

A crude model relating the evaporated mass to the energy input and material constants has been introduced by Llewellyn-Jones [7, 10], long before sparks were applied for nanoparticle production. It is based on the energy balance involving the energy flows illustrated in Fig. 5.3. With this model, it is possible to predict the eroded mass m per spark for any conductive material [8]. The equation, with some simplifications, is given by

$$m = \frac{\frac{1}{2}c_e V^2 - A\sigma\tau T_b^{\ 4} - 2(\pi A)^{0.5} t(T_b - T)k}{c_{ps}(T_m - T) + c_{pl}(T_b - T_m) + \Delta H_m + \Delta H_v}. \tag{5.1}$$

The first term in the numerator represents the spark energy available for evaporation, with V the breakdown voltage and c_e called effective gap capacitance. The total energy going into the spark can be written as $1/2CV^2$, where C is the capacitance, from which that energy is released (see the spark circuit in Fig. 2.1, Chapter 2). For the fraction of the spark energy that is available for evaporation, c_e/C, empirical values have been determined [6], but c_e is not well known. The second term assigns the heat loss by radiation from the hot spots, where T_b is the boiling point in Kelvin and τ the spark duration. As a simplification, the metal is assumed as a black body; therefore, σ is the Stefan–Boltzmann constant (5.67×10^{-8} J/(s × m × K^4). The third term denotes the heat transfer from the hot spots by conduction. k is the thermal conductivity, A the hot-spot area in m^2, and T the steady-state temperature far from the hot spot in Kelvin, in this case ambient temperature. The first and third terms of the denominator indicate the energy required to heat the metal to the melting point T_m (K), with c_{ps} the average heat capacity of the solid and ΔH_m the melting enthalpy. The second and fourth terms represent the energy required to heat the material in the liquid state to the boiling point, with c_{pl} the average heat capacity of the liquid and ΔH_v the evaporation enthalpy [2].

To compare the measured and calculated eroded mass, it is crucial to know the parameters and constants of the different alloys for the Llewellyn-Jones equation. In Table 5.2 the values for nickel, copper, and the two commercially available alloys Monel 400 and Cupronickel are listed as examples.

Table 5.2 Values for the Llewellyn-Jones equation

	Ni	Monel 400	Cupronickel	Cu
V [V]	6800	7700	6734	6800
T_b [K]	2910	2787	2665	2560
T_m [K]	1453	1325	1205	1083
c_{ps} [J/(g × K)]	0.456	0.427	0.403	0.385
c_{pl} [J/(g × K)]	0.62	0.576	0.532	0.495
ΔH_m [J/mol]	17,000	16,000	14,000	13,000
ΔH_v [J/mol]	374,000	350,000	326,000	305,000
k [W/(m × K)]	60.6	29.42	45	391.1

The breakdown voltage V is not a material constant but was measured for each experiment. The differences are most likely due to small differences in the electrode gap distances and/or electrode shapes. The material constants for pure nickel and copper are well known from the literature [9], whereas the values for the different alloys can be determined via linear interpolation. The parameters c_e, A, and τ are assumed to be independent of the electrode material and not well known. They were determined empirically by fitting the Llewellyn-Jones equation to the measured data. The fitted values are in a reasonable range compared to the literature [8, 9]. Figure 5.2 shows the model data and the measured data of the evaporation rate per spark.

The differences in the evaporation rates are mainly due to the different thermal resistances of the copper-nickel electrodes. The energy transfer through heat conduction has a greater influence than the radiation term [2]. Materials with higher thermal conductivity show less evaporation because the cooling process is more effective [12]. Accordingly, the evaporation rates of the copper-nickel alloys closely follow the trend of the thermal resistance, as seen in Fig. 5.4.

The feedstock composition influences not only the erosion rate, as shown, but also the primary particle size. With different feedstock compositions, the physical properties, such as the melting temperature, change. A lower melting point is associated with a higher diffusion coefficient of surface atoms and thus a higher tendency toward sintering and a larger primary particle size. This is in agreement with a set of qualitative rules for the primary particle size given by Feng et al. [14].

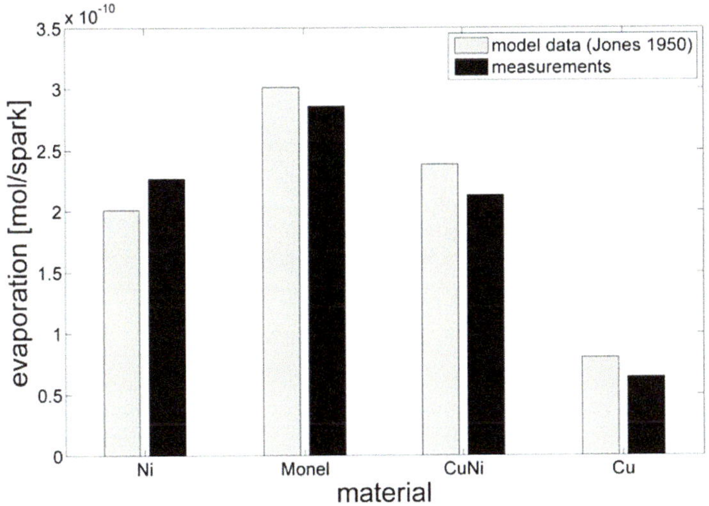

Figure 5.2 Comparison of the evaporation rate per spark between the Llewellyn-Jones model and the measured data.

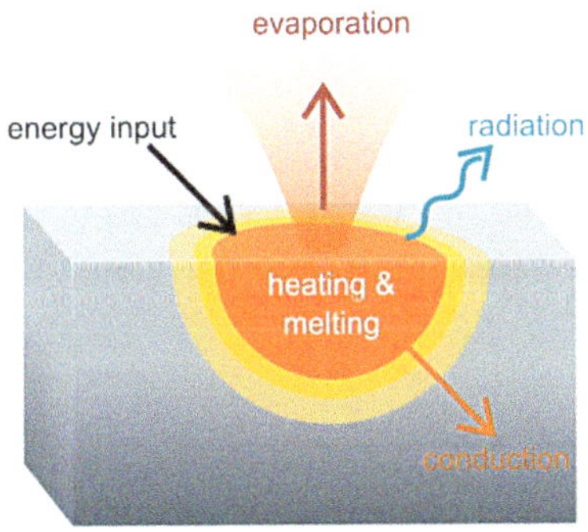

Figure 5.3 Illustration of different energy flows described by the Llewellyn-Jones model (illustration based on Llewellyn-Jones, 1950). Reprinted by permission from Springer Nature Customer Service Centre GmbH: Springer Nature, *Journal of Nanoparticle Research*, Ref. [8], copyright (2016).

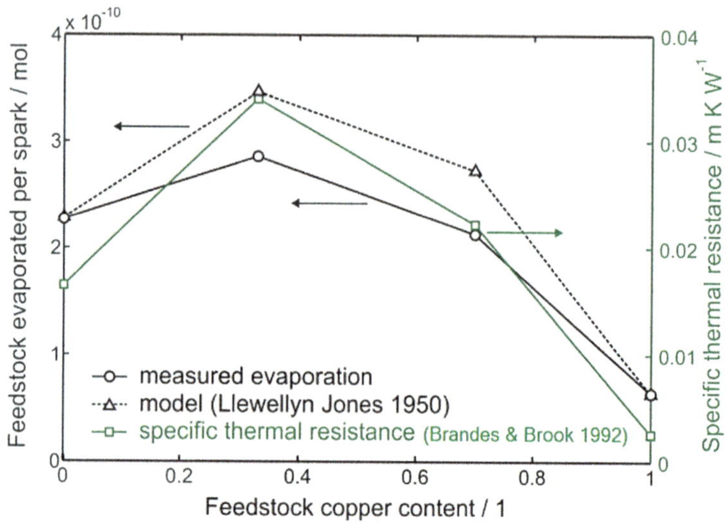

Figure 5.4 Measured evaporation rate per spark compared with the Llewellyn-Jones model and the specific thermal resistance. Reprinted by permission from Springer Nature Customer Service Centre GmbH: Springer Nature, *Journal of Nanoparticle Research*, Ref. [8], copyright (2016).

The comparison between the materials nickel, copper, Monel 400, and Cupronickel shows the trend mentioned before: Pure nickel as a feedstock electrode generates agglomerates with the smallest primary particle sizes, whereas pure copper agglomerates exhibit the largest primary particle sizes of the four compared materials. The alloys Monel 400 and Cupronickel follow the "higher melting point for smaller primary particles" rule, as seen in Fig. 5.5.

In Figure 5.6, the transmission electron microscopy (TEM) images of the corresponding materials are shown. It can clearly be seen that a higher Cu content leads to larger primary particles through enhanced sintering [8].

As demonstrated, the feedstock composition has an influence on the agglomerate structure in terms of the primary particle size and could be exploited to tailor the latter to meet geometrical requirements of the application.

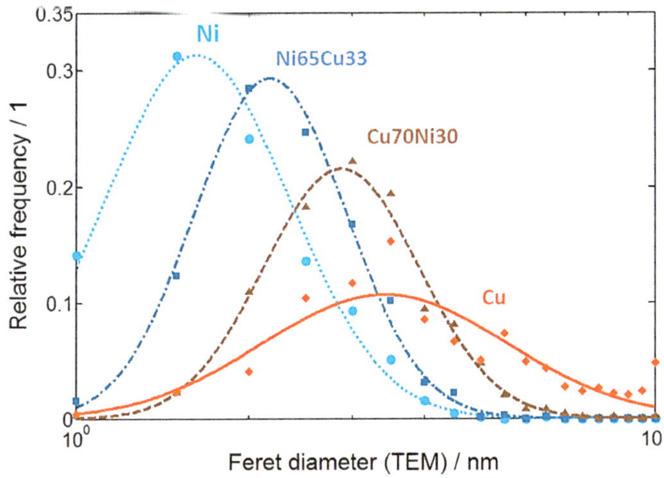

Figure 5.5 Primary particle Feret diameter of nickel, Cupronickel, Monel 400, and copper, as determined by TEM. Reprinted by permission from Springer Nature Customer Service Centre GmbH: Springer Nature, *Journal of Nanoparticle Research*, Ref. [8], copyright (2016).

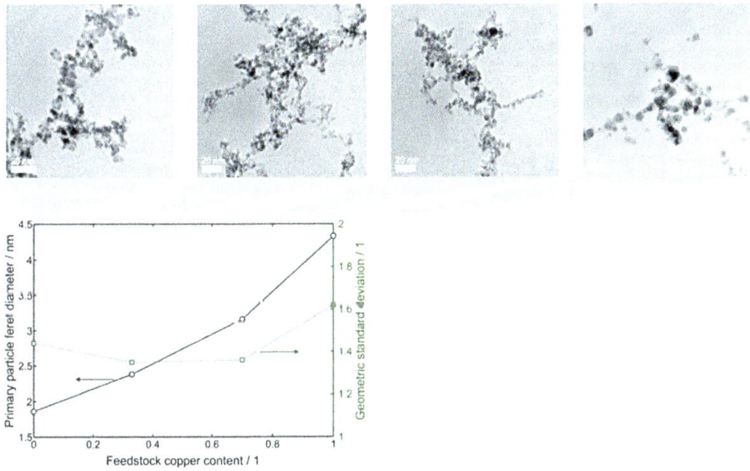

Figure 5.6 Primary particle Feret diameter in relation to the copper content and corresponding TEM images. Reprinted by permission from Springer Nature Customer Service Centre GmbH: Springer Nature, *Journal of Nanoparticle Research*, Ref. [8], copyright (2016).

5.2.1.2 The bimetallic platinum-palladium system

As thermal conductivity typically varies over a large range from metal to metal, it can be considered a dominant parameter governing the ablation rate in many cases, as in the example before. For alloys of platinum and palladium, we have a case where the pure metals as well as the alloy have similar thermal conductivities (the alloy having the smallest one) (see Table 5.3). Here, there is no qualitative correlation with the thermal conductivities, as seen in Fig. 5.7 and Table 5.3. The ablation rate of platinum, palladium, and the 50:50 alloy is evidently governed by properties other than thermal conductivity. As the differences in the other parameters (melting and boiling point enthalpies) are not large enough to explain this behavior, we conclude that there must be a significant effect not accounted for by Eq. 5.1.

Table 5.3 Comparison between erosion rate, thermal resistance, and ionization potential of platinum, palladium, and the platinum/palladium alloy

	Erosion rate [mol/spark]	Thermal resistance [(m K)/W]	Ionization potential [eV]
Pt	1.36×10^{-10}	0.0138	9.00
Pd	8.64×10^{-11}	0.0140	8.33
Pt/Pd 50/50	1.12×10^{-10}	0.0327	–

The phenomenon was already observed by Byeon et al., who speculated that there is a correlation between the generation rate of a material and the ionization potential of the atoms [4]. The results shown in Fig. 5.7 are consistent with this trend. This phenomenon cannot be seen in the copper-nickel system, with the ionization potentials of copper (7.73 eV) and nickel (7.64 eV) being similar and low compared to the palladium-platinum case. One conjecture is that for the production rate the ionization potential becomes significant when a certain value is reached. Before that, the dominant factor is the thermal resistance. Further investigations are needed to obtain a complete picture of factors influencing on the generation rate.

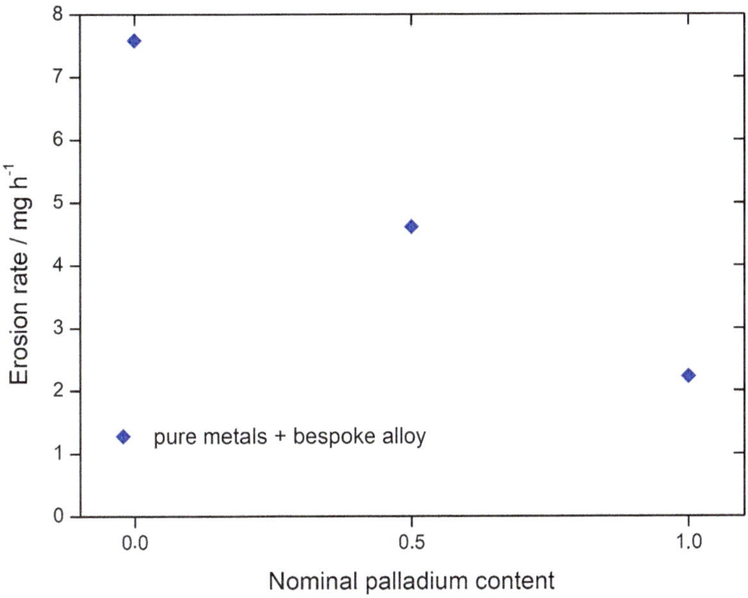

Figure 5.7 Erosion rate of platinum, palladium, and platinum-palladium alloy electrodes. Reprinted by permission from Springer Nature Customer Service Centre GmbH: Springer Nature, *Journal of Nanoparticle Research*, Ref. [8], copyright (2016).

5.2.2 Particle Composition

In the following two section (5.2.2.1 and 5.2.2.2) the reproduction of the feedstock (electrode) composition in the nanoparticles is experimentally tested, and Section 5.2.2.3 seeks explanations for this remarkable and useful phenomenon.

5.2.2.1 The bimetallic copper-nickel system

Another important variable to tailor nanoparticles for certain applications is the actual particle composition. Therefore, it is necessary to know whether the nanoparticle composition corresponds to the feedstock composition. It was already mentioned how to determine the feedstock composition (see Table 5.1).

To measure the particle composition, an elemental analysis of individual particles using energy-dispersive X-ray spectroscopy (EDX) is required. As an example, feedstock electrodes with a copper-nickel ratio of 40:60 and 70:30 are used to compare the compositions. Figure 5.8 shows the EDX results using a copper-nickel ratio of 40:60. The EDX analysis with a spot size of 0.25 nm indicates an excellent intra- and interagglomerate homogeneity: The large-area composition exhibits 40.5 atomic-% Cu and 59.5 atomic-% Ni, which correspond to a deviation of about 1%. The deviation between feedstock composition and the composition within the spot size amounts to less than ±5% for every investigated primary particle in different agglomerates.

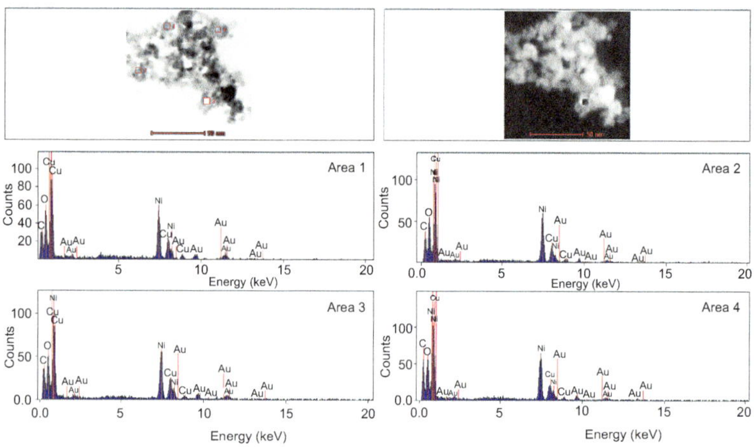

Figure 5.8 EDX spectra with corresponding TEM images using a feedstock with a copper–nickel ratio of 40:60. Reprinted by permission from Springer Nature Customer Service Centre GmbH: Springer Nature, *Journal of Nanoparticle Research*, Ref. [8], copyright (2016).

For the other feedstock compositions (70:30 copper-nickel ratio), the EDX measurements show similar results: The large-area composition exhibits 75.05 atomic-% Cu and 24.94 atomic-% Ni, corresponding to a deviation of about 7% from the specified composition (see Fig. 2.9). This deviation is assumed to be mainly due to measurement errors of EDX.

5.2.2.2 The bimetallic platinum-palladium system

Feedstock electrodes with a composition of 50:50 atomic-% were used to generate alloyed nanoparticles. Subsequently, the compositions of the produced particles were determined via EDX. In Fig. 5.10 the EDX results, as well as the corresponding TEM images, for sintered and unsintered platinum-palladium nanoparticles can be seen. The EDX results exhibit for the unsintered particles a maximum deviation from the nominal feedstock composition of ±1.5% and for the sintered particles of ±2%, which is probably due to measurement errors.

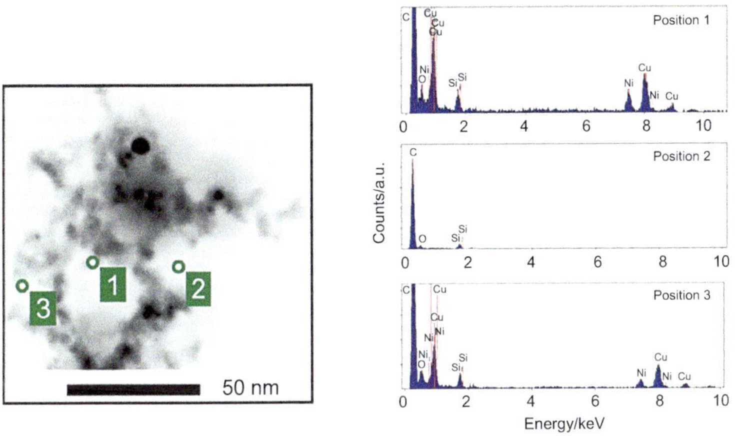

Figure 5.9 EDX analysis with the corresponding TEM image using a feedstock with a copper–nickel ratio of 70:30. Reprinted by permission from Springer Nature Customer Service Centre GmbH: Springer Nature, *Journal of Nanoparticle Research*, Ref. [8], copyright (2016).

In comparison with the analysis of copper-nickel compositions, platinum-palladium investigations show similar results: an excellent intra- and interagglomerate homogeneity and a very small variation from primary particle to primary particle with respect to the feedstock composition. The small variations seen may even be due to errors in the EDX analysis.

For the platinum-palladium particles, this similarity in the EDX spectra is conserved when inducing coalescence of the agglomerates, forming spheroid particles up to 30 nm in size. As this size is in the range of the EDX probing depth, any segregation of one of the metals to the surface would alter the mixing ratio measured by EDX. This indicates a platinum-palladium alloy or atomic-scale mixing.

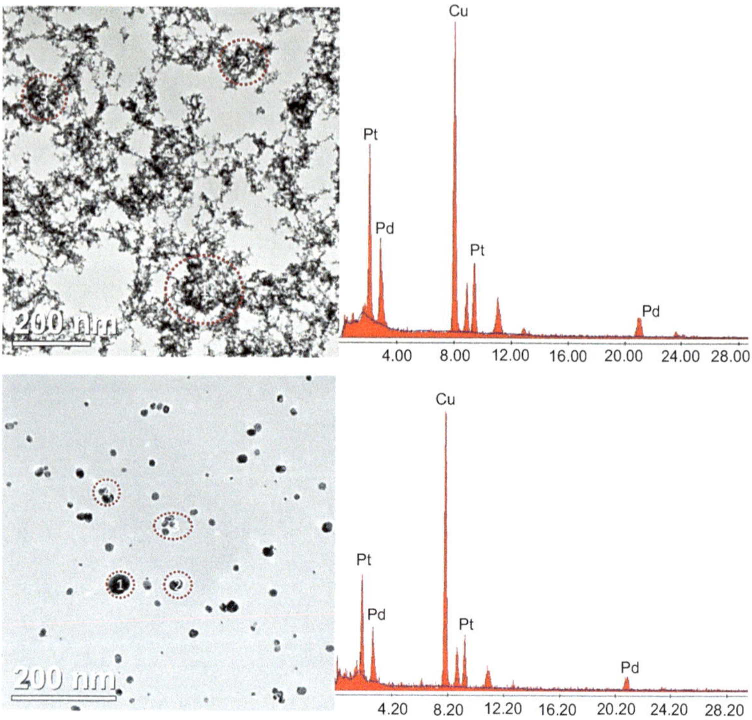

Figure 5.10 EDX analysis with corresponding TEM images using a feedstock with a platinum-palladium ratio of 50:50.

5.2.2.3 Why does spark ablation lead to particle compositions equal to the feedstock?

The TEM and EDX analyses in Sections 5.2.2.1 and 5.2.2.2 are in agreement with previous reports [2, 3], where the average particle composition has been shown to agree with the electrode composition. In addition, the results indicate that this is also true for the individual primary particles. This agreement of the particle mixing ratio with the mixing ratio in the feedstock is a remarkable feature of spark ablation, which enables predictable nanoparticle compositions for virtually any combination of materials. First, this implies that each spark forms a vapor of a homogeneous composition corresponding to the electrode composition. This means that the metal with the higher vapor pressure is not evaporated faster. Second, in the cooling process, the vapor of the higher-boiling-point metal does not condense before the one with the lower boiling point does. Otherwise, primary particles forming later in the condensation process would be enriched with the lower-boiling-point metal and the other way around.

Formation of a mixed vapor having the same mixing ratio as the electrodes is a phenomenon that has already been applied for spectroscopical analysis many years ago [15]. It is consistent with a model according to which the electrodes are ablated atomic layer by atomic layer by ion bombardment where the spark impinges on the electrode. Ion bombardment is suggested as a major ablation mechanism in Chapter 4, and it is in agreement with the observation that the (momentary) negative electrode is ablated much more strongly than the anode (see Section 5.3 below). Such cases have exclusively been reported in spark production of nanoparticles. Of course, ion bombardment also transfers heat to the layers beneath the ablated one, and formation of molten zones cannot be excluded. The occurrence of melting does not necessarily hinder emission of a homogeneously mixed vapor, though. There are indications of surface melting from visual inspection [16], and the phenomenon of occasional formation of round particles above 50 nm in size indicates formation of liquid pools, from which droplets are emitted [13], possibly aided by acoustic and electrostatic fields. Although the latter effect must be regarded as a separate particle formation mechanism of minor importance, it does indicate the tendency of some melt or liquid pool formation, probably when the spark hits the same spot multiple times. Preliminary attempts to avoid this by applying magnetic fields have been successful [17]. As in the extreme case of continuous arcs, occurrence of a molten phase cannot be avoided, and we conclude that heat transfer to the electrodes must be interrupted, and the sparks must be short enough to limit melting. Evaporation of alloys in tube ovens or continuous arcs necessarily lead to preferred evaporation of the lower boiling point (higher vapor pressure) component (distillation), and this effect is to be avoided here.

A fast condensation process is equally important. Due to the short duration of each spark, it emits a vapor cloud, about 1 mm in diameter (see Chapter 3). The inert gas flowing in rapidly mixes with these clouds, diluting the vapor and simultaneously cooling it to induce condensation. The paper of Feng et al. [14] uses a model, in which the smallness of the vapor cloud enables such extreme quenching and such high supersaturation that any collision of atoms or clusters leads to sticking from the very beginning. The step of nucleation is "omitted," and applying this picture to a binary vapor, condensation does not favor any of the two components. Provided the condensing vapor has the electrode composition, so do the particles formed by short enough sparks.

Experiments reported so far have shown that spark durations of a few microseconds, as in the experiments reported here, lead to good reproduction of the electrode composition in the nanoparticles. No systematic studies are known to us to investigate how long the spark duration may be in order to retain the composition reproduction feature. It is to be expected that the energy per spark also plays an important role here.

5.3 Spark Mixing by Using a Pair of Different Electrodes

A recent study [18] has examined mixed particle production by using different materials for the cathode and the anode. If the discharge is oscillatory, both the cathode and the anode are ablated and mixed particles are formed. These mixtures are usually on an atomic scale, and the method has been named "spark mixing" [13]. A phenomenon that is already known from thin films is that mixing gaps in the bulk tend to become narrower in nanostructures. Tabrizi et al. [3] have shown that spark mixing leads to atomically mixed crystalline nanoparticles for a number of materials that are immiscible in the bulk. Completely new materials can be made this way.

For the case of tungsten and copper, two materials showing a large resistance against mixing according to the bulk phase diagram, EDS line scans were consistent with the occurrence of 1 nm domains of pure Cu and W in a 6 nm particle [3]. Materials like this one can be considered as "cluster-assembled materials" with very special properties, and a preliminary unpublished study on electrical properties of this new material revealed a surprisingly small ohmic conductivity ($\approx 10^{-6}\ \Omega^{-1}m^{-1}$) as well as a high dielectric constant (≈ 6). However, some oxidation could not be excluded.

With an oscillatory discharge we mean a spark, during the duration τ of which the polarity is periodically reversed with a frequency $f > 1/2\tau$. The weight losses of the cathode and anode correspond to a model in which only the electrode momentarily having the negative polarity is ablated [18], at least for the conditions under which spark ablation for nanoparticle production has been carried out in the past (spark duration between 1 and 10 μs, spark energy 10–100 mJ; the circuit using a Tesla coil [19] forms an exception; see also Chapter 14). This implies a dominance of cathode ablation (cf. Chapter 4). Accordingly, spark ablation studies with electrodes of equal composition always report a stronger ablation of the cathode compared to the anode. Here the cathode and the anode are defined as the electrodes that are initially negative and positive, respectively, during the oscillatory discharge. The EDX analysis of individual primary particles indicated that the particles occur in a distribution of compositions [2], in contrast to particles from identical alloyed electrodes (see Section 5.2.2). The following sections explain how the mean particle composition as well as the broadness of the composition distribution can be influenced and predicted.

5.3.1 The Mean Composition in Spark Mixing

In most spark discharges, current and voltage perform damped oscillations because the energy released into the spark is usually stored in capacitance C_c (typically 20 nF), the electric wiring represents an inductance L_c, typically around a few μH, and the spark itself has an ohmic resistance R in the range of a few ohms. The resulting RCL circuit shown in Fig. 5.11 can be used to model the spark discharge (see Chapter 4). The damped current oscillation produced by such a circuit is described by [18]

$$I(t) = C_c U_{\max} \omega \left(1 + \frac{1}{\tau^2 \omega^2} \right) \exp\left(-\frac{t}{\tau} \right) \sin(\omega t), \tag{5.2}$$

where $U_{\max}$ is the maximum voltage, that is, the initial voltage, from which the capacitor is discharged across the gap. Under the condition of undercritical damping ($R_s^2 < 4L_c/C_c$), the damping time constant τ is given by

$$\tau = \frac{2L}{R_S}, \tag{5.3}$$

and the oscillation frequency ω is

$$\omega = \frac{1}{\sqrt{L_C C_C}} \sqrt{1 - \frac{C_C R_S^2}{4L_C}}. \tag{5.4}$$

C_C, R_S, and L_C are the capacitance, spark resistance, and circuit inductance, respectively, in the circuit of Fig. 5.11.

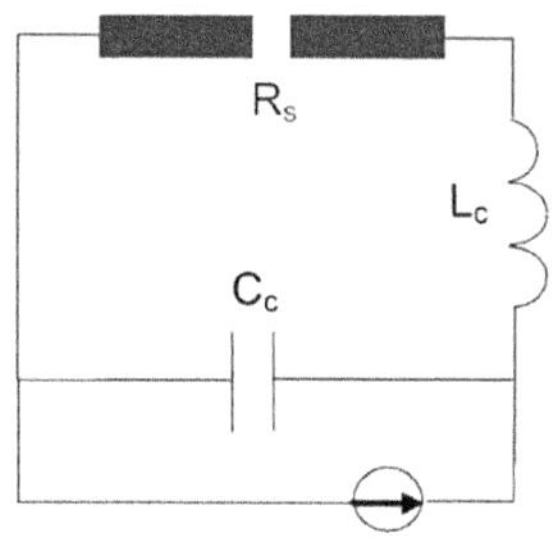

Figure 5.11 Simple model for a spark discharge circuit

Additional resistances or inductances can be added to tune the damped oscillation. Figure 5.12 shows the measured oscillations of current and voltage for a spark between Mg electrodes and Si electrodes according to Ref. [20]. The condition of undercritical damping is fulfilled for Mg, and the discharge is oscillatory. Si electrodes add a significant additional resistance, which leads to overcritical damping, where the discharge is non-oscillatory.

As only the momentarily negative electrode is ablated, the mean particle composition can evidently be tuned by choice of the circuit elements, and a resistor can be added to the circuit of Fig. 5.11, as shown in Fig. 2.1, Chapter 2. In the case of (over-)critical damping, the particles consist almost entirely of the cathode material, and for decreased damping, that is, by increasing τ according to Eq. 5.3, the anode material can increasingly be mixed in. For two materials that are ablated equally well under equal conditions (equal ablatabilities), the cathode and anode ablation ratio is approximately given by the ratio of energies during periods of negative and positive current according to

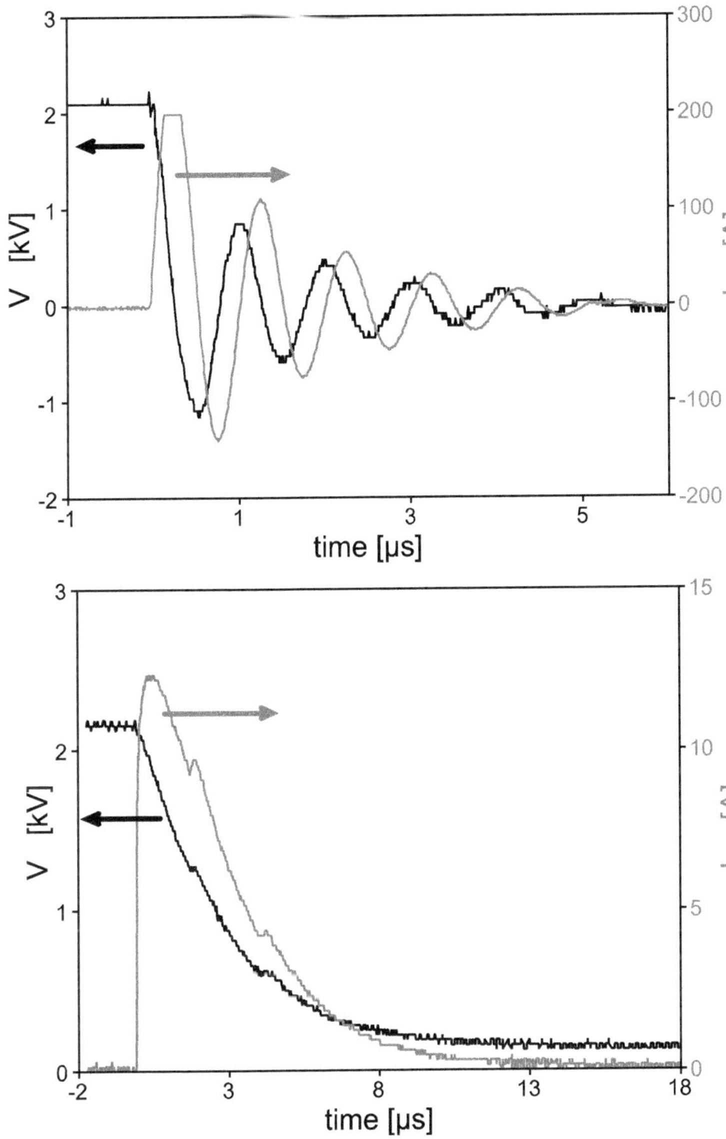

Figure 5.12 Current and voltage plots for spark between electrodes spaced 2 mm apart in Ar with C_c = 20 nF. Left: Mg; right: undoped Si. Reprinted by permission from Springer Nature Customer Service Centre GmbH: Springer Nature, *Journal of Nanoparticle Research*, Ref. [20], copyright (2011).

$$k \approx \frac{\int_0^\infty I^2 \mathrm{d}t \, (I<0)}{\int_0^\infty I^2 \mathrm{d}t \, (I>0)}. \tag{5.5}$$

From Eq. 5.2 and Eq. 5.5 it can be derived that

$$k \approx \exp\frac{\pi}{\tau\omega} \tag{5.6}$$

with τ and ω given by the circuit elements through Eqs. 5.3 and 5.4. The composition ratio of the particles (cathode material mass divided by anode material mass) is the ablation ratio of the respective electrodes, so that

$$\frac{m_\mathrm{c}}{m_\mathrm{a}}(\text{equal ablatabilities}) = k. \tag{5.7}$$

In general, different materials are ablated with different rates depending on material constants like boiling points, evaporation enthalpies, and heat capacities. Approximate measures of relative ablatabilities are given by Eq. 5.1, but experimental values are more reliable, and for specific materials they are available (see Section 2.3.1.1 and Table 2.1, Chapter 2). The mass ratio of Eq. 5.7 must be corrected by the ablatabilities ratio $\frac{C_{\mathrm{mat(Cathode)}}}{C_{\mathrm{mat(Anode)}}}$, giving

$$\frac{m_\mathrm{c}}{m_\mathrm{a}} = k\frac{C_{\mathrm{mat(Cathode)}}}{C_{\mathrm{mat(Anode)}}}. \tag{5.8}$$

Here k is given by the electrical circuit driving the spark, and the ablation ratio depends on material properties. By means of Table 5.1 and Eqs. 5.3, 5.4, 5.6, and 5.8, the relative masses ablated from the cathode and anode, $m_\mathrm{c}/m_\mathrm{a}$, equal to the mixing ratio of the particles, can be estimated as a function of the circuit elements. Here C is the storage capacitance, and L and R are best derived by recording the current or voltage oscillation with an oscilloscope and fitting the recorded curve to Eqs. 5.2–5.4, as demonstrated in Chapter 4. The range of composition ratios achievable can be extended by exchanging the cathode and anode material, but in practice, it will be difficult to reach values of k close to 1, where the numerator and the denominator in Eq. 5.5 are similar, which implies very weak damping and therewith a large τ. The decay time τ is also a measure of the spark duration, and according to the discussion in Section 5.2.2.3, problems with the production of mixed phases are to be expected from a specific spark duration upward. A recent study be Domaschke et al. [21] revealed that the absolute ablation rate of an electrode also depends on the electrode diameter (influencing heat dissipation), and this effect should make it possible to twist the ablation ratio by using an asymmetric electrode geometry (e.g., cylindrical electrodes of different diameter) in order to enable all mixing ratios.

5.3.2 The Broadness of Compositions in Spark Mixing

As seen in Fig. 5.13 (data from Tabrizi et al. [2]), spark mixing leads to a distribution of compositions. In this example, we define the composition as the mass fraction of Pd in the particles produced by spark mixing with Au or Ag:

$$\varphi_{\mathrm{Pd}} = \frac{1}{1+\dfrac{m_{\mathrm{Au,Ag}}}{m_{\mathrm{Pd}}}}. \tag{5.9}$$

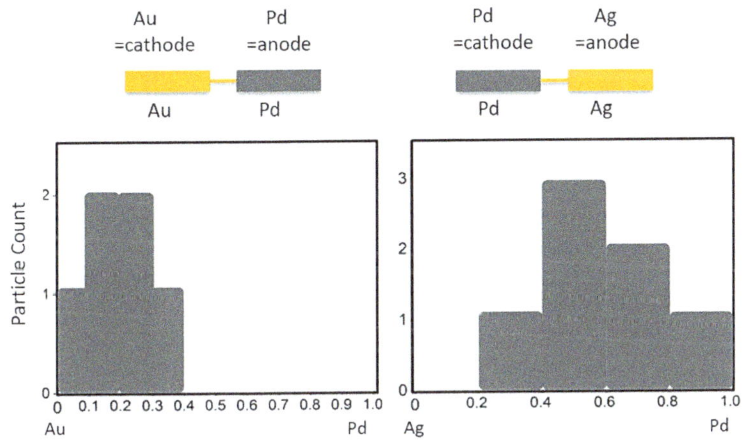

Figure 5.13 Distribution of compositions in terms of the fraction of Pd in nanoparticles produced by spark mixing with Au and Ag, φ_{Pd}, with Pd as the anode (left) and Pd as the cathode (right). Data derived from Ref. [2].

Here, m_{Pd} is the mass of Pd per particle and $m_{\mathrm{Au,Ag}}$ is the mass of Au or Ag per particle. The standard deviation of this distribution of

φ_{Pd} can be explained by the model illustrated in Fig. 5.14 [18]. The oscillating spark alternately forms vapor clouds at the cathode and the anode, which are slightly displaced from each other. Therefore there is initially some particle growth from the pure vapor, up to an "initial particle diameter" D_{pi}, before the materials completely mix in the turbulent flow. The final particles of diameter D_{pf} are thus formed by random combination and coalescence of N "initial" particles. According to Poisson statistics, Stirling's formula can be applied to describe this combinatorial process, and the standard deviation is

$$\sigma = \sqrt{\frac{\bar{\varphi}(1-\bar{\varphi})}{N}}. \tag{5.10}$$

Here $\bar{\varphi}$ is the mean composition, and N can be written as the volume ratio of the final and initial particles, so that

$$\sigma = \sqrt{\frac{\bar{\varphi}(1-\bar{\varphi})}{\left(\frac{D_{pf}}{D_{pi}}\right)^3}}. \tag{5.11}$$

Figure 5.14 Model for mixed particle formation: Pure "initial" particles (diameter D_{pi}) are first formed. They are homogeneously mixed and randomly collide and coalesce to form the "final" particles (diameter D_{pf}).

Using D_{pi} as a fitting parameter, both distributions in Fig 5.13, obtained with the same electrical spark parameters, geometry, and flow, lead to the same value of $D_{\mathrm{Pi}} \approx 3$ nm, so the composition distribution with more material combinations and final sizes under these conditions should be predictable, at least as an approximation, using that value and Eq. 5.11. Regarding the small number of particles analyzed in Fig. 5.13, this experimental data only provides rather weak support of the model (by leading to the same D_{pi}), and more studies are needed. However, the following qualitative statements according to Eq. 5.4 can be regarded as secured:

1. σ becomes smaller if the final particle gets larger.
2. σ becomes smaller if the initial particle size becomes smaller, and this can be influenced by the spark frequency, the electrode distance, or by fast turbulent mixing near the spark.
3. σ is small if one component strongly dominates—small $\bar{\varphi}$ or small $(1-\bar{\varphi})$.

5.4 Conclusions

Spark ablation easily forms mixed nanoparticles, the composition of which is basically identical with the feedstock material. Our experiments show that this is true not only for the mean composition but also for every individual (primary) particle. The simplicity of producing predictable mixtures in nanoparticles gives the spark method an extremely large potential, wherever the properties of materials or devices can be enhanced by mixing. We argue that for practically any application, mixtures must exist that are superior to the elements. This renders spark ablation an enormous potential for any nanoparticulate material.

Qualitative explanations for the reproduction of particle composition from the feedstock to the nanoparticle can be found, and the short time duration of the sparks is identified as essential.

The mass production rate was usually found to obey the Llewellyn-Jones equation, in particular where conductivity has a dominant influence. However, the case of platinum-palladium mixtures shows that this equation does not cover all effects influencing the mass production rate. The primary particle sizes of alloy particles differ from each other, where a high boiling/fusing point favors small particle sizes.

Spark mixing in oscillatory discharges leads to nanoparticles exhibiting a distribution of compositions, where the mean composition can be varied via electrical parameters of the discharge circuit. A simple model based on Poisson statistics explains the standard deviation of the composition distribution and delivers qualitative means of influencing it.

Acknowledgments

The authors would like to express their gratitude to PD Dr. rer. nat. habil. Reinhard Schneider from the Karlsruhe Institute of Technology (KIT), Prof. Guido Schmitz, and M.Sc. Yoonhee Lee from the Institute of Material Science, University of Stuttgart, for the EDX measurements. Gábor Galbács kindly acknowledges the financial support received from the Ministry of Human Capacities of Hungary through project 20391-3/2018/FEKUSTRAT.

References

1. Biskos, G., Vons, V., Yurteri, C. U. and Schmidt-Ott, A. (2008). Generation and sizing of particles for aerosol-based nanotechnology, *KONA Powder Part. J.*, **26**, pp. 13–35.
2. Tabrizi, N. S., Xu, Q., Van Der Pers, N. M., Lafont, U. and Schmidt-Ott, A. (2009). Synthesis of mixed metallic nanoparticles by spark discharge, *J. Nanopart. Res.*, **11**(5), pp. 1209–1218.
3. Tabrizi, N. S., Xu, Q., Van Der Pers, N. M. and Schmidt-Ott, A. (2010). Generation of mixed metallic nanoparticles from immiscible metals by spark discharge, *J. Nanopart. Res.*, **12**(1), pp. 247–259.
4. Byeon, J. H., Park, J. H. and Hwang, J. (2008). Spark generation of monometallic and bimetallic aerosol nanoparticles, *J. Aerosol Sci.*, **39**(10), pp. 888–896.
5. Pál, E., et al. (2012). Composition-dependent sintering behaviour of chemically synthesised CuNi nanoparticles and their application in aerosol printing for preparation of conductive microstructures, *Colloid Polym. Sci.*, **290**(10), pp. 941–952.
6. Tabrizi, N. S., Ullmann, M., Vons, V. A., Lafont, U. and Schmidt-Ott, A. (2009). Generation of nanoparticles by spark discharge, *J. Nanopart. Res.*, **11**(2), pp. 315–332.
7. Llewellyn-Jones, F. (1950). Electrode erosion by spark discharges, *Br. J. Appl. Phys.*, **1**(3), pp. 60–65.
8. Muntean, A., Wagner, M., Meyer, J. and Seipenbusch, M. (2016). Generation of copper, nickel, and CuNi alloy nanoparticles by spark discharge, *J. Nanopart. Res.*, **18**(8).
9. Brandes, E. A. and Brook, G. B. (2013) *Smithells Metals Reference Book*, 7th Ed. (Butterworth-Heinemann).

10. Llewellyn-Jones, F. (1963). The mechanism of electrode erosion in electrical discharges, *Platin. Met. Rev*, (2), pp. 58–65.

11. Lafont, U., Simonin, L., Tabrizi, N. S., Schmidt-Ott, A. and Kelder, E. M. (2009). Synthesis of nanoparticles of Cu, Sb, Sn, Sb and Cu_2Sb by densification and atomization process, *J. Nanosci. Nanotechnol.*, **9**, pp. 2546–2552.

12. Szente, R. N., Munz, R. J. and Drouet, M. G. (1994). Copper-niobium and copper-tungsten composites as plasma torch cathodes, *J. Phys. D: Appl. Phys.*, **27**(7), pp. 1443–1447.

13. Pfeiffer, T. V., Feng, J. and Schmidt-Ott, A. (2014). New developments in spark production of nanoparticles, *Adv. Powder Technol.*, **25**(1), pp. 56–70.

14. Feng, J., Biskos, G. and Schmidt-Ott, A. (2015). Toward industrial scale synthesis of ultrapure singlet nanoparticles with controllable sizes in a continuous gas-phase process, *Sci. Rep.*, **5**, art. no. 15788.

15. Scheeline, A. and Coleman, D. M. (1987). Direct solids elemental analysis: pulsed plasma sources, *Anal. Chem.*, **59**(20), pp. 1185A–1196A.

16. Kohut, A., Wagner, M., Seipenbusch, M., Geretovszky, Zs. and Galbács, G. (2018). Surface features and energy considerations related to the erosion processes of Cu and Ni electrodes in a spark discharge nanoparticle generator, *J. Aerosol Sci.*, **119**, pp. 51–61.

17. Feng, J. (2016). PhD thesis, Delft University of Technology. https://repository.tudelft.nl/islandora/object/uuid:fb6c0122-587b-471d-8009-b52ef9b69b07?collection=research.

18. Feng, J., Ramlawi, N., Biskos, G. and Schmidt-Ott, A. (2018). Internally mixed nanoparticles from oscillatory spark ablation between electrodes of different materials, *Aerosol Sci. Technol.*, **52**(5), pp. 505–514.

19. Horvath, H. and Gangl, M. (2003). A low-voltage spark generator for production of carbon particles, *J. Aerosol Sci.*, **34**, pp. 1581–1588.

20. Vons, V. A., de Smet, L. C. P. M., Munao, D., Evirgen, A., Kelder, E. M. and Schmidt-Ott, A. (2011). Silicon nanoparticles produced by spark discharge, *J. Nanopart. Res.*, **13**, pp. 4867–4879.

21. Domaschke, M., Schmidt, M. and Peukert, W. (2018). A model for the particle mass yield in the aerosol synthesis of ultrafine monometallic nanoparticles by spark ablation, *J. Aerosol Sci.*, **126**, pp. 133–142.

Chapter 6

Characterization of Nanoparticles from Spark Ablation

Maria E. Messing
Solid State Physics & NanoLund, Lund University,
Box 118, 221 00 Lund, Sweden
maria.messing@ftf.lth.se

Spark ablation basically allows production of myriads of new materials by mixing elements on the atomic scale or on the nanometer scale (see Chapters 2 and 5), with the additional flexibility in choosing an arbitrary size. To exploit these new possibilities, especially in the size regime below 20 nm, where deviations from bulk properties are guaranteed, and to discover novel features in these particles, the most advanced analysis techniques are necessary. This chapter treats techniques that are particularly suitable for characterization of these new nanoparticulate materials that can now be generated by spark ablation, and examples will be discussed. The first section will be devoted to characterization using advanced aerosol instruments, followed by a section on characterization using electron microscopes and a section on characterization by X-ray sources. Finally, possibilities of in situ characterization will briefly

Spark Ablation: Building Blocks for Nanotechnology
Edited by Andreas Schmidt-Ott

ISBN 978-981-4800-82-2 (Hardcover), 978-0-367-81709-1 (eBook)
www.jennystanford.com

be reviewed. The number of different imaginable characterization methods for aerosol nanoparticles are too numerous to be covered completely. An extensive summary can be found in the book entitled *Aerosol Measurement: Principles, Techniques, and Applications* [1]. The selected characterization methods described here are chosen because they are reasonably common, and they provide highly useful information about particle properties, including morphology and chemical composition, important for nanotechnology applications. They have all been used to characterize nanoparticles generated by spark ablation.

6.1 Introduction

Advanced characterization of nanoparticles from spark ablation or other aerosol generation methods is crucial to understand and enable the design of nanoparticles with controllable size, morphology, crystal structure, and chemical composition. If the particles will be used for specific applications, it is furthermore highly interesting to gain insight into how the mentioned particle properties affect, for example, the magnetic, catalytic, and optoelectronic behavior of new materials based on these nanoparticles.

In many aerosol generation systems, particles are already characterized in the gas phase, at least to some extent. This is referred to as online characterization before collection or deposition. It is common to have an scanning mobility particle sizer (SMPS) system or at least a neutralizer followed by a differential mobility analyzer (DMA), connected to the generation chamber, to measure the mobility diameter distribution of the produced particles. In many nanoparticle production systems there is a need to collect or deposit from a known number concentration of particles and therefore a condensation particle counter (CPC) or electrometer is also often connected to the setup. Both compact (singlet) particles and fractal-like agglomerate particles can be produced by spark ablation, depending on generation parameters (see Chapter 2). Sometimes a sintering furnace is also added to the nanoparticle production setup in order to reshape agglomerate particles into compact particles if that is requested by the intended application. To measure mobility diameter and number concentration on-line in the gas phase is a conventional procedure and will not be described in this chapter. An excellent and thorough description can be found in the textbook by Hinds [2].

If the particles are generated to target specific applications, many other properties besides the mobility diameter are interesting. The overall morphology of the particles; internal arrangement of the atoms, referred to as the crystal structure of the particle; the purity; and the chemical composition are properties that are of great interest for many applications. Most regulations of hazardous materials are based on mass concentration, but for nanoscale materials also the surface area concentration is important. Careful characterization of the mentioned particle properties is crucial, if these properties are to be linked to the behavior or performance of the particles, both from a safety aspect and for novel applications. The characterization of the performance or behavior of the particles, for example, in magnetic, catalytic, and optoelectronic applications is outside the scope of this chapter.

The properties of aerosol-generated nanoparticles can be characterized either on-line or off-line after the particles have been collected or deposited. The advantage of on-line characterization is that deposition is not needed and the particles are characterized during production, in the same gas atmosphere in which they are produced (or in vacuum). In off-line characterization the particles have often been exposed to a different gas atmosphere than during generation, which can affect their properties, and some form of deposition or collection of particles is necessary. However, some kinds of information can only be obtained by off-line characterization. To optimize off-line characterization, particles have to be deposited on an adequate substrate in a suitable amount and density, and obviously different characterization techniques have different demands, which will be discussed in each section.

The sections will start by briefly describing on-line characterization using two different advanced aerosol instruments that have been used to characterize spark-ablated particles. These systems can be connected to the nanoparticle production system. The first instrument that will be described is the aerosol particle mass (APM) analyzer that measures particle mass concentration. The second one provides chemical information about aerosol-generated particles and is called the aerosol mass spectrometer (AMS).

The next section will be devoted to off-line characterization using electron microscopy. The reason for choosing electron microscopy and not, for example, atomic probe or scanning probe microscopy is that these techniques are much more time consuming and less frequently used. Both scanning electron microscopes (SEMs) and transmission electron microscopes (TEMs) will be discussed in detail. The reason for this emphasis is that electron microscopy reveals very much information both on morphology and on chemical composition, even for particles around 1 nm in size, and is therefore often used when characterizing spark-ablated particles.

A highly useful complement to electron microscopy is based on X-rays and will be described in the subsequent section. Both X-ray diffraction (XRD) and X-ray photoelectron spectroscopy (XPS) have proven very useful to analyze spark-ablated nanoparticles and will be treated. The X-rays used for characterization can be generated both by a simple lab source or by a synchrotron. Information about particle composition and structure can often be obtained by the simple lab sources, but sometimes the synchrotron source is needed. Most often the analysis is done off-line, measuring on deposited particles, but XRD has also been performed on particles directly in the gas phase. X-ray-based methods average over a batch of particles, while electron microscopy has the capability of examining individual particles. A direct link between particle properties and behavior, for example, in catalysis, can be established by in situ characterization, which will be briefly touched upon in the final section. Both in the electron microscopy world and in the world of X-ray-based characterization, recent instrument development now enables characterization under elevated temperatures and pressures. A few in situ measurements have already been performed on spark-ablated particles.

6.2 Advanced Aerosol Instrument–Based Techniques

Traditional on-line aerosol instruments such as the SMPS system rely on the electrical mobility of particles, which is related to their size and morphology. To get information about mass, density, and composition, other methods need to be used. Highly useful for intentionally manufactured nanoparticles is the APM analyzer and the AMS to get information about the chemical content of the particles. These two instruments have both been applied to particles from spark ablation, are commercially available, and will be described in this section, starting with the APM analyzer.

6.2.1 Aerosol Particle Mass Analyzer

The principle behind the APM analyzer, which classifies aerosol particles according to their mass-to-charge ratio, was first described by Ehara et al. [3] in 1996. Since then, this instrument has been commercialized and the design has been improved to enhance both the sensitivity and the resolution. In an APM analyzer, the particle mass is measured by introducing the aerosol flow axially between an outer (r_2) and an inner (r_1) rotating cylinder, rotating at the same rotational speed ω. The outer cylinder is grounded, while a classifying voltage (V_{APM}) is applied to the inner cylinder. Only particles within a certain mass interval, for which the electrical force equals the centrifugal force, pass through the APM analyzer. The centrifugal force is mass dependent, and hence the mass can be determined according to

$$m = \frac{qE}{r\omega^2} = \frac{qV_{\mathrm{APM}}}{r^2\omega^2 \ln\left(\frac{r_2}{r_1}\right)}, \tag{6.1}$$

where r is the average radial distance to the gap between the cylinders from the axis of rotation $[(r_2 - r_1)/2]$, q the particle charge, and E the electrical field. A more detailed description of the principles behind the APM analyzer can be found in Ehara et al. [3] and McMurry et al. [4].

Since the first commercially available APM analyzer, the Kanomax APM-3600, several different designs have been suggested and other names invented, but they are all more or less based on the original design proposed by Ehara et al. and can all be referred to as particle mass classifiers (PMCs). The APM-3600 is specified to cover the classification range from about 0.01 fg (1 fg = 10^{-15} g) to 100 fg, which, for spherical particles with density 1 g/cm^3 and unit charge, corresponds to a diameter range of approximately 30–580 nm. To improve the mass sensitivity and enable measurements on smaller particles, Olfert et al. [5] suggested a design leading up to the commercially available Cambustion centrifugal particle mass analyzer and Tajima et al. [6] developed the compact APM analyzer, which is a miniaturization of the original APM analyzer design. These systems are limited to masses of about 0.5 attograms (1 ag = 10^{-18} g), corresponding to 10 nm. In 2015 Brossell et al. [7] suggested a design enabling a higher rotational speed and hence a further improved sensitivity down to 20 zg (1 zg = 10^{-21} g), which corresponds to 3.4 nm. This instrument was named the nano-PMC.

It is common to place the APM analyzer downstream of a DMA followed by a CPC to get a measurement of the mass concentration. In addition to measuring the mass of particles, the APM analyzer has also been used to measure other particle properties, such as the fractal dimension of agglomerate particles and their effective density [4], as well as to determine the number of primary particles in aggregates [8]. In the case of spark-ablated particles, the DMA-APM (Kanomax APM-3600) setup has been used to measure the mass of compact and agglomerated gold nanoparticles and to get a better estimation of their surface area concentration [9]. During the development of the nano-PMC, the mass and apparent density of coalesced and agglomerated spark-ablated silver nanoparticles were measured and the mass-mobility scaling exponent was estimated [7].

6.2.2 Aerosol Mass Spectrometer

AMSs have been used for quite some time, but the AMS continues to improve and to increase its usefulness due to improvements in the separate techniques it relies on. The basic function of an AMS is to obtain mass spectra of aerosol particles by evaporating and ionizing them, to determine their chemical composition, and to get information about their size and shape. In this chapter the AMS will be discussed very briefly, but for the interested reader the description in the book *Aerosol Measurement: Principles, Techniques, and Applications* [1] is very thorough.

The first step in the AMS is to sample the nanoparticles from the gas phase. The gas then needs to be removed, since the vaporization and ionization take place in a vacuum chamber, and the particles need to be focused into a particle beam. It is common to use one or more aerodynamic lens(es) for this purpose. Sizing of the particles may also take place in this inlet. Again, there are several different methods that can be applied in order to do this, including light scattering techniques, aerodynamic sizing, electrodynamic trapping, or particle beam chopping.

Once the particles have passed the inlet and have been focused into a beam, they are vaporized and ionized. This can be done in a single step or in two sequential steps and either continuously or in a pulsed manner. The vaporization can be either thermal or radiative. When vaporizing the particles thermally, they are continuously accelerated across the vacuum chamber to impact onto a sufficiently heated (often 600°C–1500°C) surface (filament) to flash-vaporize them. They can be ionized already at the heated surface, but it is more common to do this after vaporization by using an electron beam. Electron ionization at 70 eV is standard in an AMS and has the advantage that the obtained particle mass spectra can be compared to the spectra in reference libraries such as the National Institute of Standards and Technology (NIST).

By using a laser (or two) instead of the heated surface, the particles can be vaporized and ionized in a pulsed manner. One laser beam can be used for both vaporization and ionization, or a second laser beam can be used for the ionization. An advantage of using lasers is that they evaporate even refractory materials that cannot be vaporized by a heated surface. In the soot particle (SP)-AMS a Nd:YAG laser with a wavelength of 1064 nm is used, since soot particles absorb radiation of this wavelength.

The final step in the AMS is to send the ions into a mass spectrometer and then to use algorithms to analyze and characterize the large number of obtained spectra. The mass analyzers used in AMS are quadrupole, quadruple ion trap, or time-of-flight types. Today the time-of-flight mass analyzers are most common, since the development of computers and high-speed digitizers have made them fast enough to obtain spectra. Since thousands of spectra are acquired by mass analyzers, the processing cannot be done manually, and a number of automated methods are available that can assist in spectrum interpretation.

AMSs have been used to characterize spark-ablated particles of Ag, Au, Pd, PdAg, Fe(O), Ni(O), and CuO. To enable vaporization of these materials, the SP-AMS with a 1064 nm laser was used. The authors refer to it as a laser vaporization (LV)-AMS [10], since not only soot but also metals and some metal oxides absorb laser irradiation at this wavelength. From this study it was clear that CuO could not be vaporized, but the rest of the metals and metal oxides could and were identified in the obtained spectra. The information obtained from these AMS measurements comprised the chemical content of the particles, including impurities and metal-oxide analysis, particle mass-weighted size distributions, and some information about how compact the particles were. It was, for example, found that the amount of organic impurities on the particle surfaces decreased but that the degree of oxidation for Fe and Ni increased with increasing sintering temperature.

6.3 Electron Microscopy

Electron microscopes are truly versatile instruments, providing vast amounts of information combined with a resolution down to atomic dimensions. Often the result is presented as a set of images. The human brain is used to process images, and this may be one reason why electron microscopy is applied to engineered particles very often. In recent years, aberration-corrected electromagnetic lenses in electron microscopes have pushed the resolution limit to the atomic size range. For imaging, electrons from an electron beam hit the specimen and interact with it in different ways, either elastically, which means they maintain their energy, or inelastically, which means that they lose some of their energy to the specimen. These interactions give rise to a number of different signals that can be detected by several types of detectors in order to use them to form images, diffraction patterns, or different types of spectra. See Fig. 6.1 for an illustration of the different signals that arise from the electron–specimen interaction.

Two different types of electron microscopes exist and are used extensively to analyze spark-ablated nanoparticles: the SEM and the TEM. A SEM provides a good overview of samples containing nanoparticles and is reasonably fast and easy to operate.

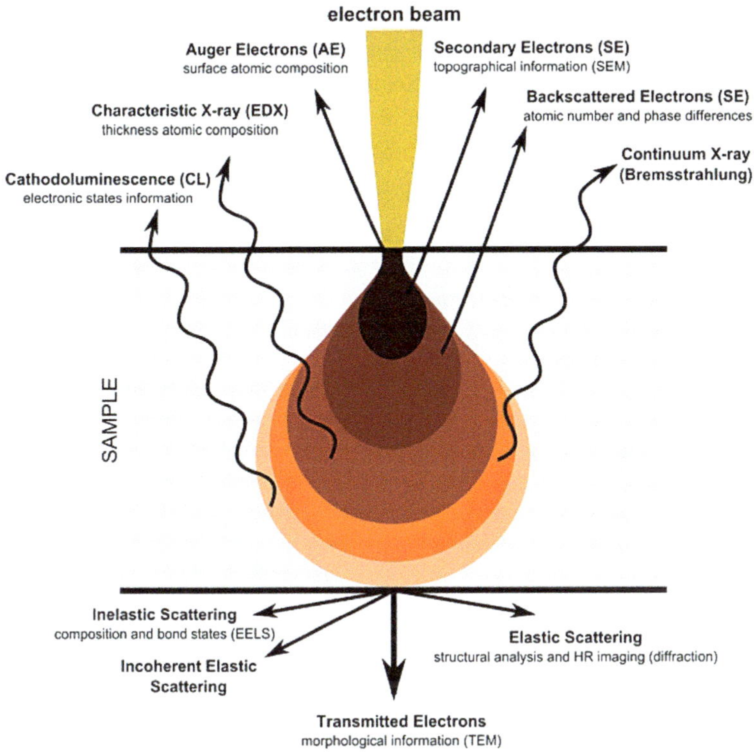

Figure 6.1 Illustration of different signals formed when an electron beam interacts with a specimen.

A TEM, on the other hand, provides detailed information about the atomic structure of individual nanoparticles but is more time consuming and rather difficult to operate. A major difference between the two methods is that in a SEM an image of the surface of the specimen is formed, whereas the images formed in a TEM contain information from the entire volume of the specimen since the electrons are *transmitted* through the sample (see Fig. 6.2).

6.3.1 Scanning Electron Microscopy

In a SEM the focused electron beam is scanned over the specimen in a raster scan pattern and the beam electrons penetrate the specimen and interact with the atoms from the so-called interaction volume, shown in Fig. 6.1. These interactions give rise to several different signals that can be detected, such as cathodoluminescence, characteristic X-rays, secondary electrons (SEs), and backscattered electrons (BSEs). Cathodoluminescence will not be discussed, since these types of detectors are quite rare. SE and BSE detectors are used to create SEM images by creating an intensity distribution map of the signal being emitted from the scanned area of the specimen. Characteristic X-rays can be used to obtain chemical information about the specimen, as will be described in Section 6.3.2.2 on analytical electron microscopy, since the mechanism is the same in SEMs and TEMs.

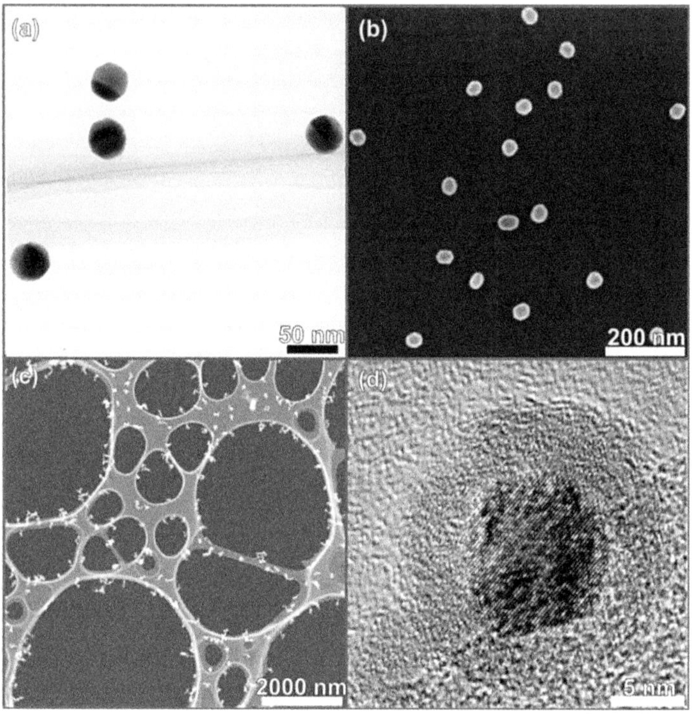

Figure 6.2 Electron microcopy images of nanoparticles. (a) TEM image of compact particles, (b) SEM image of compact particles, (c) SEM image of agglomerated particles deposited onto a TEM grid, and (d) HRTEM image of a core-shell particle with an amorphous shell and a crystalline core.

The images formed in a SEM have a wide range of magnifications, from around 10 times to more than 500,000 times, and the acceleration voltage of the beam electrons range from below 1 kV to 40 kV. To form images in a SEM, the SE and BSE are used and the images are known to have a large depth of field. SE detectors are standard in most SEMs. SEs are low-energy electrons that originate from the surface of the specimen, only from the top few nanometers. When the beam electrons interact inelastically with the atoms in the specimen, electrons from the k-shell can be emitted and these are the so-called SE. Since the SEs arise from the surface of the specimen, the SEM produces high-resolution images of the specimen surface topography. A resolution better than 1 nm can be obtained in a SEM, but a typical resolution is between 1 and 20 nm.

Images based on BSEs have a lower resolution than SE images, but contain information about the atomic number (Z) of the specimen. BSEs are high-energy electrons that have penetrated much deeper and interacted elastically with the specimen's atoms. BSE images show contrast between areas with different chemical composition since high-atomic-number atoms (heavy elements) backscatter electrons more strongly than low-atomic-number atoms (lighter elements) and therefore appear brighter in the images. In SEMs equipped with both types of detectors one can often obtain images made up by both the SE and the BSE signal.

Conventional scanning electron microscopy is performed under high vacuum, from which the restriction arises that the specimen needs to be conducting, to avoid charging effects. For spark-ablated particles, this is not a problem if they can be deposited onto a non-insulating surface. Typically, a piece of silicon substrate works fine. The solutions used to image nonconducting specimens include using a low-vacuum SEM, to operate the SEM at very low acceleration voltages (below 1 keV), or to sputter-coat the samples with a conducting material. All of these solutions come with other limitations, and it depends on the size of the spark-ablated particles whether it will be possible to image them. By operating the SEM under low vacuum, the resolution is reduced, and by operating at a low accelerating voltage the signal is lower and hence the signal-to-noise ratio is higher. Sputter-coating the specimen will work if the coating layer is much thinner than the size of the spark-ablated particles. An example of successful imaging of spark-ablated silver particles deposited on nonconducting fabric by lowering the accelerating voltage can be found in Feng et al. [11].

6.3.2 Transmission Electron Microscopy

In a TEM the electrons are transmitted through the sample and images or diffraction patterns can be directly formed in the respective image and diffraction planes. Several lenses are then used to magnify the image or diffraction pattern and to focus them onto a viewing screen or a charge-coupled device (CCD) connected to a computer. The image created by a TEM is at best a 2D shadow image of a 3D specimen averaged over the thickness of the specimen. See Fig. 6.2 for examples of TEM images. Therefore, image interpretation is far from straightforward, and a good understanding of the interactions between the electrons and the specimen is required in order to correctly interpret TEM images.

In a TEM, acceleration voltages between 80 and 300 kV are typically used, but there are examples of usage of both higher voltages, over 1 MeV, and lower voltages, down to 30 kV. In theory, a better resolution can be obtained by increasing the acceleration voltage, but then the possible beam damage to the sample is more severe. With the development of corrected microscopes, it has become more common to use lower acceleration voltages, especially for sensitive samples. Transmission electron microscopy can be performed either in the conventional or in the scanning mode, being called scanning transmission electron microscopy. In the conventional mode the sample is illuminated by a (preferably) parallel beam, whereas in STEM mode the sample is illuminated by a convergent beam; that is, a small (<10 nm) probe. To form an image in the STEM mode, the probe is scanned across the surface in a similar manner as in a SEM. This is particularly useful when images of a specific area of the sample are required, for example, to obtain chemical information from a certain location. A TEM may either be a conventional TEM, a dedicated scanning transmission electron microscope (STEM), or be able to work in both modes.

Since the electrons need to be able to penetrate through the sample, there are several strong restrictions put on the specimen to make sure it is electron-transparent. Typically, specimen preparation is a tedious and challenging task, but nanoparticles from spark ablation are an exception. This is because these particles, and if agglomerated the primary particles, are typically smaller than 20 nm in diameter. The deposition can be done by placing the TEM grid either in a deposition chamber [12] or on a filter, relying on electrostatic forces or on diffusion, respectively. Low-pressure impaction is another possibility [13]. See Chapter 2 for deposition techniques.

Once a specimen is prepared and inserted into the TEM, a large number of imaging and spectroscopy techniques exist, and huge amounts of information about the specimen can be obtained. A recommended book for more information is *Transmission Electron Microscopy* by Williams and Carter [14]. Here we will briefly describe conventional imaging, high-resolution imaging, and analytical transmission electron microscopy and how these techniques are based on different interactions between the beam electrons and the specimen.

When describing electron–specimen interactions in the TEM, the electrons can be thought of both as charged particles and as waves with a specific amplitude and phase. Electrons that enter a specimen can be scattered elastically or inelastically, either by a single atom or by an electron cloud. The contrast in TEM images is mainly due to elastically scattered electrons. When using the wave approach, the interaction of the electrons with the specimen is called diffraction. The diffraction pattern, that is, the distribution of diffracted electrons in reciprocal space, is directly linked to the crystalline structure of the specimen.

The contrast in TEM images is the result of electron–specimen interactions and is defined as the difference in intensity between two adjacent areas. When the electrons pass through the specimen, changes in both amplitude and phase may occur, giving rise to amplitude and phase contrast. A combination of amplitude and phase contrast contributes to the image, but often one type dominates. Amplitude contrast can be divided into mass-thickness contrast and diffraction contrast and is the result of differences in thickness and density or local diffraction differences in the specimen. Mass-thickness contrast depends on the atomic number, density, and thickness of the specimen. Areas of greater mass or greater thickness will show more pronounced scattering, and multiple scattering events can occur, reducing the number of electrons transmitted through the specimen in the forward direction. Diffraction contrast is related to the local variation of diffraction in the specimen. An amorphous region will scatter less than a crystalline region of the same material and will therefore appear brighter. In addition, a crystal (particle) oriented closer to a highly diffracting zone will scatter more strongly than a crystal (particle) oriented further from this zone, and hence appear darker.

Conventional transmission electron microscopy is very useful when investigating, for example, the morphology and diameter of spark-ablated particles. Some information about diameter and morphology can obviously be obtained by scanning electron microscopy; it is fairly easy to distinguish between agglomerated particles and completely compact spherical ones. If the morphology is to be investigated more accurately, for example, to follow a sintering process [15] or to find out whether particles consisting of immiscible materials do form a mixed crystal if small enough, transmission electron microscopy is a method of choice. It is also very common that the diameter distribution of spark-ablated particles is obtained from TEM images [11] and compared to the diameter given by the aerosol instruments. Finally, the primary particle size within agglomerates are often acquired from analysis of TEM images [9], because reliable on-line methods to determine it do not exist.

6.3.2.1 High-resolution transmission electron microscopy

High-resolution transmission electron microscopy (HRTEM) is used when detailed information about the structure of single particles is needed. Phase contrast is the source of contrast exploited in HRTEM images and arises when more than one beam (the direct beam and one or more diffracted beams) contributes to the image causing interference. The phase of the electron wave can change relative to the direct beam as a result of interactions with the sample causing phase contrast. The phase contrast is very sensitive to small changes in the orientation, thickness, and scattering factor of the specimen, as well as variations in the objective lens, such as focus and astigmatism. Because of this, phase contrast can be difficult to interpret, but on the other hand, it is sensitive enough to image the atomic structure of the specimen. HRTEM images contain a great deal of information, which must be correctly interpreted. Therefore, a thorough understanding of the way in which the image is formed is crucial; an excellent description can be found in Ref. [14]. In addition, HRTEM imaging should preferably be carried out together with computer simulations to ensure a correct interpretation of the images.

Another important thing to remember before carrying out HRTEM investigations is that a very limited amount of specimen is being investigated, typically one particle at a time. Therefore, one should first image the specimen at lower magnifications, and/or with a SEM, to ensure that the sample is homogenous. For example, in the case of spark mixing (see Chapter 5), a distribution of compositions is usually to be expected. Due to the diffraction contrast described earlier, making a particle closer to a highly diffracting zone appear darker, on-axis-oriented particles are more often selected for investigation than off-axis-oriented ones during HRTEM.

Imaging of the atomic structure of single particles can both be used to determine the crystal structure and to determine the chemical element of the particle. The crystal structure, that is, if a particle is amorphous, is single crystalline, or consists of different crystalline phases, can be determined directly by looking at the image (see Fig. 6.2d). To determine the chemical composition of a particle from an HRTEM image, the lattice distance, which is the distance between the atomic planes, should be measured. In a number of cases, this method delivered proof that a particle produced by spark mixing (see Chapter 5) of components that are immiscible in the bulk (i.e., in the macroscopic case) forms a crystalline alloy phase in the nanoparticulate case [16]. Figure 6.3 shows an example of a crystalline Au-Pt particle, two elements that are macroscopically immiscible in the mixing ratio produced. HRTEM has thus aided the discovery of new phases that can presently only be produced by spark ablation. In other cases, the absence of an atomic scale was revealed, where materials formed separate phases a few nanometers in size [17]. Many more examples of HRTEM imaging, and imaging by electron microscopy in general, exist in the literature on spark-ablated nanoparticles.

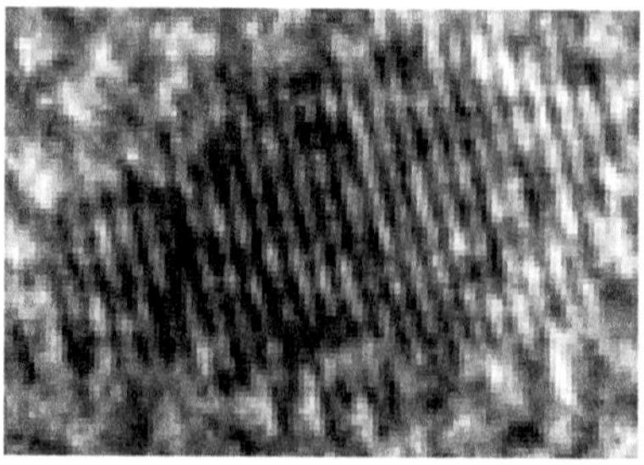

Figure 6.3 HRTEM image of a 5 nm spark-mixed Au-Pt particle with a composition of about 60:40 [24], which is in the miscibility gap of the bulk mixture. The micrograph reveals a crystalline mixture, which only exists in the nanoparticulate case with a lattice constant between the lattice constants of the pure metals.

6.3.2.2 Analytical electron microscopy

The chemical analysis of a specimen is possible in transmission electron microscopy (and scanning electron microscopy) using a technique called analytical electron microscopy, which is based on the transfer of energy from some of the electrons (those inelastically scattered) to the specimen. If an electron from a high-energy beam interacts with an inner-shell (core) electron, the latter can be ejected, leaving a hole in the inner shell. The atom will strive to return to its ground state and can do so by filling the empty inner-shell hole with an electron from an outer shell. The energy released by this transition will result in the emission of either an Auger electron or an X-ray. Both the Auger electron and the X-ray have a characteristic energy related to the energy difference between the two electron shells involved. This energy is element specific, and hence, the detection of Auger electrons or characteristic X-rays provides information about the chemical composition of the sample. The detection of characteristic X-rays is more common in transmission electron microscopy (and scanning electron microscopy) than the detection of Auger electrons, since the Auger signal is weaker. In addition, the beam electrons transmitted through the specimen lose some of their energy when exciting electrons. The energy lost by the electrons from the impinging beam can also be measured to provide information about the chemical nature of the sample.

6.3.2.2.1 *X-ray energy-dispersive spectroscopy*

Characteristic X-rays can be detected both in transmission electron microscopy and scanning electron microscopy by an X-ray energy-dispersive spectrometer. Inside the detector the incoming X-rays generate a number of electron–hole pairs, resulting in a charge pulse proportional to the X-ray energy, which can be measured. For spark-ablated nanoparticles, scanning electron microscopy X-ray energy-dispersive spectroscopy (XEDS) is not used since the resolution is too small, but for specimens with micron regions it can prove highly useful. Transmission electron microscopy XEDS, on the other hand, offers the possibility to provide information about the chemical composition inside single particles a few nanometers in diameter. The detection limit of transmission electron microscopy XEDS is often around 1–2 atomic%, and the lightest detectable element is beryllium. Quantitative information about the elements in a specimen can also be obtained by XEDS. Furthermore, if the TEM is operated in the STEM mode, information about the local chemical composition of the sample can be obtained using a probe smaller than 1 nm. This can be done at a certain point (point analysis), along a line (a line scan), or over a defined area (elemental mapping) and is particularly useful for nanoparticles consisting of more than one material, for example, core-shell particles [18]. Figure 6.4 [16] shows an example of line scans of two spark-mixed particles of Cu and W. We see that in this case two separated phases remain, which are either the pure metals or Cu-rich and W-rich. XEDS (see below) revealed peaks corresponding to the pure metals, pointing to the former.

6.3.2.2.2 *Electron energy loss spectroscopy and energy-filtered transmission electron microscopy*

A widely used technique complementary to XEDS is electron energy loss spectroscopy (EELS), in which chemical information is obtained from the transmitted beam electrons. EELS is often used for lighter elements, whereas XEDS is frequently the preferred technique for heavier elements. To measure the energy loss of the transmitted electrons a magnetic prism is used that provides a spectrum showing the distribution of intensity versus energy loss. A schematic of a typical EELS spectrum is shown in Fig. 6.5.

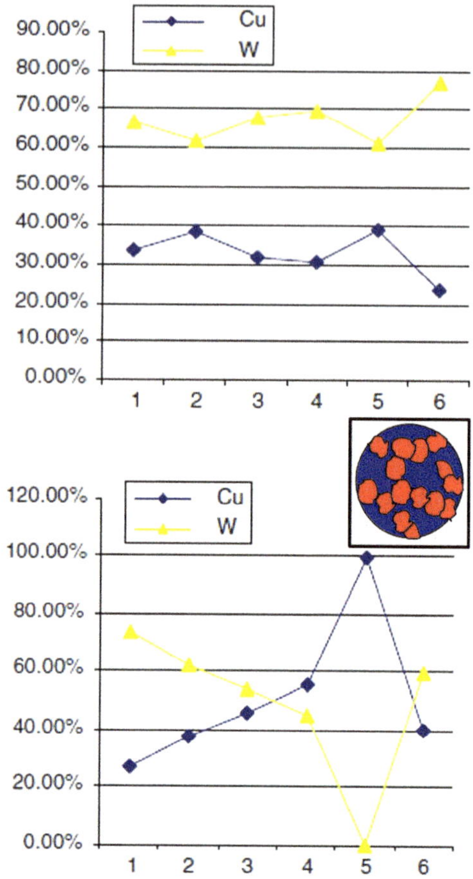

Figure 6.4 XEDS line scan through two 6 nm spark-mixed Cu-W particles [16], suggesting separate Cu and W (-rich) phases, as illustrated in the inset. Reprinted by permission from Springer Nature Customer Service Centre GmbH: Springer Nature, *Journal of Nanoparticle Research*, Ref. [16], copyright (2009).

The most intense peak is located at zero energy loss and is hence known as the zero-loss peak. This peak consists of the elastically scattered electrons. The second peak in the EELS spectrum is located between approximately 5 and 50 eV energy loss and is referred to as the plasmon peak. This peak corresponds to the interaction of the beam electrons with the weakly bound electrons in the valence or conduction bands. This interaction can result in wave-like collective oscillations of the electrons, known as plasmons, or in inter- or intraband transitions (e.g., excitation of valence electrons into the conduction band). From this low-loss region in the spectrum, information can be obtained about the thickness, the bandgap, the free electron density, and the local dielectric constant of the sample. Information like this cannot be acquired by XEDS, but the interpretation of the EELS spectrum is far from straightforward. The ionization edge, or core-loss edge, is found in the high-loss region of the spectrum (>50 eV). Similar to the peaks in the XEDS spectrum, this edge contains direct elemental information about the sample. Energies higher than the critical ionization energy can ionize atoms, and electrons with higher loss energies are seen in the spectrum as the curve slowly decreases above the ionization energy. As for XEDS, quantitative elemental information can be obtained from an EELS spectrum. In addition, information about how the ionized atom is bound, the coordination of the atom, and its density of states can be obtained from the fine structure (not resolved in Fig. 6.5), but again it is difficult to interpret.

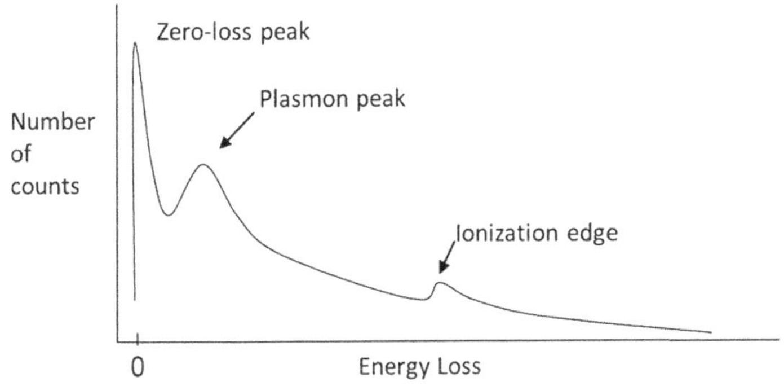

Figure 6.5 Schematic displaying an EELS spectrum (taken from Ref. [32]).

The information obtained by EELS can be used to create images displaying the chemical composition if the TEM is equipped with an energy filter. The idea behind energy-filtered transmission electron microscopy (EFTEM) is that specific energies from the EELS spectrum are selected (or filtered out) to form an image. Typically, EFTEM images are formed by selecting specific edge energies to create an elemental map of the sample. Since EELS is more suitable for lighter elements than XEDS, elemental maps of the oxygen distribution inside spark-ablated particles can be obtained [19].

6.4 X-ray-Based Characterization

Instead of using electrons as the illuminating source, X-rays can be used to obtain structural and compositional information about spark-ablated nanoparticles. A major difference compared to electron microscopy, where single particles are analyzed, is that with X-ray-based techniques measurements are performed on an ensemble of particles and the obtained information reflects an average of all of the measured particles. X-ray-based characterization can be performed with radiation both from a lab source and from a synchrotron radiation source. A synchrotron is a particle accelerator where the energy of accelerated particles is converted into electromagnetic radiation, such as hard and soft X-rays. Hard X-rays are the highest-energy X-rays, with energies greater than about 10 keV, while the lower-energy X-rays are referred to as soft X-rays. Two X-ray-based techniques have been used to analyze spark-ablated particles, namely XRD and XPS. XRD is normally a bulk analysis technique (unless a grazing incidence angle is used) where hard X-rays are normally used, whereas XPS is a highly surface-sensitive technique where soft X-rays are used, and both will be described in this section, starting with XRD.

6.4.1 X-ray Diffraction

In XRD an incoming beam of X-rays, with (typical) dimensions between 1 mm^2 and 1 μm^2, is directed onto the specimen. The incoming X-rays are elastically scattered by the electrons surrounding the atoms in the specimen. Compared to a beam of electrons directed onto a specimen and then scattered, like in a SEM, the X-rays can penetrate much deeper into the specimen, and hence, XRD is a bulk analysis technique.

To describe the mechanism behind XRD, the single crystal is a good starting point. In a single crystal the atoms are ordered in a periodically spaced lattice and the incoming X-rays are scattered at each lattice plane, resulting in a number of exiting waves. These scattered waves interfere with each other according to Bragg's law (see Fig. 6.6). Depending on the phase difference, the scattered waves can interfere destructively (zero amplitude) or constructively (twice the amplitude), producing an interference pattern with intense spots. By detecting the intensity and measuring the angle of the intense spots, information can be obtained about the symmetry and spacing of the lattice planes in the specimen. This is known as single-crystal XRD.

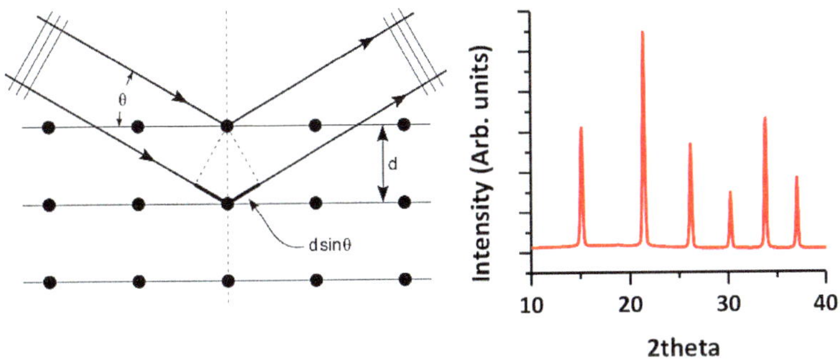

Figure 6.6 (Left) The principle behind Bragg diffraction. (Right) An example of a 2ϑ scan.

A nanoparticle with one single periodically spaced lattice can be thought of as a single crystal. However, the nanoparticle is much smaller than the X-ray beam and hence the XRD signal arises from a large number of particles. If all the nanoparticles on a specimen would be oriented to have the exact same crystalline orientation with respect to the substrate, it could be seen as a single crystal. In the case of aerosol nanoparticles, they are typically randomly oriented on the substrate, causing the diffraction planes to be randomly oriented on the specimen. This is analogous to the investigation of a polycrystalline crystal, and the technique used is called powder XRD. The interference pattern obtained consists of rings formed by the numerous different diffraction orientations, instead of bright spots. However, the lattice constant and interplane distance of the atomic planes can still be determined from these rings. One can also get information about crystallite (nanoparticle) size using the Scherrer equation [20]. A common presentation of the detected XRD data is the 2θ scan, displaying the 2θ angle of the Bragg peaks versus intensity (see Fig. 6.6). All of the XRD measurements performed on spark-ablated nanoparticles so far have been powder XRD. To learn more about powder XRD, the book entitled *Fundamentals of Powder Diffraction and Structural Characterization of Materials* by Pecharsky and Zavalij [20] is recommended.

It is fairly common to analyze spark-ablated particles using powder XRD with a lab source, but the spectra very often have low signal-to-noise ratios. An important consideration when preparing specimens for such measurements is to deposit the particles onto a substrate that interferes with the signal of the particles as little as possible, since the substrate will often dominate the measurements. To get a useful signal one needs to deposit enough material on the substrate; how much depends on the quality of the X-ray lab source and the detector used, and the particles to be analyzed. For small crystallites (nanoparticles), the small size will contribute to a broadening of the Bragg peaks, according to the Scherrer equation. This enables estimation of the particle size. Basically, this broadening also means that the smaller the particles are, the lower the signal-to-noise ratio will be. XRD has, for example, been used to demonstrate that the addition of hydrogen to the carrier gas results in the production of spark-ablated Bi particles instead of BiO particles that are produced when no hydrogen is added [15]. XRD has also been combined with small- and wide-angle X-ray scattering (SWAXS) in the analysis of magnetic spark-ablated nanoparticles [21]. Small-angle X-ray scattering (SAXS) and wide-angle X-ray scattering (WAXS) are variations of XRD techniques where the X-rays scattered to small and wide angles, respectively, are analyzed.

XRD has also been performed with a synchrotron source on spark-ablated particles deposited onto a substrate. To enable measurements on the small nanoparticles, surface X-ray diffraction (SXRD), also referred to as grazing incidence XRD, has been used. In an SXRD experiment, the sample is positioned in a way that the incoming X-ray beam hits the surface at a grazing angle. This is done to minimize the signal from the bulk, but the signal from the surface is still weak, so a high flux of incoming photons is required. A synchrotron source is therefore necessary to perform SXRD experiments. The amount of material required by SXRD is usually much smaller compared to XRD using lab sources. A surface coverage of around 5%–15% of spark-ablated particles has been found sufficient in SXRD measurements [22]. From SXRD measurements on spark-ablated palladium particles, the crystalline structure was not only confirmed, but a 2% expansion of the palladium lattice was detected due to incorporation of carbon atoms [23].

In 2018, successful XRD measurements with a synchrotron source were for the first time performed on-line on spark-ablated nanoparticles directly in the gas phase and not deposited onto any substrate. Spark-ablated platinum and gold particles were characterized using SAXS, and it was demonstrated that concentrations as low as 500,000 particles/cm^3 was detectable. Both aggregated and sintered particles were measured, and the degree of sphericity of particles after sintering, and the average diameters of the particles were determined [24].

Figure 6.7 shows an XRD spectrum of the same probe of Au-Pt particles Figs. 6.3 and 6.4 refer to. The XRD spectrum confirms that these components, immiscible in the bulk, form an alloy with a lattice constant between the pure Au and pure Pt lattice constants.

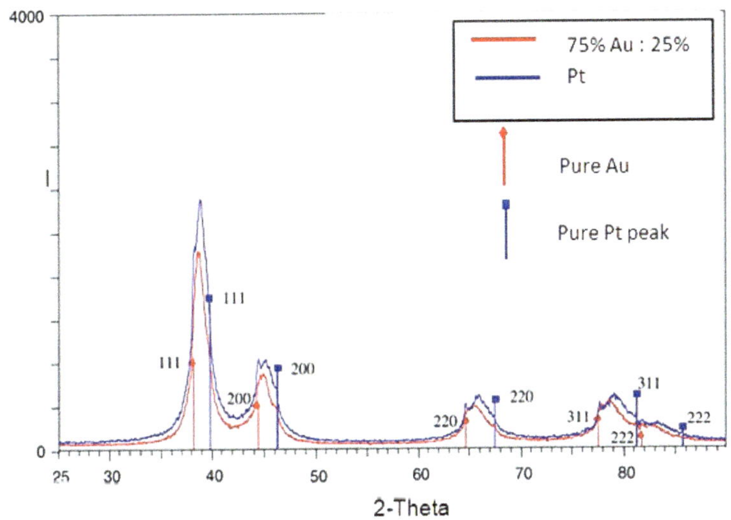

Figure 6.7 XRD spectrum of spark-mixed 5 nm Au-Pt particles [16]. The peak broadnesses are consistent with this size. The peaks for different crystal planes are always between the angles of the pure elements, indicating an alloy. Reprinted by permission from Springer Nature Customer Service Centre GmbH: Springer Nature, *Journal of Nanoparticle Research*, Ref. [16], copyright (2009).

6.4.2 X-ray Photoelectron Spectroscopy

XPS is a highly surface-sensitive technique, where core electrons are detected. These photoelectrons escape from a nanometer-thick surface layer of the specimen when it is irradiated with a beam of X-rays. The yield of emitted electrons is measured as a function of their kinetic energy. This measurement provides information about the specific elements in the specimen and their chemical surroundings. XPS can also be referred to as electron spectroscopy for chemical analysis (ESCA), which was the original name given by the group of K. Siegbahn that developed the first experimental XPS setup. XPS normally requires ultrahigh vacuum (UHV) or high-vacuum (HV) conditions (XPS at higher pressures is discussed in Section 6.5) and has been performed on spark-ablated nanoparticles using both a lab source and a synchrotron source. The requirements on specimens containing spark-ablated particles for XPS measurements are similar to those for XRD.

The principle behind XPS is that the binding energy of the core electron, E_b, is determined from the known photon energy of the incoming X-rays and the kinetic energy, E_k, of the detected photoelectrons. The binding energy is characteristic for a specific element and often also its chemical surroundings and given by

$$E_b = hv - E_k - \Phi_{spec}, \tag{6.2}$$

where, h is Planck's constant, v the frequency of the incoming light, and Φ_{spec} the work function of the spectrometer. The work function is the minimum energy required to remove an electron from the surface. The spectrometer and the specimen are connected to the same ground and align their Fermi energy levels, meaning that the final energy level the photoelectrons have to overcome to reach the detector is the vacuum level of the spectrometer.

Each peak in an XPS spectrum represents the binding energy of a core electron and thus identifies an element. Shifts of these peaks as well as their fine structures reveal information about the chemical environment of the atoms. To read more about how this information is derived and to get a thorough description of XPS, the book *Practical Surface Analysis by Auger and X-ray Photoelectron Spectroscopy* by Briggs and Seah [25] is recommended.

XPS with a lab source has been used to analyze magnetic LaFeSi nanoparticles produced by spark ablation. The XPS spectrum from a bulk sample of a LaFeSi alloy, identical with the electrodes in the spark ablation device, was compared to the XPS spectrum from the nanoparticles to confirm that the nanoparticles actually were alloys with the same composition as the electrodes [21]. XPS performed with a synchrotron source has been used to characterize Rh, Pd, and PdAg nanoparticles generated by spark ablation. In the case of Rh it was used to confirm that the particles were non-oxidized [26]; for Pd, to identify a Pd carbide phase [23]; and in the case of PdAg, to show that the particles were well-mixed alloys [27].

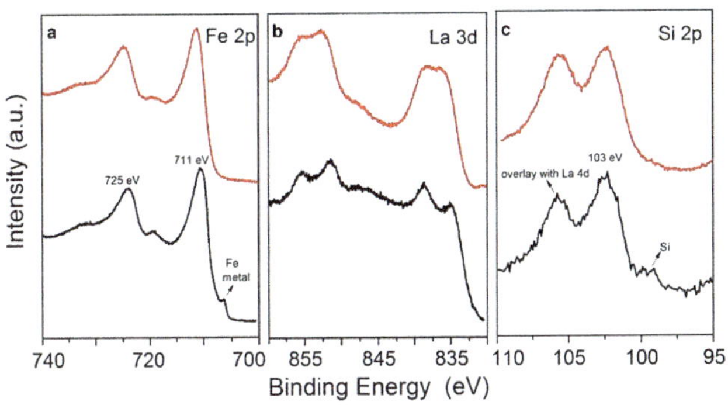

Figure 6.8 XPS spectrum of a LaFeSi nanoalloy produced by spark ablation (red) compared with the XPS spectrum of the electrodes, supporting equal compositions. Reprinted with permission from Ref. [21]. Copyright (2018) American Chemical Society. <https://pubs.acs.org/doi/10.1021/acsami.7b15441>. Further permissions related to the material excerpted should be directed to the ACS.

6.5 In situ Characterization

All of the described methods above have focused on how to characterize spark-ablated nanoparticles either after or during (on-line) generation. The focus has been on the properties of the particles as produced, but before being exposed to any particular environment (apart from air) or used in any specific application. This type of information is of course crucial and will contribute to the understanding of the behavior of specific particles in specific applications or when exposed to certain atmospheres. To advance such understanding and to improve the knowledge on how to further tailor particle properties to meet the demands, characterization of the particles during use, for example, during a catalytic reaction, can nowadays be performed. This type of characterization is referred to as in situ characterization (to distinguish it from what was here called on-line characterization).

In situ characterization is a truly emerging field both in the world of electron microscopy and in the X-ray-based characterization community. Influences on the specimen to induce the changes to be observed include an applied electric bias, a magnetic field, mechanical force, elevated temperature or pressure, or a change of the chemical environment, to give some examples. It is the technological development, such as constructions of special holders or reaction cells, that has facilitated the huge recent progress in in situ characterization. These holders or reaction cells enable measurements during exposure to a specific environment such as a reactant gas with characterization techniques that normally operate under high vacuum.

In situ electron microscopy is also called controlled atmosphere electron microscopy or, more often, environmental electron microscopy, and uses both a SEM and a TEM. In a SEM, in situ investigations of the effect on the specimen of externally applied mechanical force can be performed using modified specimen stages. Measurements under elevated temperatures and gas pressures can be performed in SEMs with special differential pumping systems allowing the gas pressure to increase around the specimen, while still keeping a high vacuum around the electron gun [28]. Customized specimen holders to create specific environments have been developed for in situ transmission electron microscopy [29]. A gas or liquid can be confined around the specimen inside a cell with electron transparent windows, for example, 10-nm-thick carbon windows. Often the elevated pressure is to be combined with an elevated temperature, and then differential pumping can be applied to restrict the gas flow above and below the specimen. Due to the development of environmental TEMs with aberration correctors and monochromators, HRTEM, and analytical electron microcopy with high spatial resolution can nowadays be used in situ. This is often employed when investigating nanoparticles exposed to gas pressures and temperatures relevant for catalysis [30]. It therefore has great potential for screening experiments aiming at discovery of new catalysts. Here spark ablation offers itself as a method of high flexibility, delivering metal and semiconductor particles of virtually any composition and size or their oxides.

In situ measurements using X-ray-based techniques can be performed both using XRD and XPS. Since X-rays are not nearly affected by surrounding gas molecules as much as electrons, XRD is possible at much higher pressures than XPS. High-pressure XRD is performed by placing the specimen in a special reaction cell, where elevated pressures and temperatures can be applied. High-pressure SXRD has been demonstrated on spark-ablated Pd particles to investigate the structural modifications of the particles upon CO oxidation [22].

In situ XPS is also known as high-pressure (HP)XPS or ambient-pressure (AP)XPS. To handle the small mean free path of the photoelectrons and to enable measurements under elevated pressures, similar as in environmental TEMs, differential pumping is used [31]. Electrostatic lenses can be applied to focus the electrons into the detector to increase the number of detectable electrons. APXPS has been used to analyze how the surface of Pd, Rh, and PdAg nanoparticles produced by spark ablation is altered when exposed to CO and oxygen. In the case of Pd nanoparticles, APXPS observed the formation of a palladium carbide phase during exposure to CO [22]. By using APXPS to compare the oxidation and reduction of a Rh single-crystal surface (a commonly used catalyst model system) and Rh nanoparticles, a significant difference in temperature dependence was found for both oxidation and reduction reactions [26]. Finally, from APXPS investigations of PdAg nanoparticles it was shown that the catalytically active Pd atoms segregated to the surface in the presence of reactant gases, opening possibilities for the design of cheaper catalysts [27].

6.6 Conclusion

As nanoparticles from spark ablation are mainly to be found in the range of the smallest lumps of matter possible (from atomic clusters to a few tens of nanometers), this fact alone calls for advanced methods of characterization. In this size range, spark ablation enables easy production of a virtually unlimited variety of new phases, but this feature is of scientific value only in combination with methods that identify these phases and characterize nanoparticulate materials in their chemical and morphological structure down to atomic dimensions.

The group of tools to be found under the name of electron microscopy, including variations such as electron diffraction, XEDS, and EELS, represents the strongest group of analytical methods for particles of these sizes. As they allow single-particle scrutiny with near-atomic or atomic resolution, they are essential to enable exploitation of the spark ablation technique for creation of new materials and devices. As the examples shown above illustrate, the complementary techniques based on X-rays as the incoming radiation, applicable to nanoparticle batches, provide more information about crystalline phases and other morphological properties.

Although in situ characterization of spark-ablated particles only has been performed in a few cases until now, spark ablation delivers highly suitable specimens for this type of characterization because of the ease with which samples of virtually any nanoparticle composition and size can be produced. It is therefore highly probable that in situ studies of nanoparticles generated by spark ablation will become very popular, in particular in view of catalyst research. Given the myriad of compositions achievable by spark mixing, each variation exhibiting different properties, spark ablation together with in situ characterization is bound to lead to the discovery of materials superior to those presently in use, especially in the field of catalysis.

References

1. Kulkarni, P., Baron, P. A. and Willeke, K. (eds.) (2011). *Aerosol Measurement: Principles, Techniques, and Applications*, 3rd Ed. (Wiley, USA).
2. Hinds, W. C. (1999). *Aerosol Technology: Properties, Behavior, and Measurement of Airborne Particles*, 2nd Ed. (Wiley, USA).
3. Ehara, K., Hagwood, C. and Coakley, K. J. (1996). Novel method to classify aerosol particles according to their mass-to-charge ratio: Aerosol particle mass analyser, *J. Aerosol Sci.*, **27**, pp. 217–234.
4. McMurry, P. H., Wang, X., Park, K. and Ehara, K. (2002). The relationship between mass and mobility for atmosphere particles: a new technique for measuring particle density, *Aerosol Sci. Technol.*, **36**, pp. 227–238.

5. Olfert, J. S., Reavell, K., St J., Rushton, M. G. and Collings, N. (2006). The experimental transfer function of the couette centrifugal particle mass analyser, *J. Aerosol Sci.*, **37**, pp. 1840–1852.

6. Tajima, N., Sakurai, H., Fukushima, N. and Ehara, K. (2013). Design Considerations and performance evaluation of a compact aerosol particle mass analyser, *Aerosol Sci. Technol.*, **47**, pp. 1152–1162.

7. Brossell, D., Valenti, M., Bezantakos, S., Schmidt-Ott, A. and Biskos, G. (2015). The nano-particle mass classifier (Nano-PMC): development, characterization, and application for determining the mass, apparent density, and shape of particles with masses down to the zeptogram range, *Aerosol Sci. Technol.*, **49**, pp. 495–507.

8. Lall, A. A., Rong, W., Madler, L. and Friedlander, S. K. (2008). Nanoparticle aggregate volume determination by electrical mobility analysis: test of idealized aggregate theory using aerosol particle mass analyzer measurements, *J. Aerosol Sci.*, **39**, pp. 403–417.

9. Messing, M. E., Svensson, C. R., Pagels, J., Meuller, B. O., Deppert, K. and Rissler, J. (2013). Gas-borne particles with tunable and highly controlled characteristics for nanotoxicology studies, *Nanotoxicology*, **7**, pp. 1052–1063.

10. Nilsson, P. T., Eriksson, A. C., Ludvigsson, L., Messing, M. E., Nordin, E. Z., Gudmundsson, A., Meuller, B. O., Deppert, K., Fortner, E. C., Onasch T. B. and Pagels, J. H. (2015). In-situ characterization of metal nanoparticles and their organic coatings using laser-vaporization aerosol mass spectrometry, *Nano Res.*, **8**, pp. 3780–3795.

11. Feng, J., Hontañoń, E., Blanes, M., Meyer, J., Guo, X., Santos, L., Paltrinieri, L., Ramlawi, N., de Smet, L. C. P. M., Nirschl, H., Kruis, F. E., Schmidt-Ott, A. and Biskos, G. (2016). Scalable and environmentally benign process for smart textile nanofinishing, *ACS Appl. Mater. Interfaces*, **8**, pp. 14756–14765.

12. Messing, M. E., Dick, K. A., Wallenberg, L. R. and Deppert K. (2009). Generation of size-selected gold nanoparticles by spark discharge - for growth of epitaxial nanowires, *Gold Bull.*, **42**, pp. 20–26.

13. Rissler, J., Messing, M. E., Malik, A. I., Nilsson, P. T., Nordin, E. Z., Bohgard, M., Sanati, M. and Pagels, J. H. (2013). Effective density characterization of soot agglomerates from various sources and comparison to aggregation theory, *Aerosol Sci. Technol.*, **47**, pp. 792–805.

14. Williams, D. B. and Carter, C. B. (2009). *Transmission Electron Microscopy: A Textbook for Materials Science*, 2nd Ed. (Springer, USA).

15. Hallberg, R. T., Ludvigsson, L., Preger, C., Meuller, B. O., Dick, K. A. and Messing, M. E. (2018). Hydrogen assisted spark discharge generated

metal nanoparticles to prevent oxide formation, *Aerosol Sci. Technol.*, **52**, pp. 347–358.

16. Tabrizi, N. S., Xu, Q., van der Pers, N. M. and Schmidt-Ott, A. (2010). Generation of mixed metallic nanoparticles from immiscible metals by spark discharge, *J. Nanopart. Res.*, **12**, pp. 247–259.
17. Pfeiffer, T. V., Kedia, P., Messing, M. E., Valvo, M. and Schmidt-Ott, A. (2015). Precursor-less coating of nanoparticles in the gas phase, *Materials*, **8**, pp. 1027–1042.
18. Wolff, M. F. H., Görlitz, D., Nielsch, K., Messing, M. E. and Deppert, K. (2011). Synthesis and magnetic characterization of MnAs nanoparticles via nano-particle conversion, *Nanotechnology*, **22**, p. 055602
19. Kreyling, W. G., Biswas, P., Messing, M. E., Gibson, N., Geiser, M., Wenk, A., Sahu, M., Deppert, K., Cydzik, I., Wigge, C., Schmid, O. and Semmler-Behnke, M. (2010). Generation and characterization of stable, highly concentrated titanium dioxide nanoparticle aerosols for rodent inhalation studies, *J. Nanopart. Res.*, **13**, pp. 511–524.
20. Pecharsky, V. and Zavalij, P. (2008). *Fundamentals of Powder Diffraction and Structural Characterization of Materials*, 2nd Ed. (Springer, USA).
21. Feng, J., Geutjens, R., Thang, N. V., Li, J., Guo, X., Keri, A., Basak, S., Galbaćs, G., Biskos, G., Nirschl, H., Zandbergen, H. W., Brück, E. and Schmidt-Ott, A. (2018). Magnetic phase transition in spark-produced ternary LaFeSi nanoalloys, *ACS Appl. Mater. Interfaces*, **10**, pp. 6073–6078.
22. Balmes, O., Resta, A., Wermeille, D., Felici, R., Messing, M. E., Deppert, K., Liu, Z., Grass, M. E., Bluhm, H., van Rijn, R., Frenken, J. W. M., Westerström, R., Blomberg, S., Andersen, J. N., Gustafson, J. and Lundgren, E. (2012). Reversible formation of a PdCx phase in Pd nanoparticles upon CO and O2 exposure, *Phys. Chem. Chem. Phys.*, **14**, pp. 4796–4801.
23. Messing M. E., Westerström, R., Meuller, B. O., Blomberg, S., Gustafson, J., Andersen, J. N., Lundgren, E., van Rijn, R., Balmes, O., Bluhm, H. and Deppert, K. (2010). Generation of Pd model catalyst nanoparticles by spark discharge, *J. Phys. Chem. C*, **114**, pp. 9257–9263.
24. McKibbin, S. R., Yngman, S., Balmes, O., Meuller, B. O., Tågerud, S., Messing, M. E., Portale, G., Sztucki, M., Deppert, K., Samuelson, L., Magnusson, M. H., Lundgren, E. and Mikkelsen, A. (2019). In situ observation of synthesized nanoparticles in ultra-dilute aerosol via X-ray scattering, *Nano Res.*, **12**, pp. 25–31.
25. Briggs, D. and Seah, M. P. (1983). *Practical Surface Analysis: By Auger and X-Ray Photoelectron Spectroscopy* (Wiley, USA).

26. Blomberg, S., Westerström, R., Martin, N. M., Lundgren, E., Andersen J. N., Messing, J. and Gustafson, J. (2014). A high pressure X-ray photoelectron spectroscopy study of oxidation and reduction of Rh(100) and Rh nanoparticles, *Surf. Sci.*, **628**, pp. 153–158.

27. Blomberg, S., Gustafson, J., Martin, N. M., Messing, M. E., Deppert, K., Liu, Z., Chang, R., Fernandes, V. R., Borg, A., Grönbeck, H. and Lundgren, E. (2013). Generation and oxidation of aerosol deposited PdAg nanoparticles, *Surf. Sci.*, **616**, pp. 186–191.

28. Donald, A. M. (2003). The use of environmental scanning electron microscopy for imaging wet and insulating materials, *Nat. Mater.*, **2**, pp. 511–516.

29. Sharma, R. (2005). An environmental transmission electron microscope for in situ synthesis and characterization of nanomaterials, *J. Mater. Res.*, **20**, pp. 1695–1707.

30. Hansen, T. W., Wagner, J. B. and Dunin-Borkowski, R. E. (2010). Aberration corrected and monochromated environmental transmission electron microscopy: challenges and prospects for materials science, *Mater. Sci. Technol.*, **26**, pp. 1338–1344.

31. Salmeron, M. and Schlögl, R. (2008). Ambient pressure photoelectron spectroscopy: a new tool for surface science and nanotechnology, *Surf. Sci. Rep.*, **63**, pp. 169–199.

32. Messing, M. E. (2011). Engineered nanoparticles generation, characterization and applications, PhD thesis, Lund University, Sweden.

Chapter 7

Atomic Clusters: Potential of Spark Discharge Generation

Anne Maisser
Energy, Environment and Water Research Center, The Cyprus Institute, Konstantinou Kavafi Street, 2121 Nicosia, Cyprus
a.maisser@cyi.ac.cy

The capability of spark discharge generation (SDG) to control the size and purity of generated particles in the size range of the very smallest materials possible carries enormous potential for various applications. Their production and processing forms an extension to the field of aerosol-based nanoengineering toward subnano dimensions. This chapter will focus on the fusion of this engineering method with the exciting field of atomic cluster research and the potential of using SDG atomic clusters as building blocks for new materials with tailored properties. It is complementary to Chapter 8, which focuses on vacuum-based technology.

Spark Ablation: Building Blocks for Nanotechnology
Edited by Andreas Schmidt-Ott

ISBN 978-981-4800-82-2 (Hardcover), 978-0-367-81709-1 (eBook)
www.jennystanford.com

7.1 Introduction

Atomic clusters typically consist of a few up to a few hundred atoms. Their physical and chemical behavior constitutes an intermediate regime that falls between bulk material and atoms [20]. Interactions between atoms may range from weak van der Waals bonds to strong ionic bonds [4]. Cluster properties strongly depend upon their size, shape, and dimensionality of their structures [1].

The effect of size on cluster properties can be classified in three distinct regimes. For particles larger than about 100 nm in diameter the properties do not exhibit any size dependency and properties can be described by the periodic table of solid-state physics. In the nanoparticle diameter range between 1 and 100 nm, cluster properties scale as a function of particle size, such as the ionization potential of metal clusters [1].

For even smaller clusters consisting of 1 to 100 atoms, properties are strongly dependent upon the number of atoms inside the cluster; however, these properties cannot be scaled as a function of size. It has been found that in this size regime they can be predicted more in analogy to the classic periodic table. In metal clusters it has been shown that cluster properties depend on the number of valence electrons in the cluster. Highly stable clusters appear at so-called magic numbers, which are related to a closed electronic shell. This behavior is described by the Jellium model [1, 3] and creates a great potential of atomic clusters as building blocks for new materials with tailored properties [5].

Cluster-assembled materials that are used for certain applications may use clusters with precisely controlled numbers of atoms (e.g., fullerenes) but also uniformly sized, larger nanoparticles [6]. In this chapter only the generation of atomic clusters with a controlled number of atoms per cluster, and their potential application as building blocks with tailored properties will be discussed.

The most widely applied atomic clusters in material design are C_{60} fullerenes. Fullerenes were first discovered in 1985 by Smalley and coworkers [23, 45, 46]. They are a special case of atomic clusters, since all their surface atoms have a full shell of valence electrons, and are thus extremely stable. Industrial-scale application is possible because fullerenes can be isolated in bulk quantities from soot [22]. Fullerenes are also capable of forming highly crystalline arrays or can be combined as hybrids with polymers. These functional materials exhibit interesting properties and are used for several applications, including, for example, solar cells [10, 17], photodetectors [15], high-capacity hydrogen storage media [53], antioxidants, and neuroprotective agents for medical applications [14, 35].

The development of fullerene materials and their applications also inspired the research and development of other cluster-assembled materials [6]. However, the greatest limitations for applications of other cluster types is that production in large industrial-scale quantities is almost impossible [20, 21].

To produce cluster-assembled materials for industrial applications, stable clusters with interesting properties need to be identified in lab experiments first. The requirements on these clusters are to be robust and likely to retain their properties in an assembled material. Another limitation to overcome is the development and establishment of a cost-efficient, high-quantity production method. Several technologies have been extensively studied over the past years. In this chapter we will describe and evaluate spark discharge generation (SDG) as a manufacturing method of atomic clusters, with the potential of being scaled up in an economical and environmentally friendly way for industrial-scale production rates.

7.2 Atomic Cluster Generation by Spark Discharge

This section focuses on the specific operation parameters of SDG to generate atomic clusters consisting of small numbers of atoms (<25). SDG was introduced in 1988 by the group of Schmidt-Ott as an aerosol generation method [43]. Ever since its introduction into the field of aerosol science and technology, the research efforts to investigate spark discharge–generated nanoparticles and to test their potential for applications have experienced rapid growth. This has been described in detail, amongst others, in a review by Meuller et al. (2012) [30], as well as by Pfeiffer et al. (2014) [37], and is subject of various chapters of this book.

Schwyn et al. showed in 1988 that SDG enables generation of small-sized nanoparticles. The detected primary particle sizes were

1.3 nm and around 5 nm for gold and carbon particles, respectively. How operational parameters of SDG relate to the produced cluster size has been thoroughly studied in the past. Tabrizi et al. carried out a very detailed study to investigate the influence of SDG operational parameters on particle size and concentration [47, 48]. The parameters that were varied were energy per spark, spark frequency, inert dilution gas, gap distance between the electrodes, dilution flow rate, etc. A scaling law predicting the generated primary particle size was later introduced by Feng et al. (2015) [11].

The results of the investigations showed that particle sizes can be decreased by increasing dilution flow rates and by lowering the energy per spark. The former measure results in faster dilution and therefore decreases collision growth as well as the time available for growth. Lowering the spark energy reduces the amount of material being evaporated by each spark. In the experiments by Feng et al. and Tabrizi et al., it has been found that by increasing the dilution flow rate the particle size decreases. While the initial decrease is rapid, the curve flattens for larger flow rates (c.f. Fig. 7.1) [11, 30, 48].

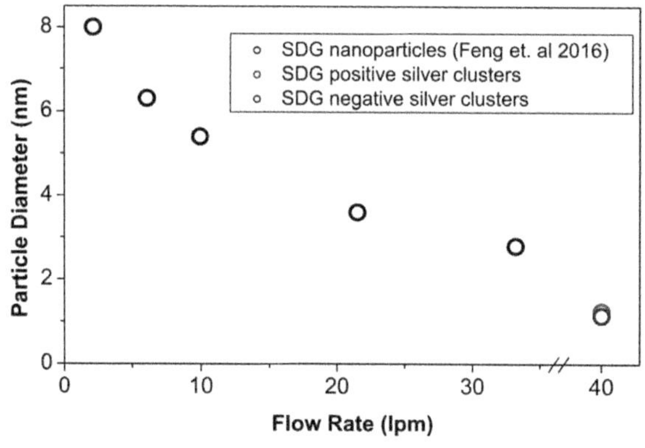

Figure 7.1 Transition between nanoparticle generation and cluster generation using SDG: Increasing the dilution flow rate while, at the same time, decreasing the energy per spark leads to a further decrease in particle size. The red and blue data points represent the mean cluster size measured for positive and negative cluster species, respectively. The data points for atomic clusters were measured at lower energy per spark compared to the black data points for larger nanoparticles. This is accounted for by the break in the *x* axis.

The major advances in improving aerosol instruments for measurements of small-sized ions down to the subnanometer size range that have taken place during recent years enabled a number of new research efforts in this size regime. Measurements were carried out to analyze air ions or to explore lower detection limits of, for example, condensation particle counters (CPCs). The possibility to generate nanomaterials with aerosol generation methods has also proven to be of interest for nanoengineering purposes. Aerosol-based nanoengineering has the advantage compared to other existing techniques that the generation and characterization stage are carried out at atmospheric pressure. The combination of the development of aerosol instruments and aerosol-based nanoengineering led to several research efforts to apply and evaluate standard aerosol generation techniques such as SDG for the production of smaller and smaller nanoparticles and clusters [26].

The first study that showed that atomic clusters could be produced and size-separated at atmospheric pressure was carried out by Schmidt-Ott and coworkers in 2009 [36] using hot-wire generation (HWG). HWG is another evaporation/condensation method to produce aerosol particles. Metallic material is evaporated from a hot wire, which is heated by passing a high electrical current through it. The material evaporated is diluted and cooled by a flowing inert gas. Similar to SDG, only the wire gets heated, which limits the sources of contamination compared to, for example, using a tube furnace as an aerosol generation method.

In comparison to SDG, the major difference in HWG is that the energy used to evaporate material is limited by the heat that is generated on the wire. In case the wire gets too hot, it will break. Other limitations of wire generation are that only a few operational parameters can be varied and that mixed-particle production is not possible. This limits the potential of optimizing the generation process and of up-scaling the amount of nanoparticles generated. This led to the development of the SDG method for atomic cluster generation.

Working at atmospheric pressure gives access to aerosol technology, in particular to separation of sizes by mobility classification. This increases the potential for industrial applications, as expensive vacuum technology can be avoided, as well as the inevitable loss of particles in the transfer from high to low pressure. A differential mobility analyzer (DMA) is used to measure size distributions of clusters generated and to size-select specific cluster species. A DMA relies on the principle of separating charged particles in a gas flow by exposing them to an electric field. The trajectories of the particles depend on their electrical mobility. In the most common configurations, a DMA consists of two concentric cylindrical electrodes. The particles are introduced close to the outer electrode. A sheath gas flow carries the particles downstream, where an outlet slit is located on the inner electrode. For a constant electrical field, only particles with a narrow electrical mobility range can exit the instrument through the outlet slit. If the particles are singly charged, the electrical mobility is related to particle size. DMAs have been developed for the analysis and size classification of larger nanoparticles. Classic instrument designs usually show poor resolution for very small particles due to their high diffusivity. This leads to diffusion broadening of the transfer function and thus decreases the resolution. However, with recent advances in designing DMAs of high resolution and transmission, atomic cluster sizes can also be measured. The highest resolutions achieved with atmospheric pressure DMAs to date are as high as 80 [2], resolution R being defined as the ratio between the measured peak position V_0 and the full width at half maximum of the peak V_{FWHM} [24].

In the measurements by Maisser et al. (see later), particles were detected using a Faraday cup electrometer (FCE), which measured the current generated by the charged particles exiting the DMA.

7.2.1 Operational Parameters of SDG for Atomic Cluster Generation

Several operational parameters of SDG were tested for their appropriateness to produce atomic clusters. The results have been published in Maisser et al. (2015) [26].

The following parameters were found to have an influence on atomic cluster generation using SDG:

- Energy per spark
- Inert gas flow
- Electrode configuration

In the following sections the specific operational parameters for cluster generation will be further discussed.

7.2.1.1 Energy per spark

As explained in Chapter 2, SDG is driven by a simple RCL (resistance, capacitance, inductance) circuit [30, 48]. A high-voltage constant current source is used to charge the capacitor and to provide the energy required for generating the spark. The resistance of the circuit is related to the gap between the electrodes, and the inductance arises from the cabling. The capacitance C is charged by the charging current I. This leads to an increase in voltage between the electrodes. When the voltage is high enough, a spark is formed, which discharges the capacitor. The process repeats itself with a certain frequency f. The frequency depends on the charging current, the capacitance, and the discharge voltage:

$$f = \frac{I}{CV_{\mathrm{d}}} \tag{7.1}$$

The discharge voltage V_{d} is related to the breakdown voltage V_{b} of the gas. The energy per spark, E, can be written as

$$E = \frac{1}{2}CV_{\mathrm{d}}^{2}. \tag{7.2}$$

The spark energy determines the amount of material evaporated from the electrode by each spark and can be controlled by the capacitance used in the electrical circuit and by the gap distance.

In comparison to when SDG is operated to produce larger nanoparticles, the spark energy for cluster generation needs to be about 1 order of magnitude lower. Another way to keep the discharge voltage low is by decreasing the gap distance. At lower gap distances the discharge will occur at lower voltages. In analogy to nanoparticle generation, it is possible to shift the mean cluster size toward larger number clusters by increasing the spark energy. This can be seen, for example, in Maisser et al. (2015), where for both positive and negative clusters an increase of capacitance and thus an increase of spark energy lead to more abundance of larger number clusters (c.f. Fig. 7.2 and Maisser et al. [26]).

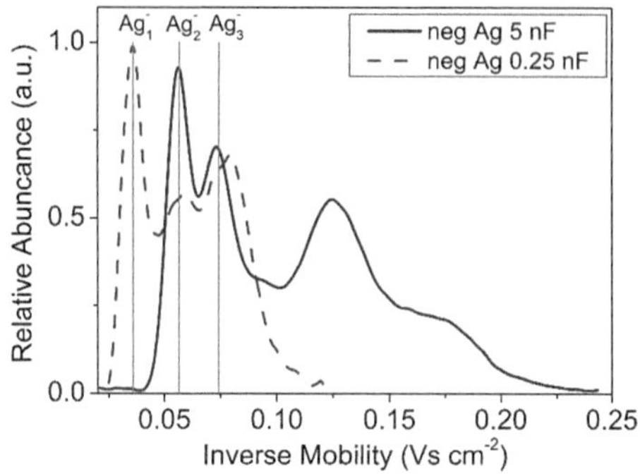

Figure 7.2 SDG negative silver atomic clusters, dependency on cluster number on spark capacitance (energy). From Ref. [26], reprinted by permission of the publisher (Taylor & Francis Ltd, http://www.tandfonline.com).

7.2.1.2 Inert gas flow

As mentioned in the previous section, the gas breakdown voltage influences the spark energy. This gives importance to the choice of gas used for a dilution flow. Initial experiments were performed using three different inert gases: argon, nitrogen, and helium. Generation of atomic clusters has been found feasible in all three gases; however, helium has some advantages compared to the other two. Using helium produces smaller particle sizes [26, 48]. This is associated with the low breakdown voltage in helium and therefore to a low discharge voltage. The low breakdown voltage in helium, therefore, enables stable operation for very low spark energies. Note that this expensive gas can be recycled, which also brings advantages with it concerning purity (see later).

To generate atomic clusters, high dilution flows are required. According to the experimental investigations, however, controlling the cluster size distribution by inert gas flow is not very practical, and adjustments of energy per spark have been found to be much easier to control in a reproducible manner.

7.2.1.3 Electrode configuration

It has been found in SDG studies that specific spark electrode shapes can help to generate smaller particles with narrower size distributions. Cole et al. (2009) [7] used sharpened electrodes to generate sub-20-nm particles. Due to the tipped electrodes, lower amounts of material are evaporated by each spark, leading to the generation of smaller particles. In another study by Han et al., a pin-to-plate configuration was used to generate small nanoparticles [16]. Compared to the rod-to-rod configuration, less agglomeration was observed, smaller particles were generated, and the size distribution measured was narrower. This could be related to a change in spark voltage associated with the change in the electric field, which would lead to lower spark frequencies and thus less material evaporated per unit time. Along the line of these findings, SDG-generated atomic clusters used a rod-to-wire electrode configuration. This enabled generation of smaller cluster sizes and stabilized cluster production over time (c.f. Fig. 7.3). However, a systematic study on the influence of electrode configuration on particle size has not been carried out to date, and it is not clear whether the rod-to-wire configuration should be recommended for atomic cluster generation in general. Further investigations are thus needed.

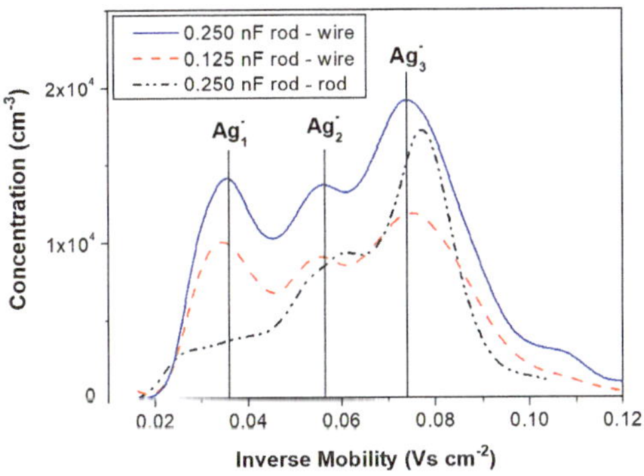

Figure 7.3 Comparison of size distributions of negatively charged silver atomic clusters using a rod–wire and a rod–rod electrode configuration.

In summary, the atomic cluster size production can be controlled by spark energy, carrier gas, electrode configuration, and dilution flow. The high number of variable operation parameters enables a high flexibility for cluster generation, where for each cluster size, material, and application, SDG can be tuned and optimized.

7.2.2 Cluster Properties

7.2.2.1 Charge state

Tabrizi et al. [47] showed that 5 nm particles from spark discharge are, at most, charged with the threefold probability compared to equilibrium bipolar charging by ion attachment, a technique usually applied in aerosol technology. The similarity with bipolar ion attachment is understandable because the spark produces ions of both polarities, which attach to the particles. Taking into account large particle losses in the bipolar charger, we can conclude from that data that the charging probability for 5 nm particles is at most a few percentage points. Because of the even smaller size of atomic clusters, it could be assumed that close to 100% of the particles carry no charge. However, earlier research on atomic clusters in vacuum systems, which applied spark discharge directly followed by expansion into vacuum (see Chapter 8 and Methling et al. [2001] [29]), unambiguously showed that up to 80% of the atomic clusters carry an elementary charge (approximately half positive and half negative). It can be concluded that metal atoms coming from the spark plasma are initially charged with a high probability. This is consistent with Chapter 4, from which we learn that close to 100% of the gas and metal atoms are ionized in the spark plasma. Outside of the plasma, where cluster and nanoparticle formation takes place, they lose their charge by recombination. At the beginning of this growth process, where atomic clusters are formed, much of the original plasma charge is still present, but diminishes in the following growth process, not being far from bipolar equilibrium charge [47] at 5 nm. In agreement with this picture, the data of Tabrizi et al. [47] indicates that the smaller the particles, the higher the charging efficiency with respect to this equilibrium.

The experiments by Maisser et al. showed that for atomic clusters, typically more positively charged clusters are produced compared to negatively charged clusters [26]. Whereas the polarities are in balance within the plasma (see Chapter 4), this is in line with results from other generation methods, where both polarities are produced. This observation can be explained by looking at the growth process downstream of the plasma, where stability characteristics are important, (metal cations are typically more stable than metal anions) [13] and by the lower mobility of positive charge carriers compared to electrons. This leads to positive charge carriers being less likely scavenged by the high electric fields generated between the two electrodes [32].

7.2.2.2 Reactivity/materials

For nanoparticle generation, and even more so for atomic cluster generation, possible sources of contamination of the generated particles need to be avoided. In SDG, the potential sources for contamination are limited to the purity of the electrodes used, and to gas purity. The measures that need to be taken to avoid impurities strongly depend on the requirements of the generated particles.

For atomic clusters, even trace amounts of impurities in the dilution gas might lead to contamination of or chemical reactions with non-inert clusters. To avoid undesired impurities or, for example, oxidation, it is recommended to use high-purity gases in combination with additional purification systems. The most abundant trace impurities are oxygen and water. For instance, it has been demonstrated that un-oxidized magnesium nanoparticles (typically highly reactive to oxygen) can be generated if all the tubing lines used are stainless steel and when the dilution gas is additionally purified using a molecular sieve and a catalyst system to remove water and oxygen impurities from the gas [30, 51]. Recycling of the carrier gas would be a measure of further reducing the contamination problem.

The high reactivity of certain cluster species with impurities also offers the opportunity to identify and even select the more stable species. Controlled amounts of impurities could be introduced into the SDG system during the generation process, while tracking changes in the produced cluster composition and distribution. Initial tests to evaluate the influence of different oxygen levels present during the SDG process have been carried out by Maisser et al. [26]. This method is known as oxygen etching. The reactivity of small atomic clusters strongly depends upon the number of atoms inside the cluster. The more reactive species react with oxygen and with each other. The stable magic number clusters, however, do not react and thus remain unchanged. Oxygen etching can be used to increase the concentration of magic number clusters and to isolate them from other cluster species generated simultaneously. This is of high practical relevance because the stable clusters are the ones of interest for applications.

7.3 Testing of the Atmospheric Pressure Cluster Source

Mobility classification is easy to carry out with inexpensive atmospheric pressure systems [2, 26, 41], and small clusters (up to ca. 10 atoms) can be separated from each other with the DMA technology presently available. A higher resolution is necessary to select larger clusters for specific applications or to obtain information on cluster purity, and this can only be achieved with conventional vacuum-based mass spectrometers, for example, time-of-flight mass spectrometers (TOF-MSs), [50] or quadrupole [29, 31]. Recently, a study has been carried out where atomic clusters produced by atmospheric pressure SDG where combined with this vacuum technology using an atmospheric pressure inlet and a time-of-flight mass spectrometer (API-TOF-MS). Initial results of this study are shown in Fig. 7.4, which revealed that the peaks of the mobility distribution, as in Fig. 7.3, actually consist of groups of clusters including oxides and hydrates, if contaminants are not sufficiently reduced. On the other hand, these measurements

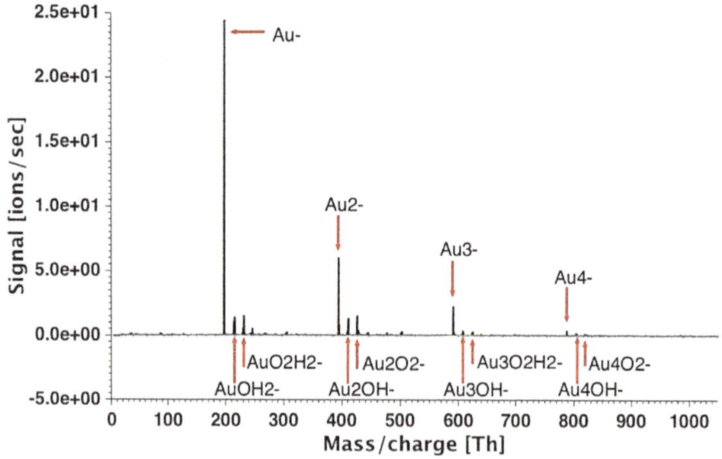

Figure 7.4 SDG gold atomic clusters measured using mass spectrometry [19].

confirm that the mobility distributions obtained in previous studies [26, 37] had mainly consisted of the pure clusters and that a simple SDG-DMA system at atmospheric pressure can be used as a source of atomic clusters [26]. In contrast to vacuum-based systems, such a source would be scalable for production of the quantities required for applications such as catalysis.

7.4 Evaluation of the Up-scaling Potential of SDG for Atomic Cluster Generation

To do cluster research or produce clusters for applications (e.g., catalysis, cluster–cluster assembled materials, etc.) there are certain requirements for the cluster generation source. The method for cluster generation should offer the option of controlling the generated cluster size and composition. For several studies it is additionally necessary to generate significant amounts of clusters, first for research purposes, where their reactivity and their potential for applications purposes are tested. Second, it is desirable that the method be even scalable to reach quantities that match industrial needs, for example, for generation of catalysts based on atomic clusters, which would induce a revolution in catalyst development, if suitable methods of immobilization are found. In the past, several techniques to generate atomic clusters have been developed and are now used as standard generation methods. As mentioned in a previous section, the greatest limitation of existing cluster methods is the difficulty to cost-efficiently up-scale output rates.

To date there exist two main approaches to generating nanoparticles and atomic clusters: wet-chemistry processes and ion beam methods. More details are discussed elsewhere [49]. However, wet-chemistry technologies have some major disadvantages. These include the slow kinetics of batch operation (which limits the up-scaling potential) and the need to use precursors, possibly resulting in contaminations, high cost, and toxicity. In addition, chemical waste is generated by wet-chemistry processes, which makes them environmentally more critical compared to gas-phase production [12, 28].

Cluster beam methods offer several advantages over other cluster generation techniques (e.g., chemical preparation techniques).

Desired cluster sizes can be selected using standard size selection tools at high resolution. The deposition energy can be controlled, which enables soft deposition to maintain the cluster morphology after the impact [39].

A number of different methods to generate ion beams exist, of which spark (or pulsed arc) ablation is one (see Chapter 8). Others are laser ablation and ion sputtering. Many of these methods are widely applied in cluster research and are typically adapted for specific experimental requirements [8]. Various cluster properties have been investigated in basic research experiments, such as electronic, optical, and magnetic properties. The dependence of certain cluster properties on size and composition has also been extensively studied in the past. Clusters have since been classified according to their ionization potential, electron affinity, dissociation energies, etc. [4]. Several very interesting properties have been found for small atomic clusters through these experiments, such as odd-even effects for cluster stability or that clusters can exhibit electronic structures similar to atoms, mimicking their properties. The term "superatoms" has been coined in connection with this [5]. Furthermore, interesting catalytic properties are predicted by, for example, Roach et al. [40] As each atom in the cluster is important, there are basically myriads of possible combinations, each exhibiting different properties, principally opening unlimited possibilities in catalysis and other fields.

There exist several excellent reviews and books discussing different cluster generation methods in great detail [8, 38, 49, 52]. Here we will only discuss the up-scaling potential of some of the existing methods in more detail.

The typical output rates of ion beam sources are in the range of nanoamps (nA) at best. To enable industrial-scale application, the output rates need to be up-scaled by orders of magnitude. Output quantities in the range of about 1–10 g of deposited clusters at 1% loading onto a suitable support material would probably be sufficiently high for some applications, such as small batch production of catalysts for pharmaceutical products [34].

In the past few years, several studies for up-scaling output rates of cluster sources have been published. The group of Palmer and coworkers presented a proof-of-principle for a source producing nA cluster beams, and the authors claim that scaling up to the mA regime would be possible This could be achieved by introducing a new kind of cluster beam source (a matrix-assembled cluster source [MACS]). In a first step, metal atoms are immobilized in a noble gas matrix by an evaporation-condensation process. The isolated atoms then undergo temperature-controlled aggregation, and finally a beam of the clusters is produced by bombarding the matrix with gas ions. This process is rather complicated, and it delivers the highest mass production rates above 1000 atoms per cluster (with nanoparticles above 5 nm in diameter). Scalability to meet industrial needs is doubtful, as the currents reported are a factor of 10^6 times lower than those targeted.

The production rate of atomic clusters generated by SDG can be optimized to some extent by adapting the operation parameters discussed in Section 7.2.1. For example, increasing the energy per spark can lead to higher production as long as the process of condensational growth is quenched sufficiently fast to avoid generating bigger particles (c.f. Fig. 7.5). However, this is not a very efficient way, and to get outputs for industrial-scale applications further engineering efforts are required.

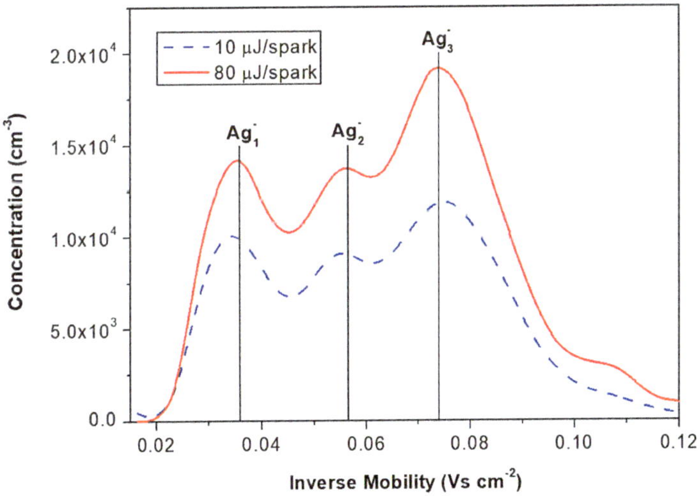

Figure 7.5 Increased production of atomic clusters by increasing energy per spark.

Recently, efforts to up-scale the output rate of nanoparticle production by SDG have also been made. A new electrical circuit to increase spark frequency has been introduced by Pfeiffer et al. [37], increasing the output rates from a few mg/hour to several g/hour compared to a classical RLC circuit. In addition to increasing the spark repetition frequency, the very simple concept of SDG also enables the operation of a large number of electrode pairs in parallel for further increasing mass output [12]. Subsequent size separation or deposition will, in any case, profit from the high percentage of charged clusters the hot spark plasma delivers. In contrast to methods that require transition into vacuum, operation at atmospheric pressure during all of the cluster preparation steps (generation, size selection, and deposition) enables collection of virtually all of the generated material in SDG.

Any process involving atomic clusters puts high demands on purity, so their specific desired properties are maintained [26]. In vacuum, continuous turbo pumping maintains the low pressure and, therefore, also reducing contaminations. In the aerosol approach, contamination can be held low enough by applying adsorbing agents. This is feasible, as shown by Vons et al. [51], where even oxidation of magnesium was avoided. However, for applications, clusters need to be removed from the clean environment. This carries the risk of clusters reacting or being contaminated during this step, as pointed out by Jena (2015) for the case of vacuum systems [20].

7.5 Conclusion and Outlook

The proof of concept and the demonstration of feasibility in using SDG for application-oriented production of atomic clusters has been provided by Maisser et al. [26]. On the basis of these very promising results we believe that the potential of SDG to become a standard technology for atomic cluster generation is significant. The high charging probability achieved by SDG for atomic clusters, stemming from the high spark temperature (see Chapters 3 and 4), is a unique feature of this method, which has not received the appreciation it deserves in view of producing sufficient quantities for applications.

However, several basic research and engineering efforts are still needed in order to transfer this technology into the area of applications. Further characterization of the SDG atomic clusters needs to be carried out with state-of-the art technologies to confirm cluster identity and purity. A detailed investigation on how changing various operation parameters affects the generated cluster sizes, concentration, and population needs to be carried out. The aerosol approach using SDG has the potential of being scaled up to produce industrial quantities of atomic clusters, but only the combination with vacuum-based analysis by MS provides a reliable characterization of the product. Combining an SDG-DMA source with MS also bears the advantage that the DMA-MS combination is structure sensitive and could provide a powerful tool for basic research. Such a cluster mobility–MS method using the newest DMA technology [9, 42] has recently been applied to determine ion structures [25, 27, 33] and is capable of resolving power in the order of 70, which is about twice the value of what is typically reported for ion mobility spectrometry–mass spectrometry (IMS-MS) measurements [18, 44].

To further study and apply the generated clusters, deposition and immobilization methods need to be found and evaluated for their application potential. We expect that the SDG-DMA source will trigger numerous investigations in the near future, and hope that SDG will fill the gap that up to now prevented atomic clusters being applied on a large scale as building blocks for new materials or as high-quality catalysts.

References

1. Arenz, M., Gilb, S. and Heiz, U. (2007). Size effects in the chemistry of small clusters, *Chem. Phys. Solid Surf.*, **12**, pp. 1–51.
2. Attoui, M., Paragano, M., Cuevas, J. and Fernandez de la Mora, J. (2013). Tandem DMA generation of strictly monomobile 1–3.5 nm particle standards, *Aerosol Sci. Technol.*, **47**, pp. 499–511.
3. Brack, M. (1993). The physics of simple metal clusters: self-consistent jellium model and semiclassical approaches, *Rev. Mod. Phys.*, **65**, pp. 677–732.
4. Castleman, A. and Jena, P. (2006). Clusters: a bridge between disciplines, *Proc. Natl. Acad. Sci. U.S.A.*, **103**, pp. 10552–10553.
5. Castleman Jr., A. and Khanna, S. (2009). Clusters, superatoms, and building blocks of new materials, *J. Phys. Chem. C*, **113**, pp. 2664–2675.

6. Claridge, S. A., Castleman Jr., A., Khanna, S. N., Murray, C. B., Sen, A. and Weiss, P. S. (2009). Cluster-assembled materials, *ACS Nano*, **3**, pp. 244–255.

7. Cole, J. J., Lin, E.-C., Barry, C. R. and Jacobs, H. O. (2009). Continuous nanoparticle generation and assembly by atmospheric pressure arc discharge, *Appl. Phys. Lett.*, **95**, p. 113101.

8. de Heer, W. A. (1993). The physics of simple metal clusters: experimental aspects and simple models, *Rev. Mod. Phys.*, **65**, pp. 611–676.

9. De La Mora, J. F., Ude, S. and Thomson, B. A. (2006). The potential of differential mobility analysis coupled to MS for the study of very large singly and multiply charged proteins and protein complexes in the gas phase, *Biotechnol. J.*, **1**, pp. 988–997.

10. Di Vece, M. (2018). Using nanoparticles as a bottom-up approach to increase solar cell efficiency, *KONA Powder Part. J.*, p. 2019005.

11. Feng, J., Huang, L., Ludvigsson, L., Messing, M. E., Maisser, A., Biskos, G. and Schmidt-Ott, A. (2015). General approach to the evolution of singlet nanoparticles from a rapidly quenched point source, *J. Phys. Chem. C*, **120**, pp. 621–630.

12. Feng, J., Guo, X., Ramlawi, N., Pfeiffer, T. V., Geutjens, R., Basak, S., Nirschl, H., Biskos, G., Zandbergen, H. W. and Schmidt-Ott, A. (2016). Green manufacturing of metallic nanoparticles: a facile and universal approach to scaling up, *J. Mater. Chem. A*, **4**, pp. 11222–11227.

13. Fernández, E. M., Soler, J. M., Garzón, I. L. and Balbás, L. C. (2004). Trends in the structure and bonding of noble metal clusters, *Phys. Rev. B*, **70**, p. 165403.

14. Georgakilas, V., Perman, J. A., Tucek, J. and Zboril, R. (2015). Broad family of carbon nanoallotropes: classification, chemistry, and applications of fullerenes, carbon dots, nanotubes, graphene, nanodiamonds, and combined superstructures, *Chem. Rev.*, **115**, pp. 4744–4822.

15. Gong, X., Tong, M., Xia, Y., Cai, W., Moon, J. S., Cao, Y., Yu, G., Shieh, C.-L., Nilsson, B. and Heeger, A. J. (2009). High-detectivity polymer photodetectors with spectral response from 300 nm to 1450 nm, *Science*, **325**, pp. 1665–1667.

16. Han, K., Kim, W., Yu, J., Lee, J., Lee, H., Woo, C. G. and Choi, M. (2012). A study of pin-to-plate type spark discharge generator for producing unagglomerated nanoaerosols, *J. Aerosol Sci.*, **52**, pp. 80–88.

17. He, Z., Zhong, C., Su, S., Xu, M., Wu, H. and Cao, Y. (2012). Enhanced power-conversion efficiency in polymer solar cells using an inverted device structure, *Nat. Photonics*, **6**, p. 591.

18. Hogan Jr., C. J., Ruotolo, B. T., Robinson, C. V. and Fernandez de la Mora, J. (2011). Tandem differential mobility analysis-mass spectrometry reveals partial gas-phase collapse of the GroEL complex, *J. Phys. Chem. B*, **115**, pp. 3614–3621.

19. Holm, S., Barmpounis, K., Attoui, M., Biskos, G. and Kangasluoma, J. (2018). *Mass Spectrometer Measurements of Spark Discharge Generated Atomic Clusters*, personal communication, Maisser, A.

20. Jena, P. (2015). Atomic clusters: opportunities in the face of challenges, *J. Phys. Chem. Lett.*, **6**(9), pp. 1549–1552.

21. Khanna, S. and Jena, P. (1995). Atomic clusters: building blocks for a class of solids, *Phys. Rev. B*, **51**, pp. 13705–13716.

22. Krätschmer, W., Lamb, L. D., Fostiropoulos, K. and Huffman, D. R. (1990). Solid C60: a new form of carbon, *Nature*, **347**, p. 354.

23. Kroto, H., Heath, J., Brien, S., Curl, R. and Smalley, R. (1985). C(60): Buckminsterfullerene, *Nature*, **318**, pp. 162–163.

24. Kulkarni, P., Baron, P. A. and Willeke, K. (2011). *Aerosol Measurement: Principles, Techniques, and Applications* (John Wiley & Sons).

25. Larriba, C. and Hogan Jr., C. J. (2013). Ion mobilities in diatomic gases: measurement versus prediction with non-specular scattering models, *J. Phys. Chem. A*, **117**, pp. 3887–3901.

26. Maisser, A., Barmpounis, K., Attoui, M., Biskos, G. and Schmidt-Ott, A. (2015). Atomic cluster generation with an atmospheric pressure spark discharge generator, *Aerosol Sci. Technol.*, **49**, pp. 886–894.

27. Maißer, A., Thomas, J. M., Larriba-Andaluz, C., He, S. and Hogan, C. J. (2015). The mass–mobility distributions of ions produced by a Po-210 source in air, *J. Aerosol Sci.*, **90**, pp. 36–50.

28. Messing, M. E., Dick, K. A., Wallenberg, L. R. and Deppert, K. (2009). Generation of size-selected gold nanoparticles by spark discharge—for growth of epitaxial nanowires, *Gold Bull.*, **42**, pp. 20–26.

29. Methling, R.-P., Senz, V., Klinkenberg, E.-D., Diederich, T., Tiggesbäumker, J., Holzhüter, G., Bansmann, J. and Meiwes-Broer, K. (2001). Magnetic studies on mass-selected iron particles, *Eur. Phys. J. D*, **16**, pp. 173–176.

30. Meuller, B. O., Messing, M. E., Engberg, D. L., Jansson, A. M., Johansson, L. I., Norlén, S. M., Tureson, N. and Deppert, K. (2012). Review of spark discharge generators for production of nanoparticle aerosols, *Aerosol Sci. Technol.*, **46**, pp. 1256–1270.

31. Miller, P. E. and Denton, M. B. (1986). The quadrupole mass filter: basic operating concepts, *J. Chem. Educ.*, **63**, p. 617.

32. Moini, M. and Eyler, J. R. (1988). Formation of small negative and positive cluster ions of gold, silver, and copper by direct laser vaporization, *J. Chem. Phys.*, **88**, pp. 5512–5515.

33. Ouyang, H., Larriba-Andaluz, C., Oberreit, D. R. and Hogan Jr., C. J. (2013). The collision cross sections of iodide salt cluster ions in air via differential mobility analysis-mass spectrometry, *J. Am. Soc. Mass Spectrom.*, **24**, pp. 1833–1847.

34. Palmer, R., Cao, L. and Yin, F. (2016). Note: Proof of principle of a new type of cluster beam source with potential for scale-up, *Rev. Sci. Instrum.*, **87**, p. 046103.

35. Partha, R. and Conyers, J. L. (2009). Biomedical applications of functionalized fullerene-based nanomaterials, *Int. J. Nanomed.*, **4**, p. 261.

36. Peineke, C., Attoui, M., Robles, R., Reber, A., Khanna, S. and Schmidt-Ott, A. (2009). Production of equal sized atomic clusters by a hot wire, *J. Aerosol Sci.*, **40**, pp. 423–430.

37. Pfeiffer, T., Feng, J. and Schmidt-Ott, A. (2014). New developments in spark production of nanoparticles, *Adv. Powder Technol.*, **25**, pp. 56–70.

38. Popok, V. N., Barke, I., Campbell, E. E. and Meiwes-Broer, K.-H. (2011). Cluster–surface interaction: from soft landing to implantation, *Surf. Sci. Rep.*, **66**, pp. 347–377.

39. Pratontep, S., Carroll, S., Xirouchaki, C., Streun, M. and Palmer, R. (2005). Size-selected cluster beam source based on radio frequency magnetron plasma sputtering and gas condensation, *Rev. Sci. Instrum.*, **76**, p. 045103.

40. Roach, P. J., Woodward, W. H., Castleman, A., Reber, A. C. and Khanna, S. N. (2009). Complementary active sites cause size-selective reactivity of aluminum cluster anions with water, *Science*, **323**, pp. 492–495.

41. Rus, J., Moro, D., Sillero, J. A., Royuela, J., Casado, A., Estevez-Molinero, F. and de la Mora, J. F. (2010). IMS–MS studies based on coupling a differential mobility analyzer (DMA) to commercial API–MS systems, *Int. J. Mass Spectrom.*, **298**, pp. 30–40.

42. Rus, J., Moro, D., Sillero, J. A., Royuela, J., Casado, A., Estevez-Molinero, F. and Fernández de la Mora, J. (2010). IMS–MS studies based on coupling a differential mobility analyzer (DMA) to commercial API–MS systems, *Int. J. Mass Spectrom.*, **298**, pp. 30–40.

43. Schwyn, S., Garwin, E. and Schmidt-Ott, A. (1988). Aerosol generation by spark discharge, *J. Aerosol Sci.*, **19**, pp. 639–642.

44. Shvartsburg, A. A. and Smith, R. D. (2008). Fundamentals of traveling wave ion mobility spectrometry, *Anal. Chem.*, **80**, pp. 9689–9699.

45. Smalley, R. E. (1992). *Doping the Fullerenes* (ACS Publications).

46. Smalley, R. E. (1997). Discovering the fullerenes, *Rev. Mod. Phys.*, **69**, p. 723.

47. Tabrizi, N., Xu, Q., Van Der Pers, N., Lafont, U. and Schmidt-Ott, A. (2009). Synthesis of mixed metallic nanoparticles by spark discharge, *J. Nanopart. Res.*, **11**, p. 1209.

48. Tabrizi, N. S., Ullmann, M., Vons, V., Lafont, U. and Schmidt-Ott, A. (2009). Generation of nanoparticles by spark discharge, *J. Nanopart. Res.*, **11**, p. 315.

49. Tyo, E. C. and Vajda, S. (2015). Catalysis by clusters with precise numbers of atoms, *Nat. Nanotechnol.*, **10**, p. 577.

50. Von Issendorff, B. and Palmer, R. (1999). A new high transmission infinite range mass selector for cluster and nanoparticle beams, *Rev. Sci. Instrum.*, **70**, pp. 4497–4501.

51. Vons, V., Anastasopol, A., Legerstee, W., Mulder, F., Eijt, S. and Schmidt-Ott, A. (2011). Low-temperature hydrogen desorption and the structural properties of spark discharge generated Mg nanoparticles, *Acta Mater.*, **59**, pp. 3070–3080.

52. Wegner, K., Piseri, P., Tafreshi, H. V. and Milani, P. (2006). Cluster beam deposition: a tool for nanoscale science and technology, *J. Phys. D: Appl. Phys.*, **39**, p. R439.

53. Yoon, M., Yang, S., Wang, E. and Zhang, Z. (2007). *Charged Fullerenes as High-Capacity Hydrogen Storage Media* (ACS Publications).

Chapter 8

The Use of Spark Ablation for Generating Cluster Beams: A Review

Claudio Piazzoni and Paolo Milani
CIMAINA and Department of Physics, University of Milano, via Celoria 16, I-21033 Milano, Italy
paolo.milani@mi.infn.it

As the use of nanostructured films and nanocomposites is becoming of interest for a wide spectrum of applications and reaching a maturity suitable for large-scale fabrication, the development of approaches able to merge parallel top-down fabrication technologies with cluster assembling constitutes an essential requisite. The present chapter deals with nanoparticle deposition based on gas-phase production by spark ablation and manipulation in effusive or supersonic beams. It is an approach particularly suitable to be matched with physical vapor deposition techniques typically used for the production of devices and smart coatings. The possibility to integrate the particle production, manipulation, and deposition stages into a continuous single process is one of the major advantages of the gas-phase route to nanostructured films.

Spark Ablation: Building Blocks for Nanotechnology
Edited by Andreas Schmidt-Ott

ISBN 978-981-4800-82-2 (Hardcover), 978-0-367-81709-1 (eBook)
www.jennystanford.com

8.1 Introduction

A wide variety of gas-phase techniques has been developed for the fabrication of nanostructured films [1]. The generation of nanostructured particulate films involves at least two steps, the formation of solid particles and the subsequent deposition on a substrate. In many processes, gas-phase modification or manipulation of the produced particles constitutes a third, intermediate step. The general term "manipulation" is used here with special reference to the following meanings: (i) the ability to sort the objects in terms of a size or geometry classification, (ii) the ability to control the position and sometimes the velocity of the nanoscale objects, and (iii) the ability to modify them physically or chemically. Depending on the specific application, a refinement of the particle population is often necessary.

Vacuum processes for solid particle deposition usually rely on expansion techniques that pass particles in a gas through an orifice or nozzle, after which expansion takes place to give a beam of particles that can be deposited onto a surface [2]. The starting vapor can be generated by a variety of means such as Joule heating of a material in a crucible, laser ablation, sputtering, or arc discharge. A broad range of materials including metals, metal oxides, and organics can be processed [2].

"Cluster sources" can be catalogued by considering the regimes governing gas introduction and extraction: continuous or pulsed, effusive or supersonic. Note that the term "cluster," as established in the community dealing with cluster/particle beams, does not necessarily refer to atomic clusters as defined in Section 1.2.1, Chapter 1, but is also used in connection with beams composed of particles considerably larger than 2 nm, which are called "nanoparticles" throughout the rest of this book. For cluster beam deposition (CBD) under vacuum conditions, a continuous gas flow must be compatible with stringent vacuum requirements and hence only effusive regimes are of practical interest. The realization and operation of pulsed cluster sources appear to be more complicated compared to continuous ones, but the reduced gas load has the advantage to allow the use of a supersonic expansion regime and the compatibility with high-vacuum (HV) and ultrahigh-vacuum (UHV) standards [2].

An advantage of all CBD techniques, as for all techniques applying spark ablation, is the separation between the particle formation region and the substrate, allowing deposition on temperature-sensitive materials, such as polymers. These aspects are of fundamental importance for applications and in particular for the compatibility of CBD processes with microfabrication and planar technologies [3].

To improve the performance of cluster sources and to develop them toward a continuous process for fabrication of nanostructured coatings, the ablation process has to be confined to a well-defined target region, thus achieving high efficiency in material vaporization. Furthermore, the carrier gas load has to be kept low to avoid large pumping systems. Source geometry and dimensions should also be optimized to produce intense and stable cluster beams [2].

The availability of efficient atomization methods is the prerequisite for the realization of an intense cluster source. Another requirement is the control of the parameters of gas aggregation determining the growth rate and hence the mass distribution of the particles. Once formed, the clusters must be extracted from the source with their original physicochemical characteristics unperturbed.

To produce nanostructured materials with tailored properties, one should also be able to control the parameters of cluster–cluster and cluster–surface interactions, which are significantly affected by the kinetic energy of the aggregates [1].

Different cluster production techniques fulfill several of the requirements cited before. However, an ideal cluster source where all the parameters can be controlled is very difficult to obtain. Among different cluster sources currently on the market, we will review in this chapter those based on spark ablation. This approach is of particular interest for the production of compact and relatively simple cluster sources able to produce intense beams of neutral and ionized particles.

In this chapter we will review three types of cluster sources based on spark ablation: the pulsed-arc cluster-ion source (PACIS), pulsed microplasma cluster source (PMCS), and arc cluster-ion source (ACIS).

8.2 PACIS

The PACIS was originally proposed for cluster beams by Meiwes-Broer and coworkers in 1990, when they introduced the concept of "a pulsed high-current arc between two electrodes in a pulse of carrier gas" [4]. The source consisted in an insulating ceramic body hosting two metal rod electrodes with a spacing of few millimeters between them. The circuit resembles the one of Fig. 2.1, Chapter 2, in that a capacitor parallel to the electrodes stores the energy released in each spark. A pulsed valve faces a channel drilled perpendicularly to the electrodes, injecting an inert carrier gas (typically He) through the gap; a suitable time matching between gas and electric pulses fires the arc and supplies the metal vapor.

The source body was coupled to a conical nozzle ("extender" [4]) and placed into HV to create the suitable conditions for cluster aggregation and seeded beam formation and transport. The design and experimental conditions were suited to the production of metal cluster ions, namely W^+_n and mixed metal–rare gas clusters (Ar_nAl^+, Kr_nAl^+) [4].

The PACIS was operated by a home-made power supply at repetition rates from 10 to 100 Hz with DC pulses 3–20 μs long, at voltages up to 1300 V and currents up to 5000 A [5, 6].

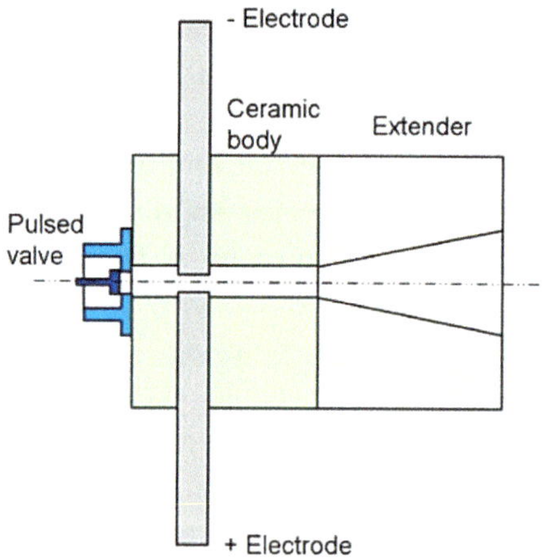

Figure 8.1 The first version of the PACIS (1990). Adapted from Ref. [4], Copyright (1990), with permission from Elsevier.

The resulting layer growth rates on the target for Ag and Pb clusters were 0.1 and 1.8Å/shot, respectively (with a shot-to-shot variation of 100%) at a distance from the source of 5 cm with a charged/neutral particles ratio of about 13 [5].

The shape and dimensions of the nozzle are crucial for the formation process of the clusters as they drive the gas dynamic and thus nucleation and neutralization processes. It was demonstrated that not only the number, dimension, and size distribution but also the charge state of the clusters was affected by this element [6]. The importance of the nozzle is witnessed by the fact that soon after the first appearance, different versions of the PACIS nozzle were proposed to improve the cluster formation capabilities and/or to adapt it to the different cathode materials used [5, 6]. Another important improvement was the introduction of more robust electric insulation (to avoid interference between the electrode discharge and the valve circuit) and of a cooling system to stabilize the source performance and improve cluster aggregation [5, 6].

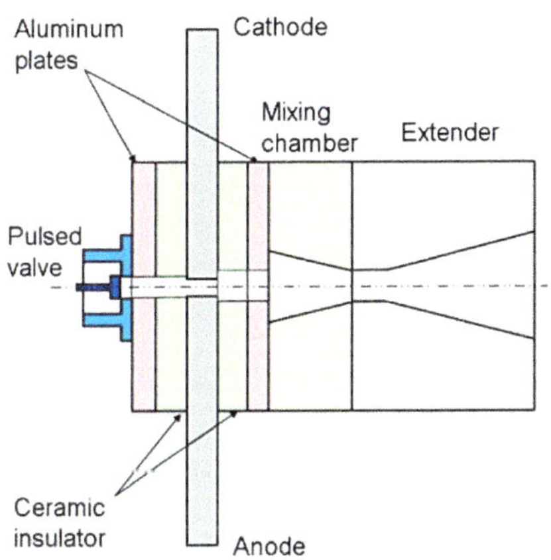

Figure 8.2 1991 PACIS version. In yellow is the so-called mixing chamber added between the discharge region and the extender and designed to facilitate cluster aggregation when the source is operated with long time pulses (~10 μs) and/or high power. Adapted by permission from Springer Nature Customer Service Centre GmbH: Springer Nature, *Zeitschrift für Physik D Atoms, Molecules and Clusters*, Ref. [5], copyright (1991).

The intrinsically charged nature of the spark-produced beam particles allows one to manipulate them via electrostatic and/or magnetic fields, placed in different vacuum chambers separated through skimmers from the source chamber. Due to the high percentage of charged clusters produced, the PACIS was mainly used for fundamental studies on free clusters, as it could be easily coupled to time-of-flight (TOF) mass spectrometers. Studies on different materials clusters, that is, Pb and Al [5], Cu and Ga [6], metal–carbon [7], metallocarbohedrene [8], and Na/Al [9], are reported in the literature. Figure 8.4 shows the TOF mass spectrum of negative Al_n^- clusters [5].

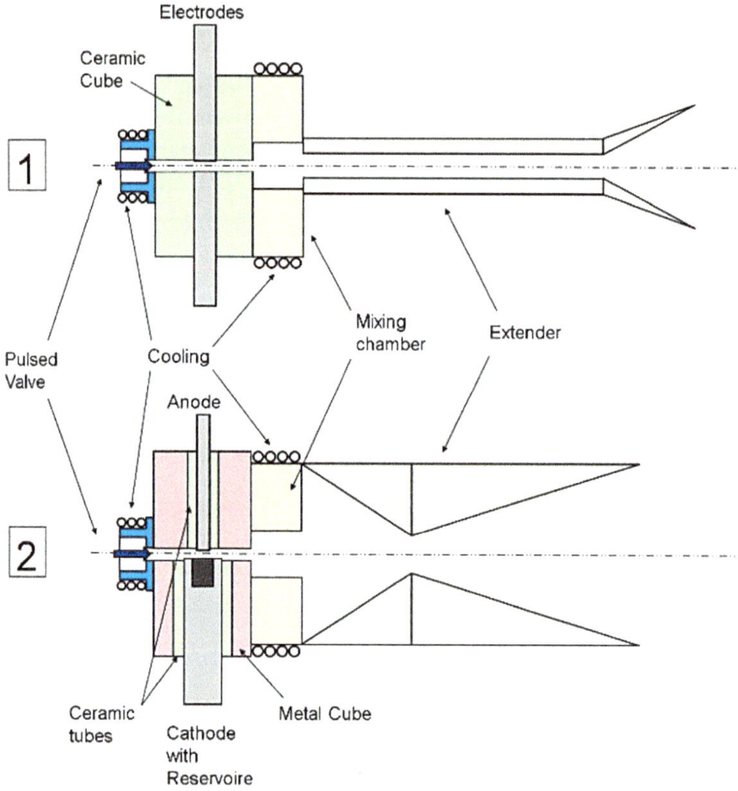

Figure 8.3 Different versions of the extender (Julich group). The two extenders (in pale yellow) can be combined with each type of the two versions of the PACIS presented here. (1) Source working with a solid cathode. (2) Source equipped with a cathode with a reservoir suitable to host liquid metals or sintered powder pellets. Adapted from Ref. [6], with the permission of AIP Publishing.

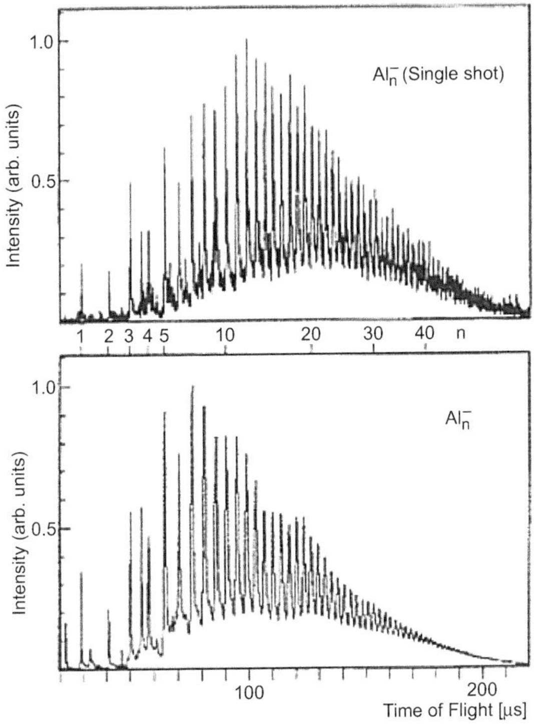

Figure 8.4 TOF spectra of Al_n^- anions. Up: Typical single-shot spectrum. Down: Averaged spectrum of about 100 shots under the same source conditions. Reprinted by permission from Springer Nature Customer Service Centre GmbH: Springer Nature, *Zeitschrift für Physik D Atoms, Molecules and Clusters*, Ref. [5], copyright (1991).

The PACIS was also used to deposit clusters on surfaces [10, 11]. Figure 8.5 reports the case of a combined TOF and scanning transmission electron microscopy (STEM) analysis of Pt^+ clusters.

The evolution of the PACIS configuration followed different directions: in the version presented in Fig. 8.6, one can recognize three pulsed valves that supply a high gas load at high repetition rates (300–1000 Hz) and a movable anode enabling to adjust the distance between the electrodes; the source was used to deposit monolayers of mass-selected Si and Al clusters [11].

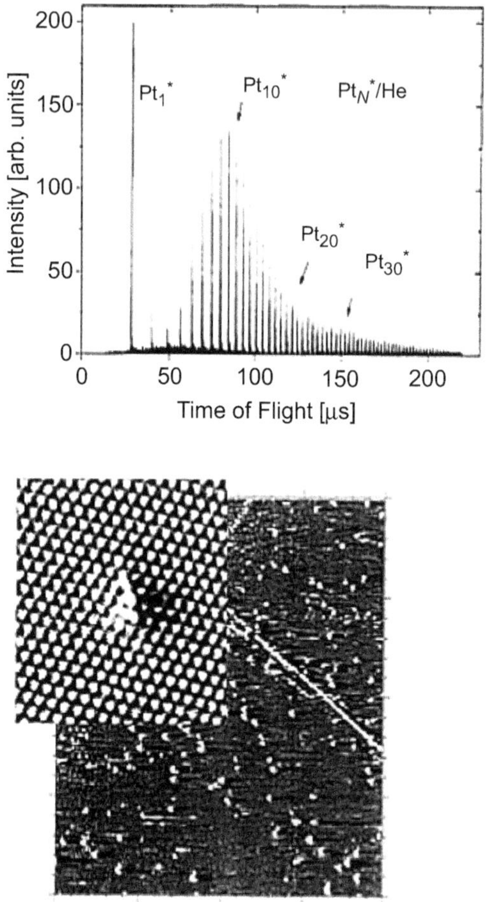

Figure 8.5 Up: Mass spectrum of positively charged platinum clusters from a PACIS. Note that the mass distribution shown here might differ from the neutral one. Down: Current image of soft-landed platinum clusters on crystalline graphite at 70 K; the typical coverage is about 2%. Inset: a single cluster on atomically resolved graphite. Reprinted from Ref. [10], Copyright (1998), with permission from Elsevier.

Figure 8.7 illustrates an example of a recent PACIS version (2015) used for the production of inorganic fullerenes (W_xS_y and Mo_xS_y) where chemical reactive species are introduced in gaseous form in a suitably designed aggregation cavity [12].

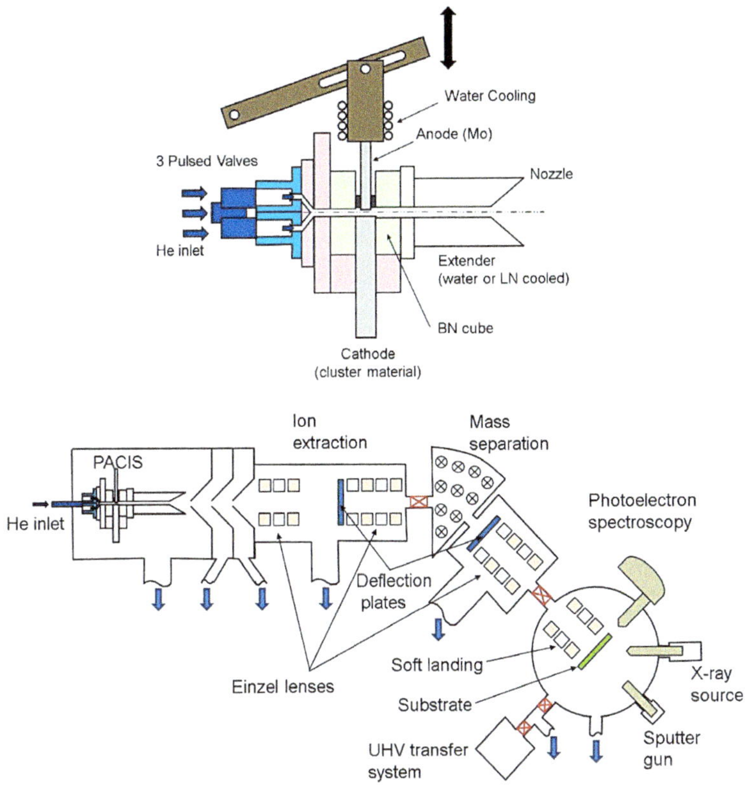

Figure 8.6 Up: Sketch of a PACIS. The cathode is eroded by a pulsed electric arc burning in the ~1 mm gap between the two electrodes. The gap can be adjusted by moving the anode, which consists of molybdenum. All metal parts are water-cooled. The three pulsed valves are operated alternatingly to achieve a maximum frequency of 1000 Hz. The length and diameter of the extender can be changed to optimize the source for the generation of smaller or larger clusters. Down: Experimental setup. The clusters are generated with a PACIS, extracted from the supersonic jet after passing three differential pumping stages and accelerated up to a kinetic energy of 500–1500 eV. The ions are mass-separated in a 45° sector magnet, guided into an ultrahigh vacuum (UHV) chamber and soft-landed on a substrate. Adapted by permission from Springer Nature Customer Service Centre GmbH: Springer Nature, *Applied Physics A: Materials Science & Processing*, Ref. [11], copyright (2001).

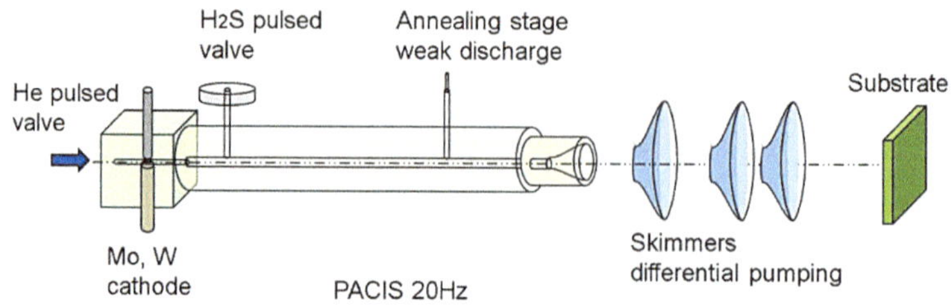

Figure 8.7 2015 PACIS version with an annealing stage and the setup for neutral cluster deposition on a transmission electron microscopy (TEM) grid. Adapted from Ref. [12], with the permission of AIP Publishing.

8.3 PMCS

The PCMS was introduced in 1997 by Milani and coworkers as an evolution of the PACIS. The first design of the PMCS was based on the firing of an electric discharge between a metallic cathode and an anode, separated by a small gap, driven by an inert gas pulse. From the technical point of view the main difference from the PACIS was the presence of a cavity with a volume of several cubic centimeters around the electrodes and the use of a cylindrical nozzle. The carrier gas is pulsed in the cavity directly on the cathode surface (instead of into the gap between the electrodes), and the cathode is continuously rotated (Fig. 8.8). This results in a different cathode sputtering mechanism that erodes it in a very confined region and produces mainly neutral clusters (positive and negative cluster ions are approximately 10% and 2% of the total, respectively) [2, 13]. Similarly to the PACIS, the PMCS was inserted in an HV chamber separated through a skimmer from following analysis and/or deposition chambers (Fig. 8.9).

The number and size distribution of the clusters is driven by the gas dynamics in the source cavity and can be tuned by changing the cavity length and/or the exit nozzle diameter, as well as through the type of carrier gas, its pressure (backing and average), the duration of the gas pulse and its timing with the electric pulse, and the electric pulse duration and intensity.

The PMCS normally produces a log-normal cluster mass distribution in a range from few tenths to several thousands of atoms per cluster with an average size at about 1000 atoms/cluster [15].

By operating the source with a pulse frequency of 5 Hz, deposition rates of 100 μm/h over a surface of 1 cm^2 placed at 500 mm from the exit of the nozzle can be achieved.

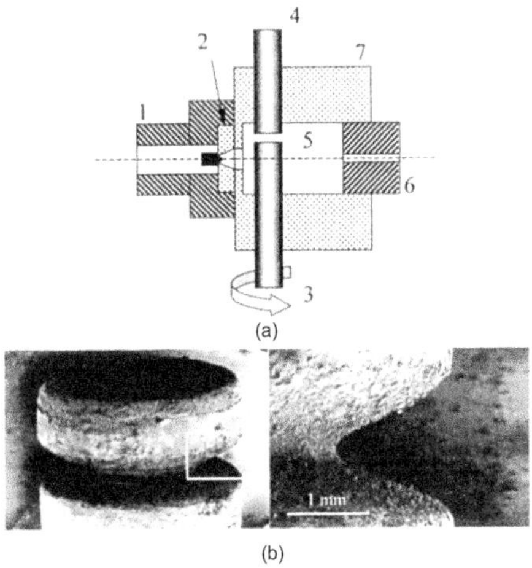

Figure 8.8 1999 first version of the PMCS. (a) Vertical cross section of the PMCS source (not to scale): (1) pulsed valve, (2) insulating valve nozzle, (3) cathode, (4) anode, (5) thermalization cavity, (6) cylindrical nozzle, and (7) ceramic main body. (b) Scanning electrode micrographs of the cathode region eroded by the plasma. The cathode is continuously rotated to prevent the formation of holes. The region in the white rectangle is shown at greater magnification (on the right). A very smooth and precise trace formed by plasma erosion is clearly distinguishable. From Ref. [14]. © IOP Publishing. Reproduced with permission. All rights reserved.

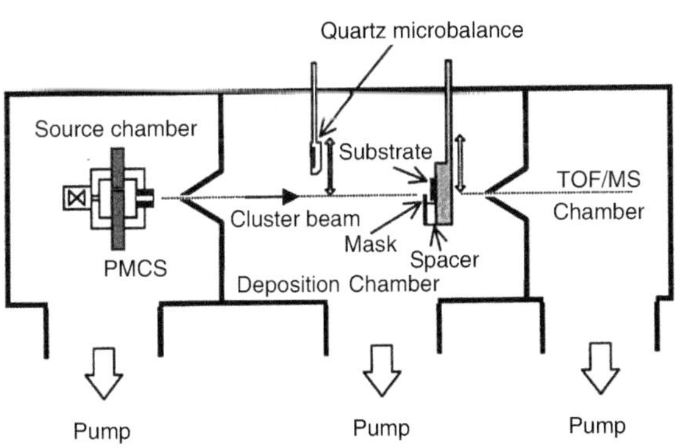

Figure 8.9 Schematic representation of the cluster beam deposition apparatus. Reprinted from Ref. [15], with the permission of AIP Publishing.

As example, in Fig. 8.10 we present an early study on carbon clusters where the particles size is analyzed in both the free and the supported case.

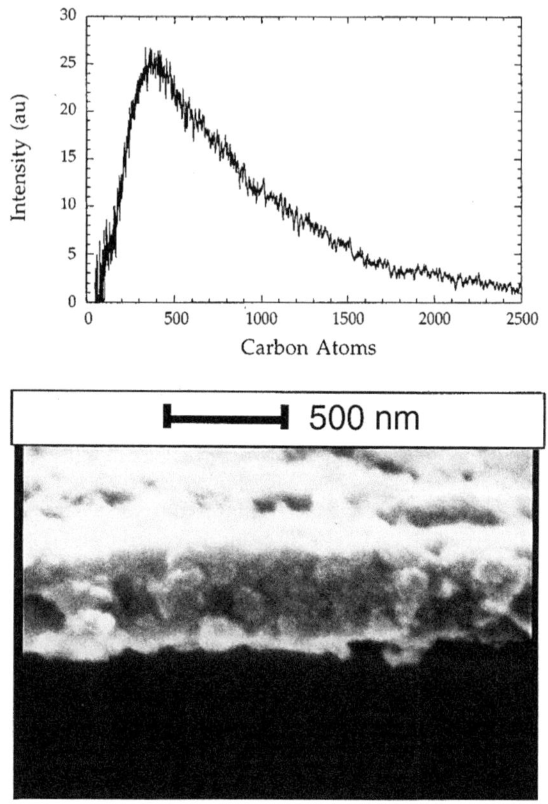

Figure 8.10 Top: Mass distribution of carbon clusters used as precursors of thin films. The TOF spectrum has been obtained by ionizing the clusters in the beam using the fourth harmonic (266 nm) of a Nd:YAG laser. Bottom: SEM micrograph of a nanocrystalline carbon thin film; high-magnification view of a section of the film: the granular structure is clearly visible. Grains as small as roughly 20 nm can be seen. Reprinted from Ref. [13], with the permission of AIP Publishing.

Figure 8.11 shows an atomic force microscopy (AFM) characterization of the evolution of the morphology of ZrO_x clusters deposited with different carrier gasses (He or Ar), with increasing quantity of deposited clusters, that is, from single particles (coverage θ = ~2%; Figs. 8.11a and 8.11b) to intermediate coverage (θ = ~50%; Figs. 8.11c and 8.11d) and continuous films (Figs. 8.11e and 8.11f). In particular, from the qualitatively point of view, Ar-deposited objects appear higher and larger than the He-deposited ones in all the three coverage conditions [16].

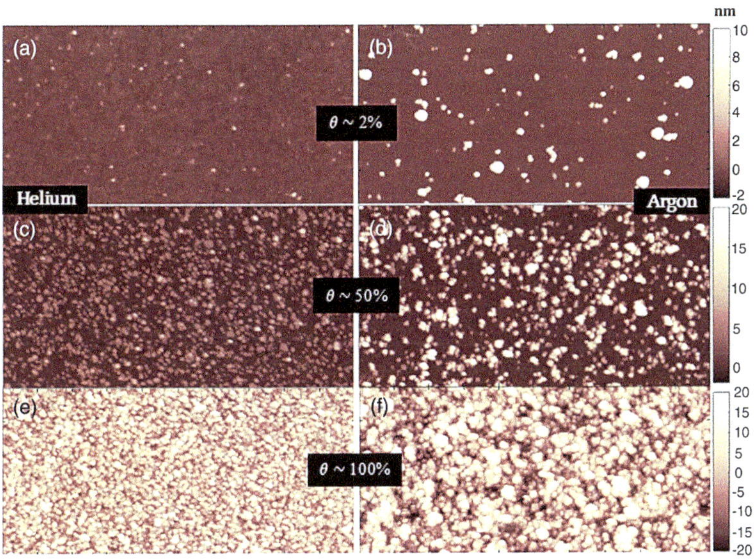

Figure 8.11 (a, b) AFM topographical maps (2 × 1 μm^2) of ns ZrO_x thin films for very low coverages deposited with helium and argon, respectively. (c, d) ns ZrO_x sample with coverage ϑ = ~50% and (e, f) ns ZrO_x continuous thin films (thickness about 50 nm). Reprinted (figure) with permission from Ref. [16]. Copyright (2018) by the American Physical Society.

Aerodynamic manipulation of the cluster beam was applied to get mass selection and focusing. Figure 8.12 shows the scheme and working principle of an 8-hole focuser used in the early stages of source evolution [15].

The high deposition rate combined with the high directionality of the cluster beam makes this source particularly suitable to the design of patterned cluster-assembled surfaces via the simple interposition of stencil masks between the source and the sample [17].

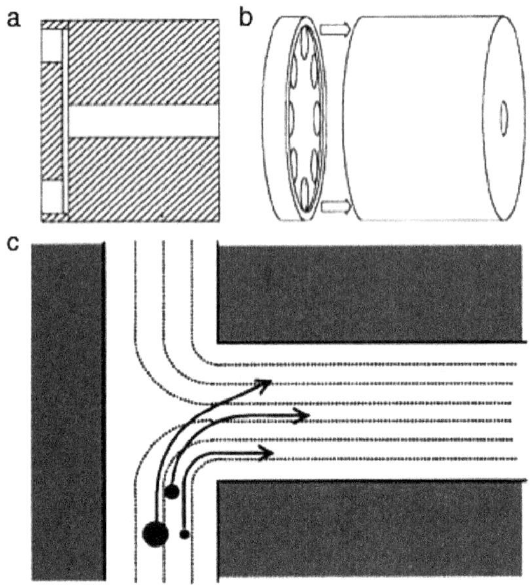

Figure 8.12 2001 8-hole focuser version. (a) Cross-sectional view and (b) perspective view of focusing nozzle assembly. (c) Focuser working principle: Heavy particles can't follow the carrier gas stream lines and hit the focuser walls, light particles follow the stream lines unperturbed, and intermediate size particles are concentrated toward the center axis. Reprinted from Ref. [15], with the permission of AIP Publishing.

Later evolutions of the source led to mounting the PMCS body outside the vacuum chamber, pumping through its exit (see Fig. 8.14) and the substitution of the focuser by aerodynamic lenses to improve cluster size selection and cluster beam directionality (see Fig. 8.15). Furthermore, a different anode configuration improved the source stability for a wide variety of conductive and semiconductive cathode materials, with the result that in contrast to the PACIS, no particular adaptation of the source design was required to work with different material cathodes. In recent years C, Ti, Ni, Fe, Si, W, Mo, MoS_2, WS_2, Au, Ag, Pt, Pd, Zr, and V clusters have been studied, either free or supported.

The introduction of precursors into the carrier gas, the use of alloy or composite electrodes (see Chapter 5), and the combination of several sources in one deposition apparatus easily allow the formation of doped or multicomponent films of controlled composition.

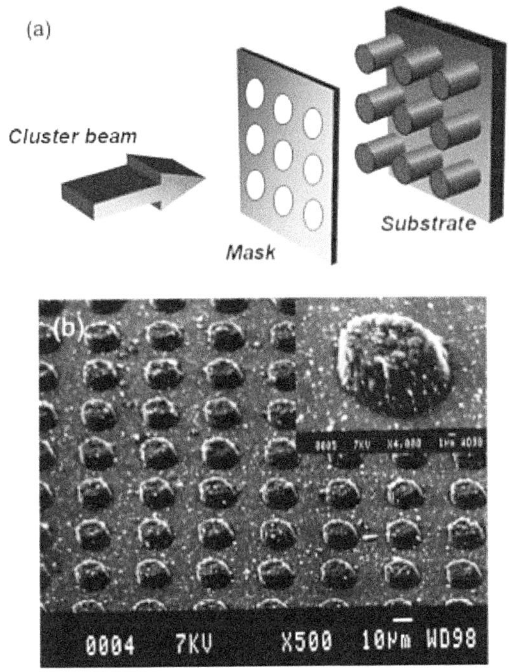

Figure 8.13 Schematic representation of the deposition of patterned films (a) by placing a hard mask into the particle beam at some distance from the substrate. The scanning electron micrograph (b) shows a carbon film obtained by this method with cluster beam deposition. From Ref. [3]. © IOP Publishing. Reproduced with permission. All rights reserved.

Figure 8.15 shows a sketch of the recent configuration of source and nozzle/focuser assembly; the enlargement shows a special composite (M-MS_2, where M can be Mo or W) cathode [18].

As an example of the large variety of materials accessible to the PMCS, Fig. 8.16 shows a high-resolution TEM image of a MoS_2/TiN composite coating deposited on a Si substrate. MoS_2 nanoparticles were deposited by PMCS CBD, equipped with the composite cathode shown in Fig. 8.15, mounting the source directly on the deposition chamber of an industrial reactive arc apparatus, which, in turn, produced the TiN hard matrix. Both MoS_2 nanoparticles and the TiN matrix are clearly visible; the diameter of the MoS_2 nanoparticles is 10–50 nm. Analysis by electron energy loss spectroscopy (EELS) confirmed the stoichiometry of the different components of the nanocomposite film [19].

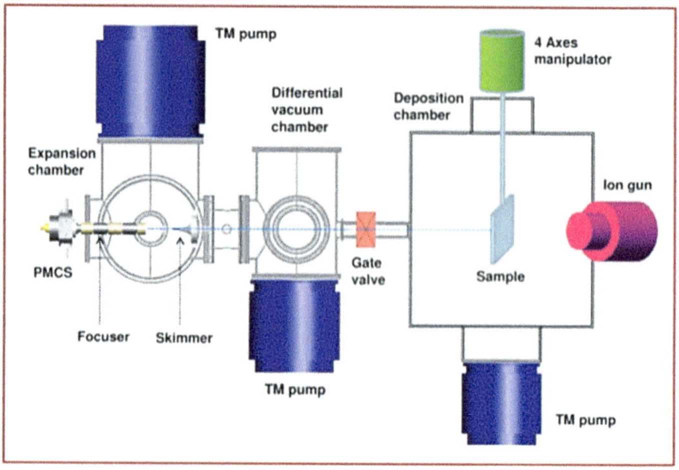

Figure 8.14 Schematic representation (not to scale) of the supersonic cluster beam deposition apparatus. The cluster source is mounted on the expansion chamber; a second source (not shown) can be mounted on the deposition chamber. From Ref. [18]. © IOP Publishing. Reproduced with permission. All rights reserved.

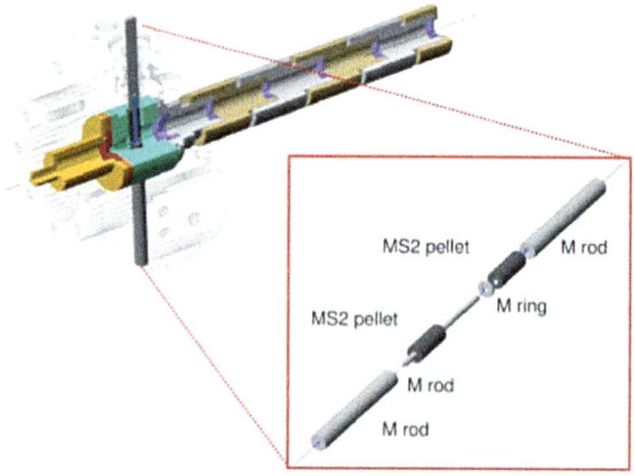

Figure 8.15 Evolution of the PMCS design. Top left: PMCS drawing. The target rod ablation takes place in the cavity of the ceramic body (green). The pulsed valve (orange/red) is mounted on the ceramic body; on the opposite side of the cavity the nozzle and the focuser (violet, gray, and bronze) are mounted. The number of aerodynamic unit elements (cavity and lens) in the focuser after the first (nozzle) can be varied from 0 to 5. Bottom right: Composed cathode exploded view. From Ref. [18]. © IOP Publishing. Reproduced with permission. All rights reserved.

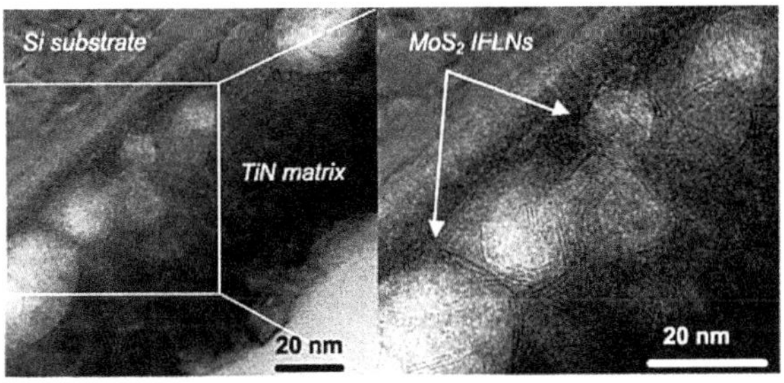

Figure 8.16 (Left) Cross-sectional TEM micrograph of a MoS_2/TiN nanocomposite. 10–50-nm-size inorganic fullerene-like nanoparticles (IFLNs) of MoS_2 are dispersed in the TiN matrix. The silicon substrate and TiN matrix can be clearly recognized. (Right) Higher-magnification view of the same region showing the fullerene-like structure of MoS_2 nanoparticles. Reprinted by permission from Springer Nature Customer Service Centre GmbH: Springer Nature, *Applied Physics A: Materials Science & Processing*, Ref. [19], copyright (2007).

The possibility of tuning the film properties in terms of morphology, crystallinity, and chemistry by changing the particle synthesis conditions, the patterning intrinsic capability, the great versatility in terms of the choice of materials and substrates, and the ease of combination with other planar deposition/fabrication techniques led to apply the PMCS to the production of different devices such as supercapacitors [20], high-throughput screening arrays for biological applications [21], microfluidics [22], flexible microelectrodes [23], gas sensors [24], low-friction films on complex mechanical components [18], soft robotics [25], etc.

We selected two applications to underline the versatility of the PMCS in terms of patterning capability/coupling with planar fabrication techniques and choice of materials/coating of complex geometries. In Fig. 8.17 we illustrate the case of gas sensors produced by supersonic CBD of different metal oxides. The micropatterned sensing arrays were deposited on an alumina wafer where a 14 × 15 array of interdigitated gold electrodes had been previously microfabricated (Fig. 8.17, up). A custom-made stencil mask was superimposed to the substrate in order to deposit the film only in the area of the interdigitated electrodes. Each microsensor device was equipped with a heater on the back face in order to control the working temperature 300°C–500°C and with interdigitated electrodes on the front face, under the sensitive layer, in order to

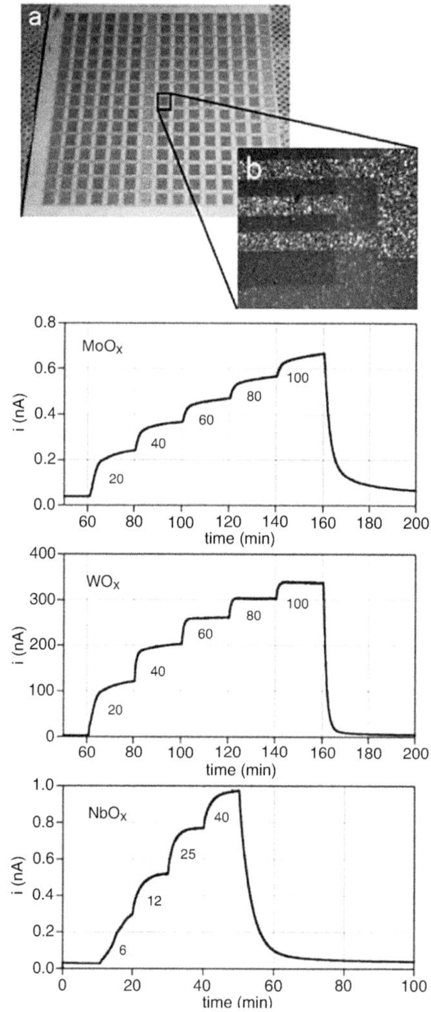

Figure 8.17 Up: (a) Array of 210 titania gas sensors made by cluster beam deposition onto an alumina wafer with interdigitated gold electrodes (b) employing hard mask patterning [Weg06]. From Ref. [3]. Down: Sensing responses of nanostructured Mo, W, and Nb oxide films, measured as the current flowing into the film upon exposure to ethanol. Ethanol concentrations are reported in the graphs. Through a huge specific surface and modified electronic structure, the nanostructure emphasizes chemical modifications occurring at the nanoparticle surface expressed in a change in film resistance (much more than tenfold respect to its baseline value in pure air). Reprinted from Ref. [24],

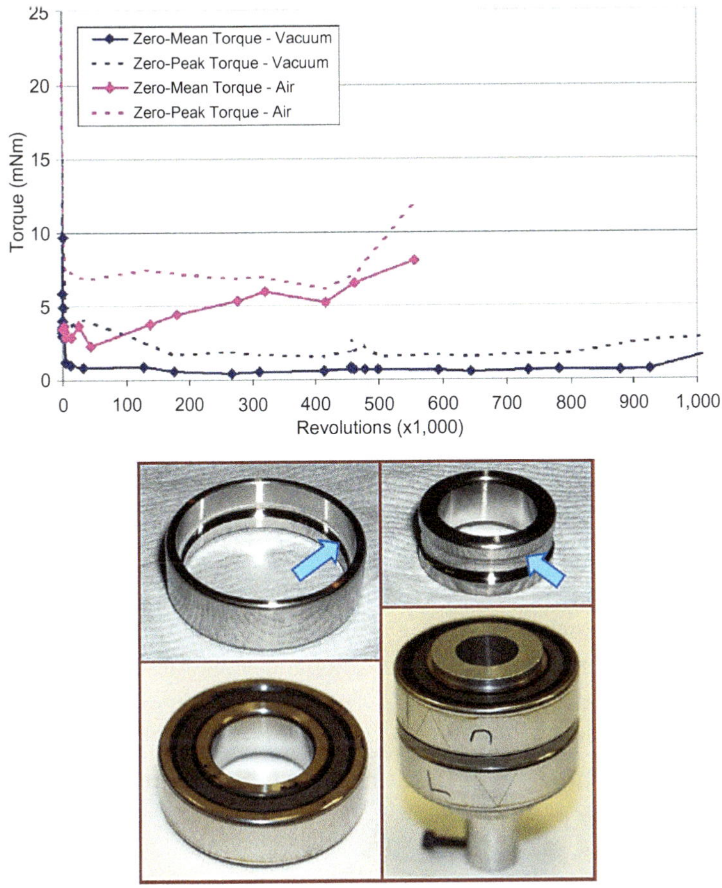

Figure 8.18 Up: Torque behavior of bearings with SCBD MoS_2 on raceways, fitted with PGM cages, running under vacuum and in air. Down: Angular contact bearings (FAG B7004), outer diameter 42 mm, inner diameter 20 mm, 11 balls 6.35 mm diameter, 15° free contact angle; Top left and top right, uncoated bearing outer and inner rings: Arrows point to the up-to-coat surfaces (races); bottom left: assembled bearing after races coating (steel spheres are encapsulated in a polymeric brown/gray PGM cage visible between the steel rings); bottom right: couple of bearings mounted in back-to-back configuration, set for bearing-rig test. From Ref. [18]. © IOP Publishing. Reproduced with permission. All rights reserved.

apply a voltage and to monitor the current during the exposure to controlled atmosphere. This fabrication principle, based on integration of PMCS patterned depositions of titanium oxide [26] and refractory metal oxides [24] with standard planar fabrication techniques, was applied to the production of chemoresistive gas sensors. Figure 8.17 (down) shows the electrical response of sensors consisting of different materials to the exposure at different concentrations of ethanol. For gas sensors, see also Chapter 9 and Chapter 11, which is dedicated to this topic.

In Fig. 8.18 we show the case of cluster-assembled MoS_2 deposited on ball bearings races to act as a solid lubricant. A custom-machined Mo/MoS_2 composite cathode was designed to allow the PMCS functioning with this fragile and nonconductive material (see Fig. 8.15). Two additional movements (i.e., satellite rotation and tilting of the sample plane) were introduced in the deposition apparatus to adapt the coating method to complex 3D surfaces of commercial objects, namely the bearing races (Fig. 8.18, down). Furthermore, ion-gun etching was used to clean the bearings surface and improve the MoS_2 coating adhesion. After coating, the bearings were assembled, mounted in couples on a test rig, and their friction reduction capability assessed either in air or in a vacuum environment; the resulting measured torque is reported in Fig. 8.18 (up) [18].

8.4 ACIS

The ACIS was introduced in 2001 by Meiwes-Broer and coworkers, and it represents a sort of step back to a continuously operated arc source characterized by a special hollow cathode configuration that ensures the suitable thermodynamic and kinetic conditions for efficient cluster formation. In the ACIS, either the cluster aggregation cavity or the exit nozzle is constituted by the cathode itself, while the inert gas (Ar) is continuously injected at a pressure of some millibars, thus sustaining the arc discharge. Due to the high current power, water cooling of the electrodes is necessary [27]. The

ACIS represents an alternative to spark sources such as the PACIS for vacuum-based cluster/nanoparticle beam setups. The flow through the continuous discharge has to be expanded directly into a low-pressure chamber to limit particle growth. The ACIS typically produces larger particles compared to the PACIS and exhibits a much higher mass production rate.

The ACIS produces intense fluxes of nanoparticles with a broad mass distribution; for example, with an iron cathode biased at 20 V and supplied with a 20 A current, typical deposition rates are some hundreds of namometers per minute over a 15 mm diameter area just after the source exit. In typical operations, up to 80% of the produced nanoparticles are charged with approximately the same content of positive and negative ions, thus enabling mass selection via electrostatic fields [27].

This source was basically conceived for the study of supported mass-selected clusters. A typical CBD apparatus based on an ACIS and equipped with an electrostatic deflector for mass selection is schematically illustrated in Fig. 8.19.

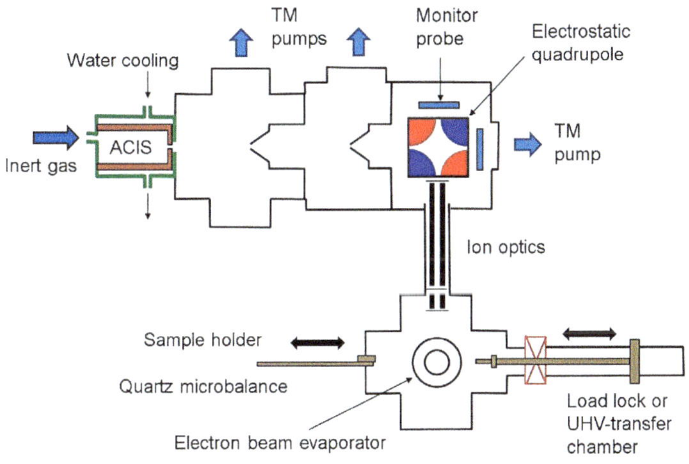

Figure 8.19 Experimental setup of a cluster beam machine based on the ACIS. Adapted by permission from Springer Nature Customer Service Centre GmbH: Springer Nature, *The European Physical Journal D: Atomic, Molecular, Optical and Plasma Physics*, Ref. [27], copyright (2001).

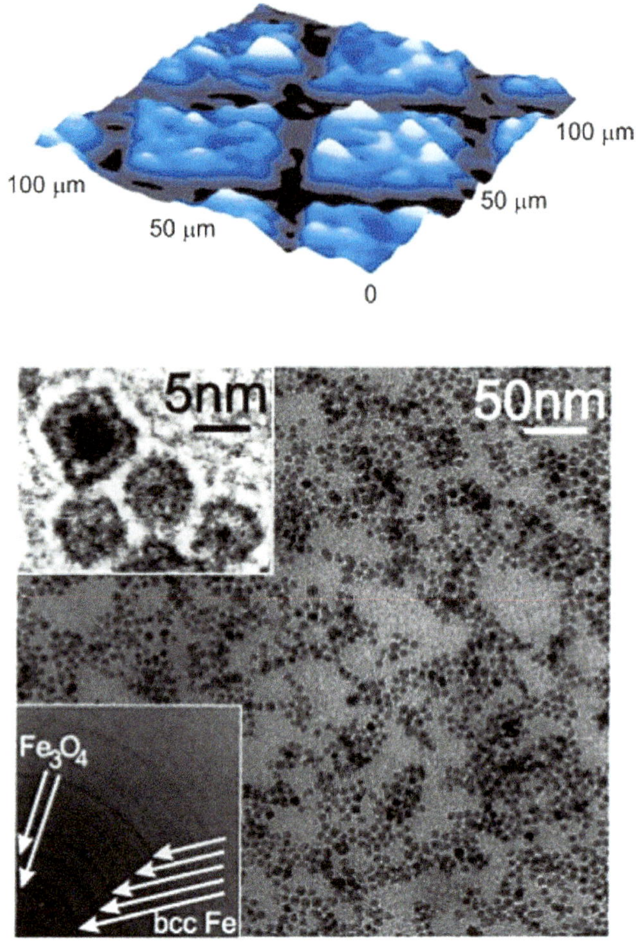

Figure 8.20 Up: AFM image of roughly size-separated (D = 10 ± 3 nm) iron clusters deposited on a graphite substrate through a mesh having a periodicity of 70 μm. The maximum height is approximately 4 μm. Down: TEM image of mass-selected iron clusters with a mean diameter of about 5 nm (cf. upper inset). Bottom inset: Beside the bcc iron diffraction pattern, additional rings resulting from, for example, Fe_3O_4 are present with lower intensity. These latter can be attributed to partial oxidation of the outer shells resulting from the transport to the microscope under ambient conditions. Reprinted by permission from Springer Nature Customer Service Centre GmbH: Springer Nature, *The European Physical Journal D: Atomic, Molecular, Optical and Plasma Physics*, Ref. [27], copyright (2001).

Figure 8.20 reports AFM and TEM analyses of mass-selected iron clusters. For AFM analysis, clusters are deposited on a graphite substrate and patterned through a mesh with a 70 μm × 70 μm periodicity, acting as a mask. The height difference between the exposed and the shielded areas results in a deposition rate of 5 nm/s. For TEM analysis mass selected iron clusters were deposited onto a commercial electron microscopy grid covered by an amorphous carbon film. The particles are well separated having a mean diameter of 5 nm in good agreement with the calculated cluster size of (5.4 ± 0.2) nm. Electron diffraction reveals that these particles consist of body-centered cubic (bcc) iron crystallites.

A second version of the ACIS was equipped with a magnetic solenoid surrounding the hollow cathode designed to control the discharge position on the cathode surface (Fig. 8.21) [28].

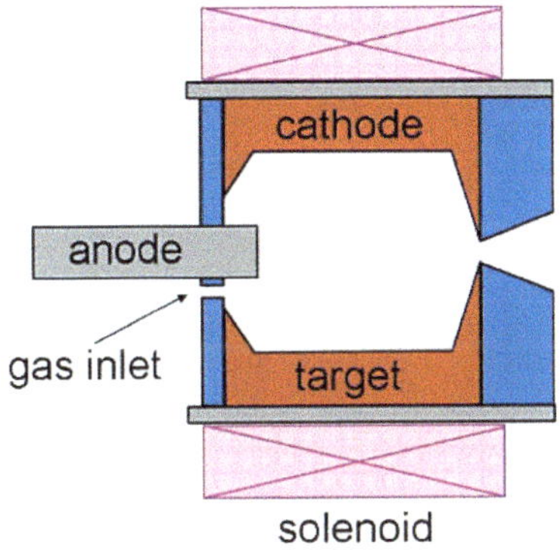

Figure 8.21 Schematic drawing of an ACIS equipped with a solenoid. Adapted by permission from Springer Nature Customer Service Centre GmbH: Springer Nature, *Applied Physics A: Materials Science & Processing*, Ref. [28], copyright (2005).

In this configuration the source was used to study Fe, Co, and FeCo [28–31] clusters and their magnetic properties.

To improve the ACIS intensity and throughput, aerodynamic focusing lenses were used (Fig. 8.22) [32].

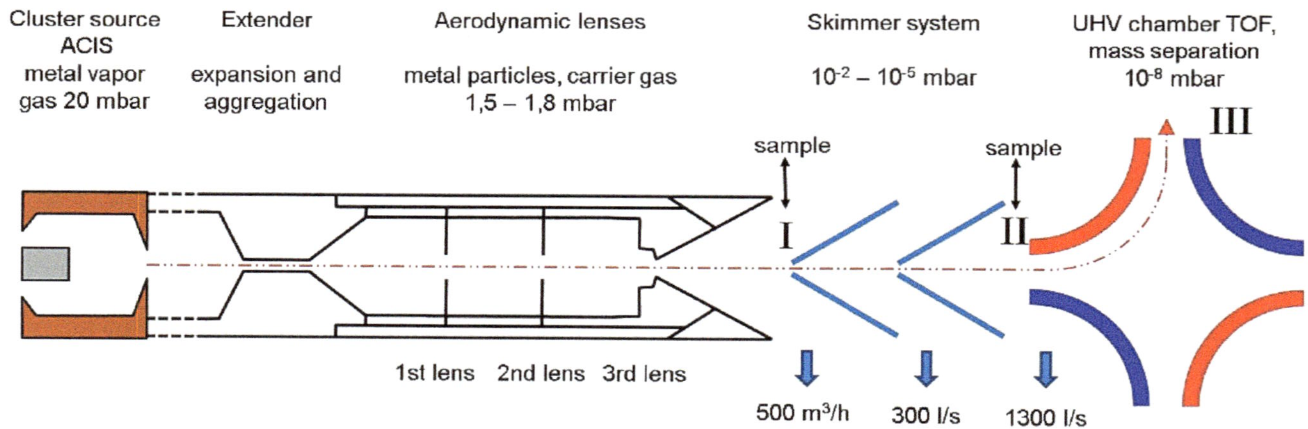

Figure 8.22 Schematic view of the experimental setup equipped with an ACIS, the aerodynamic lens system, the skimmers that separate the vacuum system into three different pumping stages, and the electrostatic quadrupole deflector for mass selection (out of scale). Adapted from Ref. [32], with the permission of AIP Publishing.

8.5 Conclusions

CBD is a tool with a degree of complexity and maturity that makes this approach very effective for functional and length-scale integration of nanoparticles in view of applications in different fields. Generation of intense cluster beams using techniques based on spark ablation is at the basis of the CBD approach. The PACIS and ACIS approaches are of interest for more fundamental studies of clusters in the gas phase and for systems where a very low density of clusters deposited on a substrate is required.

The transition from the use of a PMCS as a tool for basic research to a method for the fabrication of micro- and nanodevices is driven by the capability of manipulating neutral clusters through supersonic expansions and aerodynamic focusing. The PMCS enables the production of cluster beams that meet the requisites necessary for applications: high intensity, stability, repeatability, high lateral resolution, and parallel processing. The use of a PMCS is compatible with planar fabrication technologies typical of thin-film deposition and microelectromechanical system (MEMS) fabrication.

References

1. Huttel, Y., Martínez, L., Mayoral, A. and Fernández, I. (2018). *MRS Commun.*, **8**(3), pp. 947–954.
2. Milani P. and Iannotta S. (1999). *Cluster Beam Synthesis of Nanostructured Materials* (Springer).
3. Wegner, K., Piseri, P., Vahedi Tafreshi, H. and Milani, P. (2006). *J. Phys. D: Appl. Phys.*, **39**, pp. R439–R459.
4. Ganteför, G., Siekmann, H. R., Lutz, H. O. and Meiwes-Broer, K. H. (1990). *Chem. Phys. Lett.*, **165**, p. 293.
5. Siekmann, H. R., Lüder, C., Faehrmann, J., Lutz, H. O. and Meiwes-Broer, K. H. (1991). *Z. Phys. D*, **20**, p. 417.
6. Cha, C. Y., Ganteför, G. and Eberhardt, W. (1992). *Rev. Sci. Instrum.*, **63**, p. 5661.
7. Lu, W., Huang, R., Ding, J. and Yang, S. (1996). *J. Chem. Phys.*, **104**(17).
8. Blessing, N., Burkart, S. and Ganteför, G. (2001). *Eur. Phys. J. D*, **17**, pp. 37–41.

9. Wang, H., Zhang, X., Ko, Y. J., Grubisic, A., Li, X., Ganteför, G., Schnöckel, H., Eichhorn, B. W., Lee, M.-S., Jena, P., Kandalam, A. K., Kiran, B. and Bowen, K. H. (2014). *J. Chem. Phys.*, **140**, p. 054301.

10. Bettac, A., Köller, L., Rank, V. and Meiwes-Broer, K. H. (1998). *Surf. Sci.*, **402–404**, pp. 475–479.

11. Klipp, B., Grass, M., Müller, J., Stolcic, D., Lutz, U., Ganteför, G., Schlenker, T., Boneberg, J. and Leiderer, P. (2001). *Appl. Phys. A*, **73**, pp. 547–554.

12. Pietsch, S., Dollinger, A., Strobel, C. H., Park, E. J., Ganteför, G., Seo, H. O., Kim, Y. D., Idrobo, J. C. and Pennycook, S. J. (2015). *J. Appl. Phys.*, **118**, p. 134302.

13. Milani, P., Ferretti, M., Piseri, P., Bottani, C. E., Ferrari, A., Li Bassi, A., Guzzetti, G. and Patrini, M. (1997). *J. Appl. Phys.*, **82**, p. 5793.

14. Barborini, E., Piseri, P. and Milani, P. (1999). *J. Phys. D: Appl. Phys.*, **32**, pp. L105–L109.

15. Piseri, P., Podestà, A., Barborini, E. and Milani, P. (2001). *Rev. Sci. Instrum.*, **72**, pp. 2261–2267.

16. Borghi, F., Podestà, A., Piazzoni, C. and Milani, P. (2018). *Phys. Rev. Appl.*, **9**, p. 044016.

17. Barborini, E., Piseri, P., Podestà, A. and Milani, P. (2010). *Appl. Phys. Lett.*, **77**, pp. 1059–1061.

18. Piazzoni, C., Buttery, M., Hampson, M. R., Roberts, E. W., Ducati, C., Lenardi, C., Cavaliere, F., Piseri, P. and Milani, P. (2015). *J. Phys. D: Appl. Phys.*, **48**.

19. Piazzoni, C., Blomqvist, M., Podestà, A., Bardizza, G., Bonati, M., Piseri, P., Milani, P., Davies, C., Hatto, P., Ducati, C., Sedláčková, K. and Radnóczi, G. (2008). *Appl. Phys. A*, **90**, pp. 101–104.

20. Bongiorno, G., Podestà, A., Ravagnan, L., Piseri, P., Milani, P., Lenardi, C., Miglio, S., Bruzzi, M. and Ducati, C. (2006). *J. Mater. Sci.: Mater. Electron.*, **17**, pp. 427–441.

21. Scopelliti, P. E., Borgonovo, A., Indrieri, M., Giorgetti, L., Bongiorno, G., Carbone, R., Podestà, A. and Milani, P. (2010). *PLoS One*, **5**(7), p. e11862.

22. Zanardi, A., Bandiera, D., Bertolini, F., Corsini, C. A., Gregato, G., Milani, P., Barborini, E. and Carbone, R. (2010). *BioTechniques*, **49**, pp. 497–504.

23. Marelli, M., Divitini, G., Collini, C., Ravagnan, L., Corbelli, C., Ghisleri, C., Gianfelice, A., Lenardi, C., Lorenzelli, L. and Milani, P. (2011). *J. Micromech. Microeng.*, **21**, p. 045013.

24. Barborini, E., Ducati, C., Leccardi, M., Bertolini, G., Repetto, P. and Milani, P. (2011). *Jpn. J. Appl. Phys.*, **50**, p. 01AK01.

25. Santaniello, T., Migliorini, L., Yan, Y., Lenardi, C. and Milani, P. (2018). *J. Nanopart. Res.*, **20**, p. 250.

26. Mazza, T., Barborini, E., Kholmanov, I. N., Piseri, P., Bongiorno, G., Vinati, S., Milani, P., Ducati, C., Cattaneo, D., Li Bassi, A., Bottani, C. E., Taurino, A. M. and Siciliano, P. (2005). *Appl. Phys. Lett.*, **87**, p. 103108.

27. Methling, R., Senz, V., Klinkenberg, E. D., Diederich, T., Tiggesbäumker, J., Holzhüter, G., Bansmann, J. and Meiwes-Broer, K. H. (2001). *Eur. Phys. J. D*, **16**, p. 173.

28. Bansmann, J. and Kleibert, A. (2005). *Appl. Phys. A*, **80**, p. 957.

29. Kleibert, A., Passig, J., Meiwes-Broer, K. H., Getzlaff, M. and Bansmann, J. (2007). *J. Appl. Phys.*, **101**, p. 114318.

30. Fraile-Rodriguez, A., Nolting, F., Bansmann, J., Kleibert, A. and Heyderman, L. J. (2007). *J. Magn. Magn. Mater.*, **316**, pp. 426–428.

31. Rosellen, W., Bettermann, H., Veltum, T. and Getzlaff, M. (2012). *Physica E*, **44**, pp. 1683–1686.

32. Passig, J., Meiwes-Broer, K.-H. and Tiggesbäumker, J. (2006). *Rev. Sci. Instrum.*, **77**, p. 093304.

Chapter 9

Patterned 3D Nanostructure Arrays from Charged Aerosols

Jicheng Feng, Kiwoong Lee, and Mansoo Choi
Global Frontier Center for Multiscale Energy Systems,
Seoul National University, Seoul 08826, Republic of Korea
Department of Mechanical and Aerospace Engineering
Seoul National University, Seoul 08826, Republic of Korea
mchoi@snu.ac.kr

Three-dimensional (3D) nano-/microstructures open up new possibilities for developing nanodevices used in sensors, electronics/optoelectronics, and energy conversion systems. In this chapter, we review a nanoscale patterning technology, coined as "electric-field-assisted aerosol lithography" (EAAL), for making arrays of complex 3D nano-/microstructures under atmospheric conditions. In EAAL, electrical charging of prepatterns leads to focusing nanoscopic electrostatic lenses that steer the deposition of charged aerosols into precisely addressable locations on a surface. The construction of 3D micro-/nanostructures requires a consistent supply of building blocks in the form of charged aerosol particles a few nanometers in size. Spark ablation is picked to create such building blocks, and associated modifications are also introduced here.

Spark Ablation: Building Blocks for Nanotechnology
Edited by Andreas Schmidt-Ott

ISBN 978-981-4800-82-2 (Hardcover), 978-0-367-81709-1 (eBook)
www.jennystanford.com

9.1 Introduction

Well-ordered multidimensional multiscale nano-/microstructures provide a platform for creating novel nanodevices, including sensors, electronic/magnetic, and photonic devices (e.g., solar cells). Although various techniques for the fabrication of 3D micro-/nanostructures have been developed, particularly in the past decade, each existing method has its own limitation that currently hinders mass production of 3D micro-/nanostructures. For example, well-established nanofabrication techniques utilizing electron beams ([1–3]), focused ion beams [4], and laser beams [5, 6] are inherently a serial approach that requires a high-vacuum environment. Therefore, these approaches have been largely confined to laboratory scale, not available to scale up to mass production. Microcontact printing [7] that utilizes nanopatterned stamps is a parallel approach; however, apparently it has difficulty in fabricating 3D structures. The capillary force could be utilized effectively to introduce nanoparticles (NPs) into the prepatterned trench area to make well-ordered NP arrays [8, 9]. Still, the method does not favor the fabrication of 3D structures. Although 3D printing technology has been extensively explored over the past years [10–13], its resolution does not yet reach to the nanoscale. These observations suggest a strong need to develop a new way of constructing 3D nano-/microstructures in a practical manner. The preferred approach should work under atmospheric conditions and be operated in a parallel process (vs. serial direct-write or serial wire-bonding methods) to produce 3D nanostructures on a large surface area with nanometer resolution.

In this chapter, we review a new approach utilizing aerosol technology to develop 3D nano-/microstructure arrays in a practical way. The method utilizes charged aerosols that are precisely positioned into the desired location on a substrate using a process called "ion-assisted aerosol lithography" (IAAL) or "electric-field-assisted aerosol lithography" (EAAL). Throughout this chapter, we stick to the term "EAAL." Making nanoscale-resolution 3D structures requires smaller NP building blocks down to a size of sub-10 nm. It is also important to provide unagglomerated particles to assure compact 3D structures. To achieve that, we will present a number of recent developments in spark discharge. The parallel patterning process can make complex 3D nanostructures on a surface from microscale to macroscale and simply works under atmospheric conditions. The 3D assembly module via this method will be explained and various applications of the resultant 3D nanostructures will be introduced.

9.2 Spark Discharge Techniques for Generating a Few-Nanometer-Size Unagglomerated Charged Aerosols

Spark discharge was introduced to make aerosol NPs by Schwyn et al. [14]. It consists of a pair of electrodes connected to a resistance-inductance-capacitance (RLC) circuit, as shown in Fig. 9.1. The electrodes face each other and are separated by a gap (ca. 1 mm), within which an inert gas flushes through. A constant current source continuously charges the capacitance until reaching a breakdown voltage and forms a spark discharge in the interelectrode gap. After discharging, the capacitance is recharged for preparing a consecutive spark. The process repeats itself, while the variation of the breakdown voltage appears problematic [15]. Spark energy, which is an important parameter to determine the particle properties, is proportional to the square of the breakdown voltage. The voltage variation therefore leads to non-uniformity of the generated aerosols. To keep a constant spark energy, Pfeiffer et al. and Feng et al. developed a switching RLC circuit (cf. Fig. 9.1) [16–18], which always applies the same voltage to ignite sparks. Despite the various developments, this section mainly focuses on the generation of desired NP building blocks in order to make patterned 3D nanostructures.

To make patterned 3D nanostructures with high controllability/reproducibility/scalability/resolution, we first modify size distributions of NPs generated by the spark ablation. Besides size dependence on operating parameters [18–20], electrode configurations were found important to control the particle size. Then, we increase the portion of positively charged particles to encourage the 3D nanostructure growth, since they are attracted to a negatively biased substrate and subsequently deposited there, or both polarities can be reversed accordingly. Although the NP structures were shown to reach nanoscale resolution, parallel patterning on a large surface area of macroscale remains a challenge. To overcome that, we will also detail the development for sparking multiple pairs of electrodes in parallel. Finally, we will show how to maintain the crystallinity and purity of NPs, which make the final 3D nanostructures also preserve good crystallinity and purity, as required for certain applications.

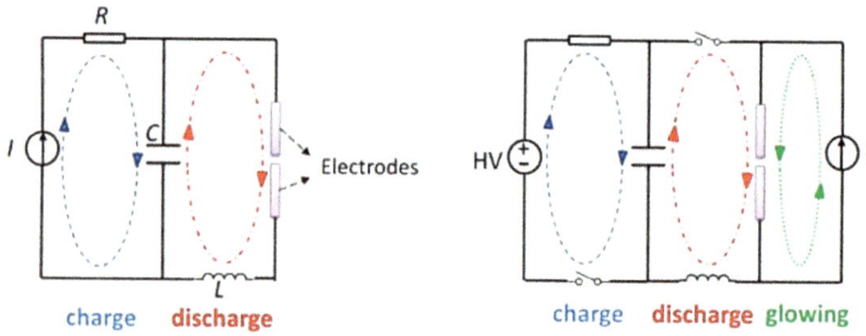

Figure 9.1 Electrical circuit for creating spark discharges. Left: Classical RLC spark circuit. Right: Switching circuit that can reach a spark repetition frequency as high as 25 kHz [16]. The switching circuit contains a continuous current source that sustains a low glowing discharge to facilitate spark ignition, too low in power to ablate the electrodes. From Ref. [17]. Reproduced by permission of The Royal Society of Chemistry.

9.2.1 Pin-to-Plate and Wire-in-Hole Configuration Spark Discharge

In the classical setup of spark ablation, electrode configuration is mainly based on rod-to-rod (cf. Fig. 9.2a) [21]. We found out that changing this configuration to pin-to-plate (cf. Fig. 9.2b) enables the production of unagglomerated NPs with a narrower size distribution. We attribute this to the reduction of the particle duration in the aerosol form (i.e., residence time or coagulation volume) and quickly capture them onto a substrate. Although this pin-to-plate-type spark discharge was successful in generating a few nanometer unagglomerated particles, we identified a problem of consistent production. In the case of long-term sparking, the sharp pin was significantly ablated to result in a broader size distribution than that obtained initially. Motivated by this, we developed a wire-in-hole-type electrode configuration (Fig. 9.2c). The wire-in-hole configuration ensures a longer and more stable spark operation, because a large surface area of the wire cylinder is exposed to sparks. As a result, this greatly improves the consistency of particle size distribution with time. Making smaller unagglomerated NPs through this way is in line with the findings from Feng et al. [19, 20], who shortened particle duration in aerosol by increasing the gas flow rate. In contrast to that, this way to make sub-10-nm particles is beneficial for printing 3D nanostructures (detailed reasons are shown later) [21, 22].

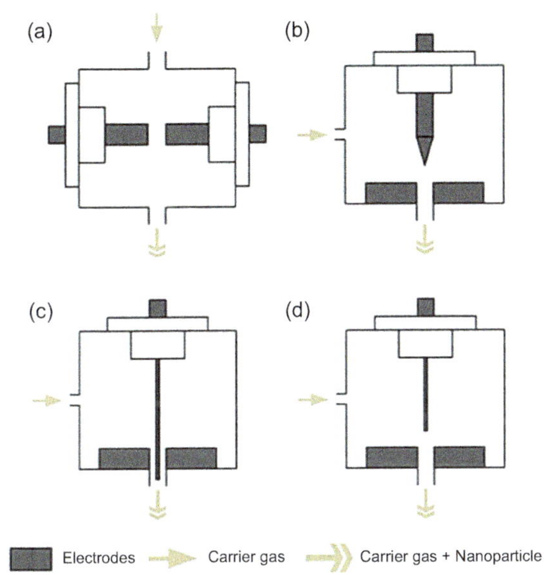

Figure 9.2 Electrode configurations in a spark chamber. (a) Rod-to-rod, (b) pin-to-plate, (c) wire-in hole, and (d) wire-to-plate. From Ref. [21], reprinted by permission of the publisher (Taylor & Francis Ltd, http://www.tandfonline.com).

9.2.2 Enhancing NP Throughput

Given a spark energy, the NP mass production rate is proportional to the spark repetition frequency [17]. In the classical RLC circuit for initiating sparks, increasing frequency leads to premature spark discharges associated with the residual spark plasma. To minimize these premature discharges, wire-to-plate electrodes were used for a fast removal of the residual plasma. As compared to other types of electrode configuration, such an arrangement results in an increase of not only local velocity of the carrier gas but also electrical field intensity around the spark domain [23]. On the other hand, raising the spark repetition frequency in the classical RLC circuit can develop continuous arc discharges, signaling that an upper operating frequency threshold exists in the system. Exceeding the threshold, linearity does not hold between the NP mass production rate and the frequency [17]. Much larger particles (hundreds of nanometers to a few microns) are produced in arc discharges (hot plasma temperature), thereby giving a higher mass production rate of these undesirably large particles. In the recent works of Feng et al. [17] and Pfeiffer et al. [16], they decoupled the charge and discharge cycle in a switching RLC circuit. Using that circuit, the highest possible spark frequency can reach 25 kHz. As a result, the mass production rate is enhanced by about 2 orders of magnitude, while maintaining the production of sub-10-nm particles and the virtually unlimited mixing possibilities [17].

Besides increasing NP throughput in a single electrode pair [23], we developed multiple pins to holes on a plate electrode for making parallel sparks in a chamber. Doing so, we also gained the benefit of patterning uniform NP structures on an extended surface area of 50 mm × 50 mm [24]. The uniformity of patterned 3D nanostructures was improved by a factor of 3 compared to the single-pin system (i.e., a point particle source). In the single-pin setup, the concentration of charged aerosols is inhomogeneously distributed above a large substrate, leading to taller (shorter) nanostructures at the center (edge) of the substrate [24]. When using multiple pins, the charged aerosol cloud was enlarged to well cover the entire substrate. In addition, proximity collection (i.e., the substrate was placed inside the spark chamber instead of using an extended tubing system) was applied for assembling very tiny NPs/clusters formed at the initial stage [19, 20], reflected by denser nanostructures. Similar results were obtained by increasing the flow rate. In addition, proximity collection largely preserves the charged species (e.g., charged aerosols and ions), the combination of which is essential to obtain a good focusing effect. This work therefore brings our technology into a practical use for the resulting 3D nanostructures on a macroscale.

9.2.3 Increasing the Portion of the Positively Charged Particles

Spark ablation generally produces a higher fraction of charged particles (mainly carrying one elementary charge) without the need of any charging device [25]. This fulfills the requirements of EAAL [26], which relies on a localized electric field to precisely direct charged NPs for the construction of 3D nanostructures. Evidently, increasing the portion of charged particles is the sole approach to promoting the growth of 3D nanostructures. In spark ablation, particle charging mainly stems from ion attachment [27]. To enhance the particle–ion collisions, positive ions were continuously added into the aerosols by directly connecting a second voltage source that allowed a corona discharge mode before breakdown. The amount of positively charged particles was increased by a factor of ca. 2, while those of negatively charged ones diminished by about half.

9.2.4 Generation of Highly Crystalline Silicon NPs

Spark ablation does not employ any liquid precursors/stabilizers, thereby bringing a highly pure NP surface. Considering also the high reactivity of NPs with the trace amount of reagents (e.g., O_2, probably during the production or sample transport), the NPs collected are normally covered with an oxide layer (the thickness depends on the handling process). Apart from that, oxidation of NPs induces the formation of partial amorphous structures [28, 29], which are undesirable to some applications (e.g., solar cells, quantum dots). To address this problem, hydrogen was added to the carrier gas for passivating the surface of NPs via binding hydrogen atoms there [28, 29]. As an example, the good crystallinity of produced Si NPs was confirmed by transmission electron microscopy (TEM), X-ray diffraction (XRD), and Raman spectra, whereas Fourier transform infrared (FTIR) spectroscopy measurements showed a reduction of NP oxidation. The work provides a general approach to handling "clean" NP building blocks for 3D nanostructure arrays and toward other applications strictly requiring a highly pure NP surface.

9.3 Formation of 3D Nanostructure Arrays

3D nanostructure arrays can be made by delivering NP building blocks into precisely addressable locations predetermined on a substrate. This means that some kind of force must drive the particles to the location, and different forces were already exemplified in Section 9.1. In this chapter, we combine the Columbic force-dominated 3D assembly of charged aerosol NPs with topographically patterned substrates formed by conventional lithography. In this process, the reshaped electric field streamlines steer the transport of charged aerosols to precisely addressable locations. A prolonged deposition piles up charged NPs to form 3D nanostructures where charge dissipation can occur. This way of making 3D structure arrays remarkably reaches nanoscale resolution and operates under atmospheric conditions.

9.3.1 Electric-Field-Assisted Aerosol Lithography

Choi's group introduced the use of EAAL for patterning 3D nanostructure arrays on a surface [26]. In the method, charged aerosols together with ions of the same polarity are injected into a chamber where an electric field is applied. In the chamber, a negative voltage is applied to a substrate prepatterned by a dielectric layer (e.g., photoresist [PR]). Considering the mobility difference between ions and charged aerosols, the former move faster to accumulate on the surface of the dielectric prepatterns. Other ions are deposited onto openings of the substrate and are neutralized immediately. The ions on the surface of the prepatterns distort the originally almost flat equipotential plane into a convex-shaped equipotential plane that acts as a nanoscopic electrostatic lens around each prepattern. Later-arriving charged aerosols follow the localized electric field line near the substrate and are deposited into the openings (cf. Fig. 9.3). The degree of focusing can be manipulated by the amount of ions injected. In 2006, this method was, for the first time, explained in detail and used for parallel patterning of charged aerosol NPs of various metals [26]. It was found that the width of deposited NP pillars was about 1/6 of the width of opening patterns. As already mentioned, this focused deposition is attributed to the electrostatic lens effect induced by either predeposited ions or a potential on the prepattern surface.

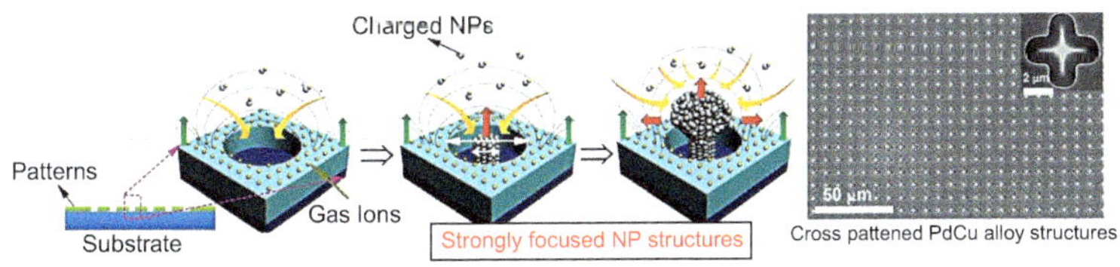

Figure 9.3 Schematic of the assembly progress of charged aerosols via EAAL. The red, green, and yellow arrows represent the growing direction of nanostructures, repelling electric forces, and direction of coming aerosols, respectively. Adapted from Ref. [30]. © IOP Publishing. Reproduced with permission. All rights reserved.

From a theoretical perspective, the NP deposition process in EAAL can be described by a Lagrangian particle trajectory [31]. Provided that the Poisson equation was used to obtain the electric field, the particle trajectory can be calculated through the Langevin equation:

$$m_{\mathrm{p}} \frac{dv_{\mathrm{p}}}{dt} = F_{\mathrm{D}} + F_{\mathrm{B}} + F_{\mathrm{C}} + F_{\mathrm{vdW}} \tag{9.1}$$

It contains these terms: the fluid drag force F_{D}, the Brownian force F_{B}, the Coulomb force F_{C}, and the van der Waals force F_{vdW} [32–35], where m_{p}, t, and v_{p} refer to the mass of a particle, the time, and the particle velocity, respectively.

To clarify the generality of EAAL, next we discuss particle material, the number of particle charges, and the type of substrate hosting the 3D nanostructures. The particle material, in principle, can be any, depending on which aerosol technology is used for making the charged aerosol NPs. It was found that EAAL favors singly charged NPs in terms of making uniform/compact 3D nanostructures and ensuring a clean dielectric surface [36]. With respect to the substrate, conductivity is important to frame localized "funnel like" electrical field streamlines and to prevent charge accumulation during deposition. Notably, we also showed the possibility of making 3D nanostructures on a nonconductive substrate [33].

To broaden the scope of particle materials, Choi's group also used EAAL to successfully pattern protein and polystyrene NP structures [36, 37]. By electrospraying protein molecules to form charged aerosols, the patterned micro-/nanoarrays had 50–130 nm features, even deposited into the valleys within the deep microchannels [37]

In addition to polystyrene NP patterns, polymeric particle patterns were made from electrospray deposition through a mask placed on a substrate [38]. Differently, the focusing effect was produced by setting a potential difference to a conductive mask and a dielectric substrate (opposite to our case of using a dielectric mask and a conductive substrate).

Using electrospraying for producing charged aerosols, the ill-defined number of charges on particles was proven to exacerbate the focusing effect during deposition [36]. The reason for this is that more charges on a single particle [39] make it not fully follow the electric field lines due to the effect of inertia (cf. Fig. 9.4). Similar to impaction, these NPs (gaining a higher velocity than those with one elementary charge) easily deviate from the electric field lines, thus depositing onto the mask surface or the mask openings randomly. With the help of a neutralizer, the focused deposition was eventually recaptured but with inevitable particle agglomeration and a significant loss of charged particles. It is worthwhile to note that spark ablation produces a higher percentage of charged NPs, which only take one elementary charge without the need of any secondary charging device [40]. This turns out to be a key advantage for highly focused NP deposition, closely correlated to uniform/compact 3D nanostructures and a clean dielectric surface.

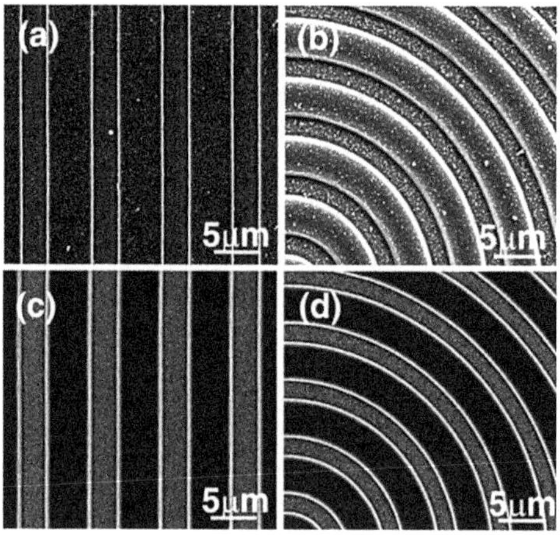

Figure 9.4 Patterned deposition for ill-defined (a, b) and well-defined (c, d) number of charges of polystyrene NPs produced by electrospraying. Reprinted from Ref. [36], with the permission of AIP Publishing.

9.3.2 3D Assembly of NPs via EAAL

A conductive substrate (e.g., a highly doped silicon wafer) is covered by a layer of dielectric materials having prepatterned open spacings. Figure 9.3a schematically shows a single-hole PR pattern (color-coded mint) covering a Si substrate (brilliant blue). Upon applying a voltage to the substrate, the charged aerosols are guided along the electric field line and subsequently deposited on the exposed conductive region in a converging manner (cf. Fig. 9.3b). Furthermore, Fig. 9.3c shows that the NP deposition develops to 3D nanostructures arrays (more details will be discussed later), followed by removal of the dielectric layer (dipping in acetone or O_2 plasma etching). Figure 9.5 provides a direct visual of cross PR patterns on a Si wafer.

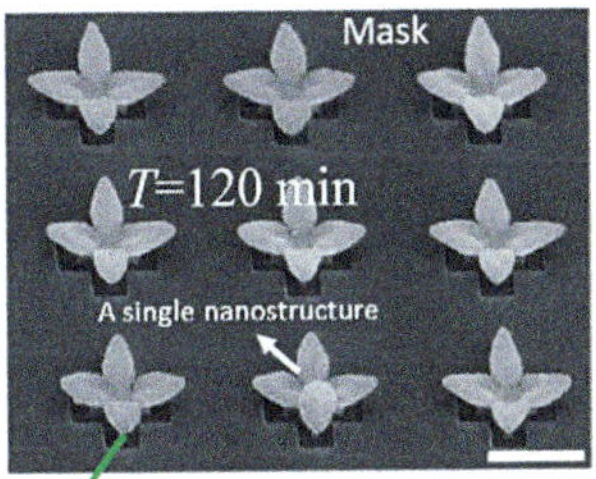

Figure 9.5 SEM image of the prepatterned substrate (with a cross pattern, made of PR; scale bar = 1.5 μm). Reprinted with permission from Ref. [41]. Copyright (2011) American Chemical Society.

As the nanostructures grow taller than the thickness of the dielectric layer, their growth splits into lateral and vertical directions because of the strong repelling electric field from the PR surface (cf. Fig. 9.3c). To give a direct visual, we provide scanning electron microscopy (SEM) images (cf. Fig. 9.6) of arrays of 3D nanostructures consisting of Cu NPs on the PR prepatterned Si substrate [41, 42]. Figures 9.6a and 9.6b show the resulting 3D nanostructures when using square arrays of 1.5 μm long cross PR patterns for 90 and 120 min deposition, respectively. In this case, 3D nanostructures give fourfold flower-like shapes. Changing to square prepatterned substrates, Figs. 9.6c and 9.6d represent the resulting

3D nanostructures for NP deposition of 40 and 90 min, respectively. Therefore, it can be observed that PR patterns play an essential role in determining the shapes of the final 3D nanostructures.

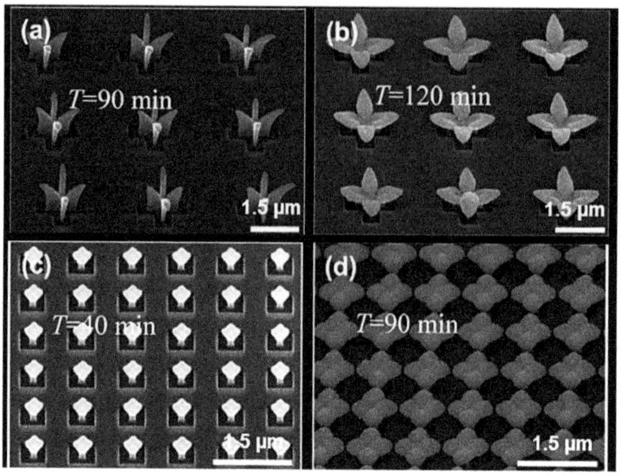

Figure 9.6 (a, b) SEM images of 3D nanostructures on PR cross patterns for 90 and 120 min deposition and (c, d) on PR square patterns for 40 and 90 min deposition. Reprinted with permission from Ref. [41]. Copyright (2011) American Chemical Society.

The final 3D nanostructures, however, can be completely different from the opening patterns (cf. Fig. 9.6). For example, a round opening can make a mushroom-like nanostructure, while a square and a cross pattern allow to make clovers and lotus petals [41]. The shape of 3D nanostructures depends on several factors, such as the arrangement of adjacent hole patterns, NP size, deposition time, surface charge density on the dielectric surface, and flow rate of charged aerosols. As shown in Fig. 9.6b,d, 3D nanostructures grow toward their neighborhoods and eventually connect to each other, regardless of the shape of the prepatterns. Such a multifurcating mechanism has been recently elaborated [34]. When nanostructures grow upward and become higher than the PR layer (cf. Fig. 9.7a), they start leaning toward the nearest structure. This is because the potential field acts like an electrical trap. At different heights from the surface of the substrate (e.g., Z = 1.5 and 2 μm relatively far from the pattern with a thickness of ca. 0.3 μm; cf. Fig. 9.7b), the potential fields are interconnected, explaining the formation of bridges among the nanostructures. In the case that the adjacent nanostructures are sufficiently distant, one can see in Fig. 9.7c that the shape of PR patterns does determine the shape of the resulting nanostructures.

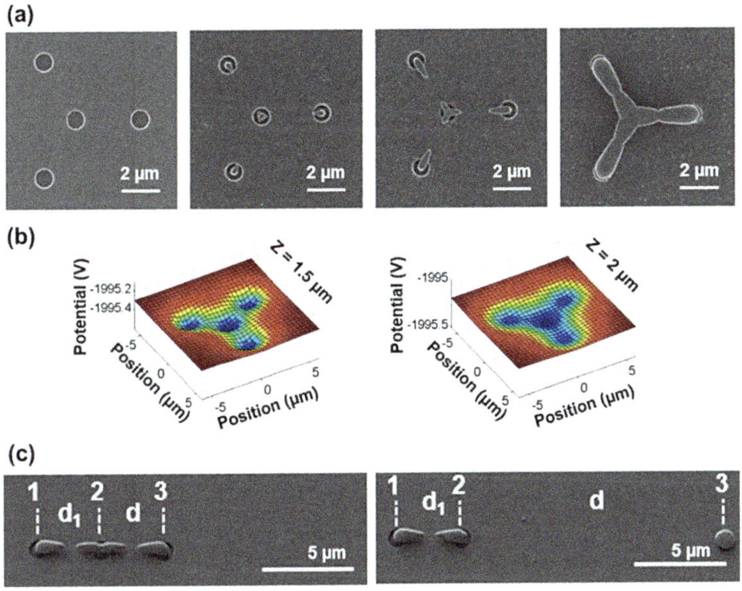

Figure 9.7 (a) SEM images of Cu nanostructures on 1 μm hole patterns at different deposition times (0, 5, 15, and 25 min). (b) Electric potential calculated at different heights of Z = 1.5 and 2 μm. (c) SEM images of three Cu nanostructures at different hole locations. d_1 = 3.5 μm, d = 3.5, and d = 14 μm. Reprinted with permission from Ref. [34]. Copyright (2016) Wiley-VCH.

To this end, we describe that EAAL can make a matrix of complex 3D structures with a single layer. The following part explores the possibility of stacking multilayers of 3D nanostructures, even feasible with different materials and dimensions. Using a prepatterned substrate may limit a vertical growth of consecutive 3D nanostructures. The height of the 3D nanostructures is more or less in the order of 1 μm. To stack a second layer vertically, we managed to deposit the NPs on top of the existing 3D nanostructures from the first deposition (cf. Fig. 9.8) [35]. Both depositions occurred within identical cross-opening prepatterns and an identical relative thickness of the PR layer. In other words, the first deposition used a PR layer with a thickness of 100 nm, while the second layer deposition had a distance of 100 nm between the top surface of the PR layer and the head of the 3D nanostructures.

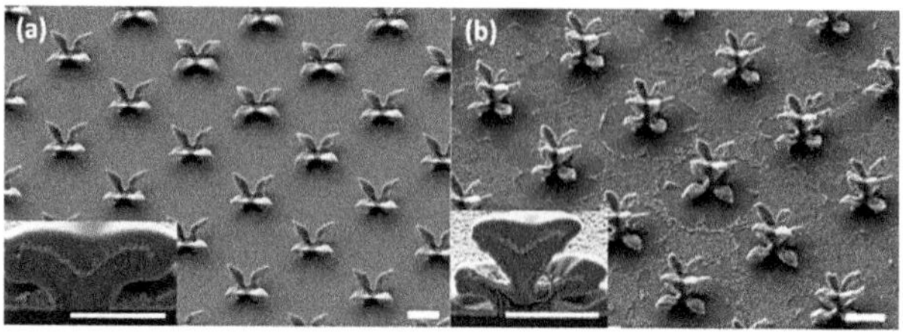

Figure 9.8 SEM images of monolayer and double-layer 3D nanostructures (scale bar = 1 μm). Reprinted from Ref. [35]. © IOP Publishing. Reproduced with permission. All rights reserved.

9.3.3 NP Focusing Mask

In 2010, Choi's group extended EAAL by eliminating the step of PR prepatterning via a so-called focusing mask concept [33]. The method employed a nonconducting stencil mask having openings, which adheres to the same principle of ion-induced focusing. To formulate this concept, they successfully achieved the focused patterning of NPs on the surface of thick glass and a polymer substrate (cf. Fig. 9.9) [33]. However, such nonconductive substrates always lead to charge accumulations (from deposited charged NPs and ions) within the nanostructures. The consequence of this was that the incoming charged NPs were repelled and scattered away, resulting in self-termination growth and unwanted deposition of charged NPs onto the mask surface. Therefore, patterning on nonconductive substrates (e.g., thick glass, polymer film) requires means of conducting away the abundant charges. This can be achieved by covering the nonconductive substrate (beneath the mask; cf. Fig. 9.10) with a thin conducting liquid before deposition and evaporate it afterward [33].

Previously, a plastic microstencil mask was already utilized for making a metal pattern with thermal deposition [43]. However, the resolution was limited to tens of microns. Later, Kim et al. proposed a combination of electrospraying with a freestanding mask for patterning thin films of organic or biomacromolecules, with the highest resolution of 2 μm line and space [44]. The weak focusing effect was attributed to the charged particles on the microstencil mask. Similarly, they reported that the aperture patterns of the mask determine the deposited patterns.

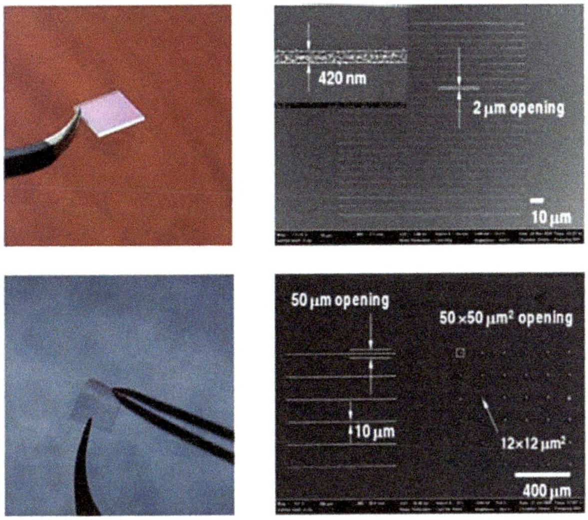

Figure 9.9 Optical and SEM images on the first row are taken from a 0.7-mm-thick glass substrate, while the bottom ones are based on a 0.1-mm-thick polymeric substrate. Reprinted with permission from Ref. [33]. Copyright (2010) Wiley-VCH.

Further enhancing the flexibility of EAAL, Choi's group also used a floating mask or freestanding mask instead of directly placing a mask on a substrate (cf. Fig. 9.10) [45]. Such modification not only simplifies the patterning procedures but also enables the increase of array density and patterning of different material NPs on the same substrate by sequential deposition, synchronized with lateral mask translation [33, 45]. Therefore, the floating mask imparts the most important feature of the noncontact printing technique that may work for complex surfaces of substrates. A detailed study on this by Choi's group is ongoing [46].

In addition, the freestanding mask allows sputtering a metallic film (100 nm thick), where an electric potential difference can be made to a negatively biased substrate (cf. Fig. 9.11). Doing so was proven to also reshape the electric field streamlines [45], similar to the ion-induced electrostatic lens effect. Raising the potential difference was shown to scale down the nanostructure patterns [45].

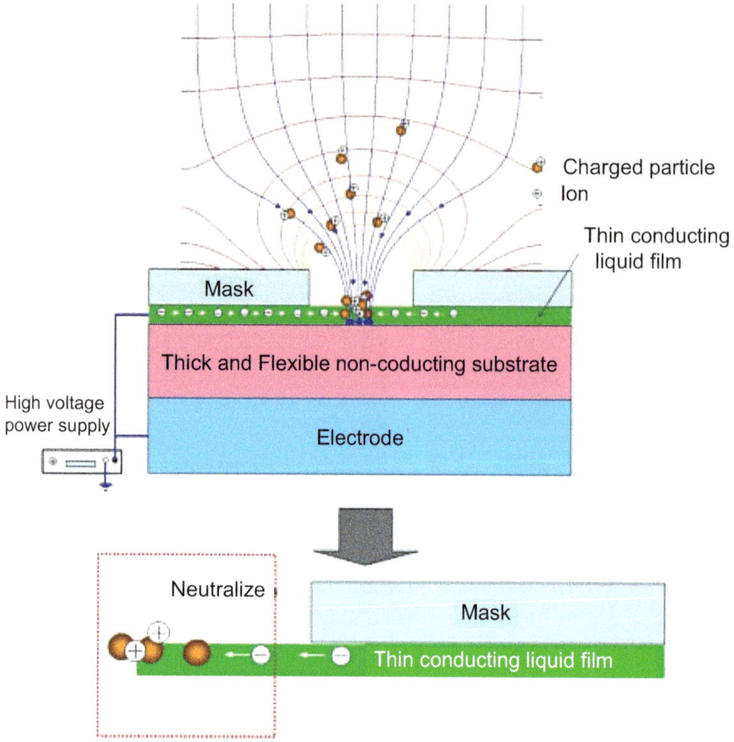

Figure 9.10 Conducting liquid film coating on a nonconductive substrate and charge neutralization beneath the mask. Reprinted with permission from Ref. [33]. Copyright (2010) Wiley-VCH.

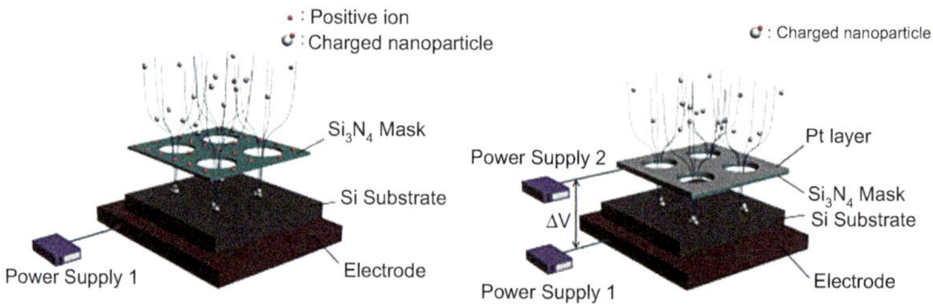

Figure 9.11 Comparison of the focusing effect induced by ion accumulation (left) and by the potential difference between the floating mask and the substrate (right). The stencil mask and the Si substrate were separated by 4 μm using a nanopositioning stage. Reprinted from Ref. [45]. Copyright (2015), with permission from Elsevier.

9.3.4 Parallel Patterning of 3D Nanostructures on a Large Surface Area

To lodge practical applications, the most important goal is to enlarge the patterned surface area with great uniformity. Choi's group already successfully scaled up the patterning of 3D nanostructures on a surface from the order of 100 μm × 100 μm to that of 50,000 μm × 50,000 μm [24]. The core concept was to enlarge the cloud of charged aerosols above the surface of a substrate, circumventing the spatially non-uniform concentration profile from a "point particle source." To address that, they developed multiple pins to holes on a plate electrode (cf. Fig. 9.12a). Using this setup, the nanostructure arrays show a constant height and diameter at a surface area of 50,000 μm × 50,000 μm (cf. Fig. 9.12b,c) [24].

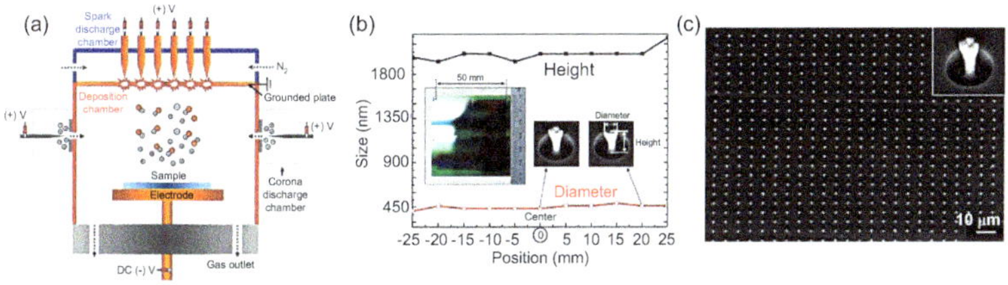

Figure 9.12 Patterning 3D nanostructure arrays by using spark discharge equipped with multiple pins to holes on a plate electrode (a) and demonstration of the parallelism of the nanostructure arrays (b, c). Reprinted from Ref. [24]. © IOP Publishing. Reproduced with permission. All rights reserved.

9.3.5 Remarks on Spark-Produced NPs for Patterning 3D Nanostructures

The charging probability for spark-produced NPs is roughly about 5% [19], much higher than in most other aerosol generation methods. Simultaneously, the spark plasma also contains gas ions. Making such combined charged species is a unique advantage for focused deposition of charged NPs to form 3D nanostructures. Even more, the NP size generally lies below 10 nm (also having a narrow size distribution) [17], thus only carrying one elementary charge and effectively eliminating inertial impaction [36]. In this regard, the motion of singly charged NPs is strictly dominated by the external electrical field, and a process for reducing particle charge is not necessary [36]. A third unique feature is that spark-produced NPs can be highly pure, promoting local particle coalescence upon deposition within the NP structures. Such local coalescence helps to compact the nanostructures and thus to improve their conductivity. Finally, the unlimited mixing combinations give great versatility to the materials of 3D nanostructures, broadening future applications.

9.4 Applications

9.4.1 Gas Sensors

Gas molecules diffuse to nanopatterns consisting of NPs, which leads to a change in electrical or optical properties, from which an electrical signal can be derived. This signal is correlated to the concentration of the gas (e.g., CO and H_2). The sensing device based on an NP line pattern was reported to be capable of detecting the CO concentration down to 1 ppm [33]. Its sensitivity, in principle, can further be enhanced by changing the width of the patterned lines, in accordance with the opening space of the patterned mask.

Apart from the line patterns for gas sensors [33], EAAL was also applied to make 3D microstructures that are mutually bridged but floating over a substrate [34, 41]. This enables further advancing interconnects used in sensors [34], nanoelectronics [47], etc. This point is illustrated with an example drawn from previous work [34]. Figure 9.7a shows a dielectric layer having three holes positioned at the vertices and one hole located at the orthocenter of an equilateral triangle (edge length ca. 5.8 μm; hole diameter 1 μm). The charged NPs were initially focused into the center of each opening. The structures grew upward to a comparable thickness of the patterns, terminating their vertical growth while transforming into lateral growth toward the center. The nanostructure at the center hole opens three bridges toward the nanostructures grown from the three holes at the vertices. These nanostructures were eventually connected, as shown in the rightmost panel in Fig. 9.7a. Applying this principle, a CO gas sensor was made based on the tetra-furcated Cu NP structure, and its response was enhanced by a factor of more than 2 for CO at 100 ppm compared to the NP film [34]. The asymmetric

NP structures can be further modified to establish multidirectional networks and junctions suitable for nanoelectronics, sensors, etc.

Depositing different functional NPs either above or below an NP film was reported to considerably improve the conductivity of NP structures, thereby delivering exceptionally high sensitivity in solid-state gas sensors [48]. This concept can be integrated into EAAL for patterning NP structures used in sensing devices or other nanodevices.

9.4.2 Plasmonic Sensors

Plasmonics offers a key to improve the efficiency in widespread applications (see also Chapter 12) [49]. Storing part of their energy in the electron gas of a metal, surface plasmons are capable of creating strongly localized and intense optical fields, which consequently enhance a range of optical processes. Known examples are surface-enhanced Raman scattering (SERS), metal-enhanced fluorescence (MEF), and plasmonic solar cells. MEF has been proven to improve fluorescence intensities of luminescent probes [50], useful in sensitive detection of biomolecules (cf. Chapter 13).

MEF stems from the plasmon resonance of metal NPs (e.g., silver or gold) that interact with a fluorophore. This leads to enhancement of excitation rates through local field amplification and to an increase in emission efficiency by enhanced radiative decay rates, both of which contribute to an overall increase in fluorescent emission intensities (see Chapter 13 for biomedical applications of this effect). The increase of luminescence up-conversion was theoretically predicted to follow a fourth-power dependence on local field enhancement, but the typical gains are weaker than in SERS [51]. According to this theory, human immunoglobulin G–fluorescein isothiocyanate (IgG-FITC; Sigma, F9636) deposited on top of Ag nanostructures was shown to emit a significantly improved fluorescence signal (cf. Fig. 9.13) [37]. The work distinguishes from a traditional dipping process, as it makes a desired locus through stacking different materials layer by layer [37]. In the traditional process, it relies on dipping a metal-prepatterned substrate into a protein solution, followed by washing away the proteins absorbed on the nonmetallic parts. Still, some residual proteins are absorbed, thus producing an unwanted background signal.

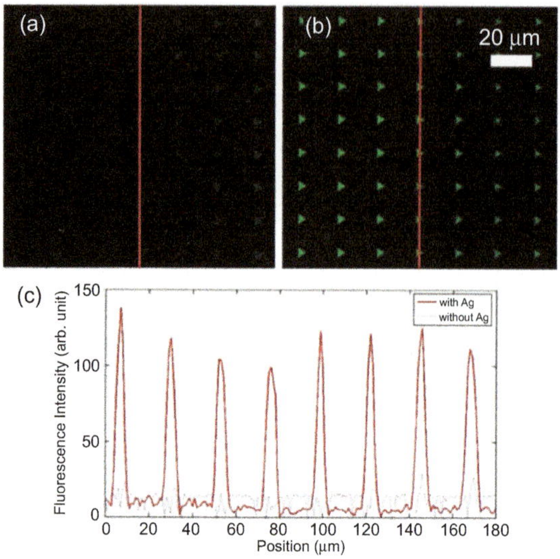

Figure 9.13 Considerably improved fluorescence signal. Confocal microscopic images are corresponding to (a) IgG-FITC and (3) IgG-FITC/Ag nanostructrues. (c) The fluorescence signal is shown to be enhanced by a factor of roughly 6. Reprinted with permission from Ref. [37]. Copyright (2011) Wiley-VCH.

Raman scattering by a single molecule has been reported, with an intensified signal up to the order of 10^{14} [52], due to the hot spot (produced by Ag/Au nanostructures)-induced local field enhancement. One may notice that the enhancement here is much stronger than the case of fluorescence. This is because Raman scattering is a coherent process in contrast to the typically incoherent absorption-emission fluorescence (excited-state absorption and nonradiative energy transfer differ from Raman scattering) [51].

Flower 3D nanostructures were shown to have electromagnetic hot spots at the 3D nanogaps between adjacent petals. Choi's group has shown that electromagnetic hot spots can be engineered by the number of petals (cf. Fig. 9.14a,b) [53]. Figure 9.14c,d shows that an increase in the number of petals leads to an increased number of hot spots, thereby enhancing the field intensity significantly. In comparison with the particle film, such 3D nanostructures were shown to improve the intensity of Raman scattering by around 2 orders of magnitude [53].

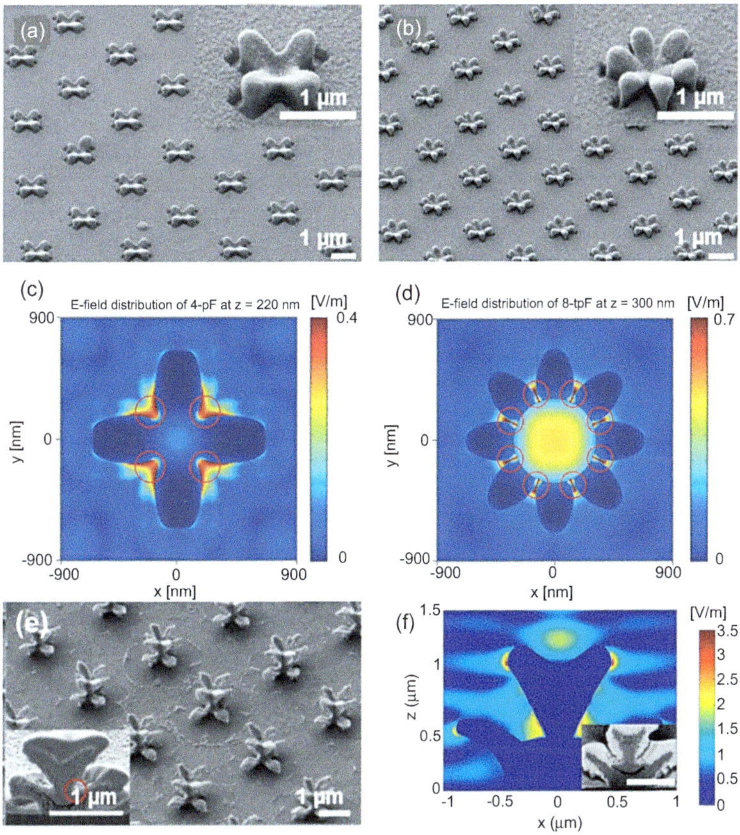

Figure 9.14 (a) SEM image of 3D nanostructures with four petals. (b) SEM image of 3D nanostructures with eight petals. (c) Calculated electric field intensity distribution of 3D nanostructure with four petals at z = 220 nm. (d) Calculated electric field intensity distribution of 3D nanostructure with eight petals at z = 300 nm. Reprinted with permission from Ref. [53]. Copyright (2014) Wiley-VCH. (e) SEM image of stacked bilayer 3D nanostructures. (f) Calculated electric field intensity distribution of vertical cross section. Reprinted from Ref. [35]. © IOP Publishing. Reproduced with permission. All rights reserved.

Furthermore, Lee et al. demonstrated that additional vertical hot spots can be created by vertical-stacking 3D nanostructures through consecutive execution of EAAL and e-beam lithography (Fig. 9.14e) [35]. They discovered that even though the number of hot spots was increased twice, SERS peak intensity was enhanced by more than fivefold. Such disproportional enhancement was attributed to additional vertical hot spots created near the pillar connecting the upper and lower 3D nanostructures, as shown in Fig. 9.14f.

9.4.3 Solar Cells

Solar cells are another attractive field for exploiting 3D nanostructures, which have optical advantages such as electromagnetic hot spots or cavities. 3D nanostructures made by EAAL were shown to improve the efficiency and performance of solar cells. Ha et al. demonstrated that the power conversion efficiency of nc-Si:H thin-film solar cells can be improved by up to 22% by incorporating 3D nanostructures (Fig. 9.15a) [54]. This enhancement was attributed to the surface plasmon resonances surrounding the structures at multiple frequencies, which increase the light scattering and the absorption

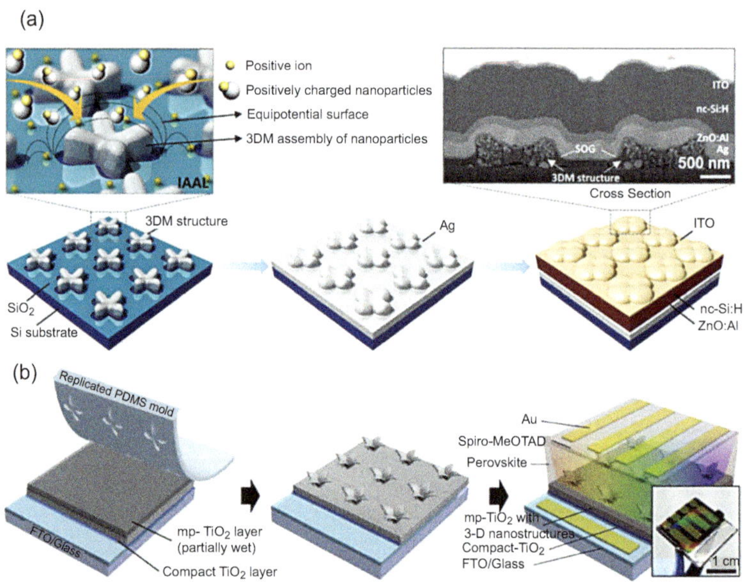

Figure 9.15 (a) Schematic illustration for the preparation of an nc-Si:H thin-film solar cell using 3D nanostructures via ion-assisted aerosol lithography (IAAL). Reprinted from Ref. [54]. © IOP Publishing. Reproduced with permission. All rights reserved. (b) Schematic illustration of patterning of TiO_2 using a PDMS mold and fabrication of a perovskite solar cell incorporating 3D nanostructures. Reprinted from Ref. [55], Copyright (2016), with permission from Elsevier.

efficiency (see also Chapter 12). Moreover, Jang et al. used a new technology combining 3D assembly via EAAL and soft lithography to fabricate 3D-nanopatterned TiO_2 electrodes [55]. As shown in Fig. 9.15b, they made 3D-patterned TiO_2 electrodes by stamping 3D patterned polydimethylsiloxane (PDMS) molds replicated from 3D nanostructures via EAAL. By employing 3D-patterned TiO_2 electrodes, perovskite solar cells can exhibit improved optoelectronic performance and enhanced charge collection due to effective light harvesting. Their work improved the approximate average values of short-circuit current density by 5%, the open-circuit voltage by 2%, the fill factor by 3%, and the photon conversion efficiency by 10%, in reference to those with flat TiO_2.

9.5 Conclusions

We have described the implementation of aerosol technologies for EAAL to print 3D nanostructure arrays, together with some applications. In this method, the transport of charged aerosols is directed to precisely addressable/programmable locations with nanometer resolution on a surface, eventually structuring to arrays of complex 3D nano-/micropatterns. In principle, this process should work with any material that can be converted to charged aerosols. In this respect, spark ablation is a perfect match for EAAL in terms of making uniform/compact 3D nanostructure arrays, even on a macroscale surface. For example, it can be modified to supply sub-10-nm building blocks consistently, and this consistent production was maintained when sparking multiple pairs of electrodes in parallel. Considering all the developments (particularly the concept of a floating mask), EAAL bears great potential for further development to a noncontact 3D nanoprinting technology that enables the manufacturing of 3D nano-/microstructures on a mass/macroscale.

References

1. Su, Y. W., Wu, C. S., Chen, C. C. and Chen, C. D. (2003). Fabrication of two-dimensional arrays of CdSe pillars using E-beam lithography and electrochemical deposition, *Adv. Mater.*, **15**(1), pp. 49–51.

2. Yamazaki, K. and Yamaguchi, H. (2012). Electron beam lithography on vertical side faces of micrometer-order Si block, *J. Vac. Sci. Technol., B*, **30**(4), p. 041601.
3. Feng, J., Tuominen, M. T. and Rothstein, J. P. (2011). Hierarchical superhydrophobic surfaces fabricated by dual-scale electron-beam-lithography with well-ordered secondary nanostructures, *Adv. Funct. Mater.*, **21**(19), pp. 3715–3722.
4. Keskinbora, K., Grevent, C., Hirscher, M., Weigand, M. and Schutz, G. (2015). Single-step 3D nanofabrication of kinoform optics via gray-scale focused ion beam lithography for efficient x-ray focusing, *Adv. Opt. Mater.*, **3**(6), pp. 792–800.
5. Li, Z., Siklos, M., Pucher, N., Cicha, K., Ajami, A., Husinsky, W., Rosspeintner, A., Vauthey, E., Gescheidt, G., Stampfl, J. and Liska, R. (2011). Synthesis and structure-activity relationship of several aromatic ketone-based two-photon initiators, *J. Polym. Sci., Part A: Polym. Chem.*, **49**(17), pp. 3688–3699.
6. Ovsianikov, A., Viertl, J., Chichkov, B., Oubaha, M., MacCraith, B., Sakellari, I., Giakoumaki, A., Gray, D., Vamvakaki, M., Farsari, M. and Fotakis, C. (2008). Ultra-low shrinkage hybrid photosensitive material for two-photon polymerization microfabrication, *ACS Nano*, **2**(11), pp. 2257–2262.
7. Wilbur, J. L., Kumar, A., Biebuyck, H. A., Kim, E. and Whitesides, G. M. (1996). Microcontact printing of self-assembled monolayers: applications in microfabrication, *Nanotechnology*, **7**(4), pp. 452–457.
8. Cui, Y., Björk, M. T., Liddle, J. A., Sönnichsen, C., Boussert, B. and Alivisatos, A. P. (2004). Integration of colloidal nanocrystals into lithographically patterned devices, *Nano Lett.*, **4**(6), pp. 1093–1098.
9. Kraus, T., Malaquin, L., Schmid, H., Riess, W., Spencer, N. D. and Wolf, H. (2007). Nanoparticle printing with single-particle resolution, *Nat. Nanotechnol.*, **2**(9), pp. 570–576.
10. Park, J. U., Hardy, M., Kang, S. J., Barton, K., Adair, K., Mukhopadhyay, D. K., Lee, C. Y., Strano, M. S., Alleyne, A. G., Georgiadis, J. G., Ferreira, P. M. and Rogers, J. A. (2007). High-resolution electrohydrodynamic jet printing, *Nat. Mater.*, **6**(10), pp. 782–789.
11. Galliker, P., Schneider, J., Eghlidi, H., Kress, S., Sandoghdar, V. and Poulikakos, D. (2012). Direct printing of nanostructures by electrostatic autofocussing of ink nanodroplets, *Nat. Commun.*, **3**, p. 890.
12. Tarafder, S., Balla, V. K., Davies, N. M., Bandyopadhyay, A. and Bose, S. (2013). Microwave-sintered 3D printed tricalcium phosphate scaffolds

for bone tissue engineering, *J. Tissue Eng. Regen. Med.*, **7**(8), pp. 631–641.

13. Khalyfa, A., Vogt, S., Weisser, J., Grimm, G., Rechtenbach, A., Meyer, W. and Schnabelrauch, M. (2007). Development of a new calcium phosphate powder-binder system for the 3D printing of patient specific implants, *J. Mater. Sci. Mater. Med.*, **18**(5), pp. 909–916.
14. Schwyn, S., Garwin, E. and Schmidt-Ott, A. (1988). Aerosol generation by spark discharge, *J. Aerosol Sci.*, **19**(5), pp. 639–642.
15. Tabrizi, N. S., Ullmann, M., Vons, V. A., Lafont, U. and Schmidt-Ott, A. (2009). Generation of nanoparticles by spark discharge, *J. Nanopart. Res.*, **11**, pp. 315–332.
16. Pfeiffer, T. V., Feng, J. and Schmidt-Ott, A. (2014). New developments in spark production of nanoparticles, *Adv. Powder Technol.*, **25**(1), pp. 56–70.
17. Feng, J., Guo, X., Ramlawi, N., Pfeiffer, T. V., Geutjens, R., Basak, S., Nirschl, H., Biskos, G., Zandbergen, H. W. and Schmidt-ott, A. (2016). Green manufacturing of metallic nanoparticles: a facile and universal approach to scaling up, *J. Mater. Chem. A*, **4**, pp. 11222–11227.
18. Feng, J. (2016). Scalable spark ablation synthesis of nanoparticles: fundamental considerations and application in textile nanofinishing, TU Delft.
19. Feng, J., Biskos, G. and Schmidt-Ott, A. (2015). Toward industrial scale synthesis of ultrapure singlet nanoparticles with controllable sizes in a continuous gas-phase process, *Sci. Rep.*, **5**, p. 15788.
20. Feng, J., Huang, L., Ludvigsson, L., Messing, M. E., Maisser, A., Biskos, G. and Schmidt-Ott, A. (2016). General approach to the evolution of singlet nanoparticles from a rapidly quenched point source, *J. Phys. Chem. C*, **120**(1), pp. 621–630.
21. Chae, S., Lee, D., Kim, M.-C., Kim, D. S. and Choi, M. (2015). Wire in-hole-type spark discharge generator for long-time consistent generation of unagglomerated nanoparticles, *Aerosol Sci. Technol.*, **49**(7), pp. 463–471.
22. Han, K., Kim, W., Yu, J., Lee, J., Lee, H., Gyu Woo, C. and Choi, M. (2012). A study of pin-to-plate type spark discharge generator for producing unagglomerated nanoaerosols, *J. Aerosol Sci.*, **52**, pp. 80–88.
23. Noh, S. R., Lee, D., Park, S. J., Kim, D. S. and Choi, M. (2017). High throughput nanoparticle generation utilizing high-frequency spark discharges via rapid spark plasma removal, *Aerosol Sci. Technol.*, **51**(1), pp. 116–122.

24. Ha, K., Choi, H., Jung, K., Han, K., Lee, J.-K., Ahn, K. and Choi, M. (2014). Large-area assembly of three-dimensional nanoparticle structures via ion assisted aerosol lithography with a multi-pin spark discharge generator, *Nanotechnology*, **25**(22), p. 225302.

25. Noh, S. R., Kim, D. S., Park, S. J. and Choi, M. (2016). Enhanced yield of positively charged particles from a spark discharge generator via in situ corona discharges, *J. Aerosol Sci.*, **101**, pp. 188–195.

26. Kim, H., Kim, J., Yang, H., Suh, J., Kim, T., Han, B., Kim, S., Kim, D. S., Pikhitsa, P. V. and Choi, M. (2006). Parallel patterning of nanoparticles via electrodynamic focusing of charged aerosols, *Nat. Nanotechnol.*, **1**(2), pp. 117–121.

27. Alonso, M., Hernandez-Sierra, A. and Alguacil, F. J. (2003). Electrical charging of aerosol nanoparticles and some practical applications, *Rev. Metal.*, **39**, pp. 41–57.

28. Lee, D., Lee, K., Kim, D. S., Lee, J.-K., Park, S. J. and Choi, M. (2017). Hydrogen-assisted spark discharge generation of highly crystalline and surface-passivated silicon nanoparticles, *J. Aerosol Sci.*, **114**, pp. 139–145.

29. Feng, J., Geutjens, R., Thang, N. V., Li, J., Guo, X., Kéri, A., Basak, S., Galbács, G., Biskos, G., Nirschl, H., Zandbergen, H. W; Brück, E. and Schmidt-Ott, A. (2018). Magnetic phase transition in spark-produced ternary LaFeSi nanoalloys, *ACS Appl. Mater. Interfaces*, **10**(7), pp. 6073–6078.

30. Ha, K., Choi, H., Jung, K., Han, K., Lee, J. K., Ahn, K. and Choi, M. (2014). Large-area assembly of three-dimensional nanoparticle structures via ion assisted aerosol lithography with a multi-pin spark discharge generator, *Nanotechnology*, **25**(22), p. 225302.

31. Krinke, T. J., Deppert, K., Magnusson, M. H. and Fissan, H. (2002). Nanostructured deposition of nanoparticles from the gas phase, *Part. Part. Syst. Char.*, **19**(5), pp. 321–326.

32. You, S. and Choi, M. (2007). Numerical simulation of microscopic motion and deposition of nanoparticles via electrodynamic focusing, *Aerosol Sci.*, **38**, pp. 1140–1149.

33. You, S., Han, K., Kim, H., Lee, H., Woo, C. G., Jeong, C., Nam, W. and Choi, M. (2010). High-resolution, parallel patterning of nanoparticles via an ion-induced focusing mask, *Small*, **6**(19), pp. 2146–2152.

34. Bae, Y., Pikhitsa, P. V., Cho, H. and Choi, M. (2017). Multifurcation assembly of charged aerosols and its application to 3D structured gas sensors, *Adv. Mater.*, **29**(2), p. 1604159.

35. Lee, K., Choi, H., Kim, D. S., Jang, M. S. and Choi, M. (2017). Vertical stacking of three-dimensional nanostructures via an aerosol lithography for advanced optical applications, *Nanotechnology*, **28**(47), p. 475302.

36. Lee, H., You, S., Woo, C. G., Lim, K., Jun, K. and Choi, M. (2009). Focused patterning of nanoparticles by controlling electric field induced particle motion, *Appl. Phys. Lett.*, **94**(5), p. 053104.

37. Woo, C. G., Shin, H., Jeong, C., Jun, K., Lee, J., Lee, J.-R., Lee, H., You, S., Son, Y. and Choi, M. (2011). Selective nanopatterning of protein via ion-induced focusing and its application to metal-enhanced fluorescence, *Small*, **7**(13), pp. 1790–1794.

38. Xie, J., Rezvanpour, A., Wang, C.-H. and Hua, J. (2010). Electric field controlled electrospray deposition for precise particle pattern and cell pattern formation, *AIChE J.*, **56**(10), pp. 2607–2621.

39. Suh, J., Han, B., Okuyama, K. and Choi, M. (2005). Highly charging of nanoparticles through electrospray of nanoparticle suspension, *J. Colloid Interface Sci.*, **287**(1), pp. 135–140.

40. Hoppel, W. A. and Frick, G. M. (1986). Ion—aerosol attachment coefficients and the steady-state charge distribution on aerosols in a bipolar ion environment, *Aerosol Sci. Technol.*, **5**(1), pp. 1–21.

41. Lee, H., You, S., Pikhitsa, P. V; Kim, J., Kwon, S., Woo, C. G. and Choi, M. (2011). Three-dimensional assembly of nanoparticles from charged aerosols, *Nano Lett.*, **11**(1), pp. 119–124.

42. Choi, M., Lee, H., You, S. and Woo, C. (2008). Array formation of 3-D nanostructure of nanoparticle via electrodynamic focusing of charged aerosols, *AAAR 27th Annual Conference*, Florida, p. 55.

43. Kim, G., Kim, B. and Brugger, J. (2003). All-photoplastic microstencil with self-alignment for multiple layer shadow-mask patterning, *Sens. Actuators, A*, **107**(2), pp. 132–136.

44. Kim, J. W., Yamagata, Y., Kim, B. J. and Higuchi, T. (2009). Direct and dry micro-patterning of nano-particles by electrospray deposition through a micro-stencil mask, *J. Micromech. Microeng.*, **19**(2), p. 025021.

45. Choi, H., Kang, S., Jung, W., Jung, Y. ho; Park, S. J., Kim, D. S. and Choi, M. (2015). Controlled electrostatic focusing of charged aerosol nanoparticles via an electrified mask, *J. Aerosol Sci.*, **88**, pp. 90–97.

46. Choi, M. (2018). Three dimensional nanoprinting via aerosol technology, *Aerosol Technology*, Bilbao.

47. Atabaki, A. H., Moazeni, S., Pavanello, F., Gevorgyan, H., Notaros, J., Alloatti, L., Wade, M. T., Sun, C., Kruger, S. A., Meng, H., Qubaisi, K. A.,

Wang, I; Zhang, B., Khilo, A., Baiocco, C. V., Popović, M. A., Stojanović, V. M. and Ram, R. J. (2018). Integrating photonics with silicon nanoelectronics for the next generation of systems on a chip, *Nature*, **556**(7701), pp. 349–354.

48. Tricoli, A. and Pratsinis, S. E. (2010). Dispersed nanoelectrode devices, *Nat. Nanotechnol.*, **5**(1), pp. 54–60.

49. Aslam, U., Chavez, S. and Linic, S. (2017). Controlling energy flow in multimetallic nanostructures for plasmonic catalysis, *Nat. Nanotechnol.*, **12**(10), pp. 1000–1005.

50. Yin, D., Wang, C., Ouyang, J., Zhang, X., Jiao, Z., Feng, Y., Song, K., Liu, B., Cao, X., Zhang, L., Han, Y. and Wu, M. (2014). Synthesis of a novel core–shell nanocomposite Ag@SiO_2 @Lu_2O_3:Gd/Yb/Er for large enhancing upconversion luminescence and bioimaging, *ACS Appl. Mater. Interfaces*, **6**(21), pp. 18480–18488.

51. Esteban, R., Laroche, M. and Greffet, J. J. (2009). Influence of metallic nanoparticles on upconversion processes, *J. Appl. Phys.*, **105**(3), p. 033107.

52. Lu, D., Cho, S. K., Ahn, S., Brun, L., Summers, C. J. and Park, W. (2014). Plasmon enhancement mechanism for the upconversion processes in $NaYF_4$:Yb^{3+},Er^{3+} nanoparticles: Maxwell *versus* Förster, *ACS Nano*, **8**(8), pp. 7780–7792.

53. Jung, K., Hahn, J., In, S., Bae, Y., Lee, H., Pikhitsa, P. V., Ahn, K., Ha, K., Lee, J.-K., Park, N. and Choi, M. (2014). Hotspot-engineered 3D multipetal flower assemblies for surface-enhanced Raman spectroscopy, *Adv. Mater.*, **26**(34), pp. 5924–5929.

54. Ha, K., Jang, E., Jang, S., Lee, J. K., Jang, M. S., Choi, H., Cho, J. S. and Choi, M. (2016). A light-trapping strategy for nanocrystalline silicon thin-film solar cells using three-dimensionally assembled nanoparticle structures, *Nanotechnology*, ***27***(5), p. 055403.

55. Jang, S., Yoon, J., Ha, K., Kim, M., Kim, D. H., Kim, S. M., Kang, S. M., Park, S. J., Jung, H. S. and Choi, M. (2016). Facile fabrication of three-dimensional TiO_2 structures for highly efficient perovskite solar cells, *Nano Energy*, **22**, pp. 499–506.

Chapter 10

Catalytic Applications of Nanoparticles Produced by Spark Ablation

Alfred P. Weber
Institute of Particle Technology, Clausthal University of Technology, Leibnizstrasse 19, D-38678 Clausthal-Zellerfeld, Germany
alfred.weber@mvt.tu-clausthal.de

In this chapter the applications of metal and metal oxide nanoparticles generated by spark discharges for catalytic reactions ranging from methanation over hydrogen oxidation to the use of the particles as photoinitiators for polymerization are presented. Besides the generation of particles by spark ablation, the studies have in common that all processes are performed in the aerosol state. This provides the advantage that the influence of the particle size, structure, and surface state on reactions can be investigated without interference of a support material. A universal problem of fragile dendritic agglomerates, which are formed from very small primary particles, is their inherent instability against thermal and mechanical stress. Enhanced by the reduced melting point of the primary particles, these agglomerates undergo fast coalescence and restructuring during catalytic reactions at elevated temperatures,

Spark Ablation: Building Blocks for Nanotechnology
Edited by Andreas Schmidt-Ott

ISBN 978-981-4800-82-2 (Hardcover), 978-0-367-81709-1 (eBook)
www.jennystanford.com

significantly reducing the available surface area. Therefore, a new aerosol approach will also be presented to improve the thermal and mechanical stability of the agglomerates by precisely depositing silica only in the necks between the primary particles.

10.1 Introduction

In heterogeneous catalysis the active component consists preferentially of very small particles in order to obtain a high surface area [1]. However, the high fraction of surface atoms reduces the melting point and decreases the stability of the particles against sintering [2]. Therefore, to obtain and keep good dispersion, the active particles are usually deposited onto a support material with a high specific surface area. For the catalytic reaction the substrate plays an important role, not only by enhancing the stability of the active component, but also by providing particle–support interactions, which, in turn, may change the reaction pathway [3]. Thus, the specific influence of different particle properties on the reaction under investigation may be masked due to mutual interactions and are hard to be separated.

On the one hand, in this chapter aerosol catalysis is introduced where catalytic reactions take place on gas-borne nanoparticles, called aerosol catalysis [4]. Aerosol catalysis does not suffer from the problems mentioned above and represents an ideal tool to study the behavior of catalyst particles with regard to their characteristic properties such as size, morphology, and surface conditions. Besides the prevention of the support interference, aerosol catalysis can also circumvent the problem of educt transport limitations, offering direct access to the reaction kinetics. In addition, short residence times lead to low conversions so that the reaction conditions do not change and reaction-specific characteristics such as reaction orders of different educt gases are not needed to quantify the catalytic activity of the particles under investigation.

On the other hand, catalyst particles with highly defined properties facilitate the investigation of their suitability and activity for a given catalytic reaction. Here spark ablation offers the synthesis of particles of high purity [5]. This is a consequence of particle formation where a metal vapor is produced from the ablated surface from which nanoparticles form by nucleation and condensation [6]. Moreover, the carrier gas can be purified to an extent that impurity levels in the parts per billion (ppb) range are reached [7]. In this way, non-oxidized Mg nanoparticles have been produced [7]. Once the primary particles have formed, the morphology of the subsequent particles can be adjusted by tailoring coalescence and collision kinetics, as shown later. Finally, also the particle surface can be modified, within certain limits, employing gas-phase reactions [8].

10.2 Particle Generation and Conditioning

Metal and metal oxide nanoparticles were produced by spark ablation. Two opposing high-purity electrodes of the material of interest were charged in parallel to a capacitor (with a capacitance of 20 nF) using a constant current until the breakdown voltage was reached. The discharge frequency was controlled with the loading current, and the discharge energy was varied by adjusting the breakdown voltage related to the distance between the electrodes.

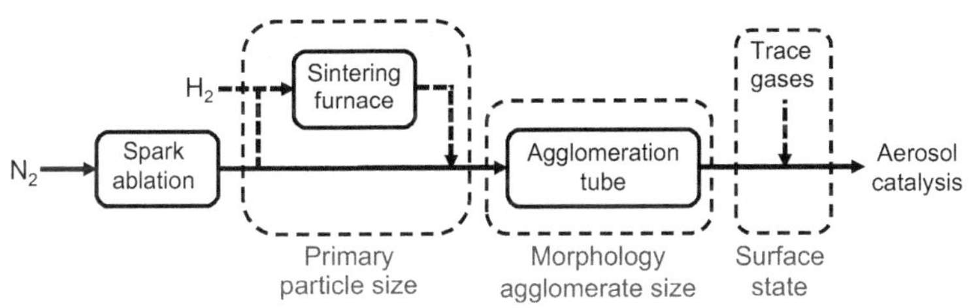

Figure 10.1 Schematic diagram of the experimental setup to produce defined catalyst nanoparticles by spark ablation: The primary particle size can be adjusted by sintering (possibly facilitated by hydrogen addition), agglomerate size and structure are determined by spark ablation parameters (see Chapter 2) and coagulation time, and the particle surface may be modified with the addition of reactive trace gases.

The ensuing spark vaporizes small but fairly constant amounts of metal, which then form nanoparticles by homogeneous nucleation [9]. These so-called primary particles undergo collisions, resulting in chainlike agglomerates. Using thermally induced restructuring, these agglomerates can become more compact in packing all the way up to complete coalescence when passing through a sintering furnace (Fig. 10.1). By controlling the number concentration and the residence time before sintering, the primary particle size can be varied over a certain range (from a few nanometers up to several tens of nanometers) [10]. Since partially oxidized surfaces may be hard to sinter completely, the addition of a few percentage points of hydrogen, reducing oxidation, can accelerate the sintering kinetics [8].

10.3 Aerosol Catalysis and Analytics

For the aerosol catalysis experiments, preconditioned nanoparticles were mixed with educt gases and introduced as aerosols into the reactor. Depending on the reaction under consideration, the reaction started at the mixing point or had to be initiated by increasing the temperature in a tube furnace (Fig. 10.2).

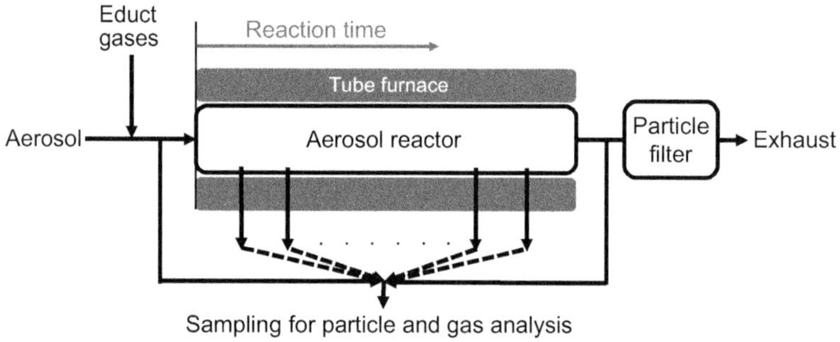

Figure 10.2 Schematic diagram of the experimental setup to perform aerosol catalysis experiments: The gasborne catalyst particles are mixed with the educt gases and introduced into the reactor, where samples are taken at different residence times for particle characterization (SMPS, TEM, BET, TGA, APES, mass concentration) and for gas analysis (FTIR).

Samples for particle characterization and gas analysis were taken before and during the reaction along the reactor. Particles were measured with respect to agglomerate size distribution, primary particle size, and morphology and surface state using scanning mobility particle sizer (SMPS), transmission electron microscopy (TEM), and aerosol photoemission spectroscopy (APES). The photoemission can be described by the Fowler–Nordheim equation [11]:

$$Y = c * (h\nu - \phi)^m, \tag{10.1}$$

where Y is the photoelectric yield defined as the probability of the emission of an electron per incident photon of energy $h\nu$, ϕ is the material-dependent work function, c is the photoemission constant, and m an exponent, which was found to be 2 for metals and graphite [12].

In addition, the specific surface area was determined with nitrogen adsorption (Brunauer–Emmett–Teller [BET]) and carbon deposits on the particle surface were quantified using thermogravimetric analysis (TGA) in air. The gas composition was followed with Fourier transform infrared (FTIR) spectrometry using a home-built aerosol cell. Before releasing the exhaust gas to the hood, the nanoparticles were removed from the aerosol with a particle filter. The residence time in the reactor was varied between a few tens of milliseconds up to several tens of seconds by exchanging the quartz glass reaction tube.

Since the conversion was always very low (typically below the thousandth level), it was safe to assume that the composition of the educt gases did not change as the reaction proceeded. Therefore, details of the reaction such as the reaction orders of the different educts gases had not to be taken into account. By relating the amount of product gases to the total available surface area the turnover rate (TOR) could be directly determined, which is a measure for the production rate of product molecules per active surface site:

$$\text{TOR} = \#\ \text{product\_molecules}/(\#\ \text{active\_surf\_sites}^*\text{time}) \quad (10.2)$$

However, as will be shown below, for fast reactions only a fraction of the total surface area as measured with BET participates in the reaction, which will be called active surface in what follows.

10.4 Nickel Nanoparticles for Methane Formation

The goal of this chapter is to show the particular catalytic properties of aerosol nanoparticles produced with spark ablation (cf. insert in TEM micrograph in Fig. 10.3). Therefore, the formation of methane from carbon monoxide and hydrogen over a Ni catalyst was chosen as a simple and thoroughly studied reaction offering ample data on classical supported catalysts. This so-called methanation can be represented by [13]

$$CO + 3H_2 \rightarrow CH_4 + H_2O\,. \tag{10.3}$$

According to the volcano curve, nickel is one of the best catalysts for the methanation reaction [14]. Although the adsorption of CO on the particle surface as one of the main reaction steps is supposed to be size dependent, it was observed that the methanation was quite insensitive to a particle size below 25 nm [15]. This contradiction was explained by a so-called compensation effect in hydrogen and carbon monoxide adsorption at active sites [16]. For the overall reaction, the rate-limiting step seems to be the hydrogenation of adsorbed CH on the particle surface [17]. In systems with high external and internal diffusion characteristics (such as aerosol particles) the overall reaction rate can be approximated by the surface reaction step of $CH^*+H^* \rightarrow CH_2^*$ (all species in the adsorbed state). Since in the methanation over Ni catalysts methane is formed almost exclusively (>98%), selectivity issues have not to be considered and the on-line measured evolution of the methane concentration can directly be used to determine the overall reaction rate.

10.4.1 Comparison of Aerosol Catalysis with Classical Methanation

To compare the basic reactions characteristics of aerosol catalysis with the classical methanantion on extended catalysts, activation energies and reaction orders were determined. The reaction rate r_{CH4} for the formation of methane according to Eq. 10.3 is given by

$$r_{CH4} = k_0 \times \exp(-E_{act}/RT) \times p_{H2}{}^{\alpha H2} \times p_{CO}{}^{\alpha CO}\,, \tag{10.4}$$

where k_0 and E_{act} are the pre-exponential factor and the activation energy for methane formation, respectively, and p_{H2} and p_{CO} are the partial pressures of hydrogen and carbon monoxide, respectively. The exponents α_{H2} and α_{CO} are the reaction orders of the methanation for hydrogen and carbon monoxide, respectively.

Since α_{H2} is known to be close to unity [18], only the reaction order of carbon monoxide was determined by evaluating r_{CH4} versus p_{CO} in a double logarithmic plot. When changing the partial pressure of CO from 5250 Pa to 11,010 Pa, the reaction order changed from −0.15 to −0.52, which falls within the range published in the literature (−1 to +0.5) [18]. Also the activation energy found for aerosol catalysis with a value of 96 kJ/mol agreed well with literature values for temperatures of up to 350°C (103–109 kJ/mol) for single-crystal nickel catalysts [19, 20] and nickel foils [21].

Above 350°C the reaction rate increases with a reduced slope compared to the lower-temperature behavior, resulting in a lower activation energy of 65 kJ/mol. This effect is due to the accelerated formation rate of adsorbed carbon atoms on the catalysts surface. If the hydrogenation rate cannot keep pace with the CO dissociation rate at higher temperatures, carbon deposits build up on the particle surface. Goodman found that the formation of relatively inactive graphite on a single-crystal nickel catalyst requires a certain surface carbon level and a critical temperature of 375°C [20]. Therefore, this poisoning of the aerosol nickel particles observed above 350°C is attributed to the formation of graphite on the particle surface, as further confirmed by aerosol photoemission measurements presented later.

It can be concluded that the key features of the methanation (i.e., overall activation energy, reaction order, and poisoning effects) are similar for aerosol and supported catalysts. However, some substantial differences regarding size effects and especially turnover rates will be presented and discussed in the following section.

10.4.2 Size Effects

In the parlance of catalyst behavior, structure-sensitive and structure-insensitive reactions are differentiated [22]. For a structure-sensitive reaction the catalyst shows an optimum size, even when corrected for the surface area increase with decreasing particle size (specific surface area effect). For metallic nanoparticles the optimal diameter lies in the range of a few nanometers. The reason for the size effect is still under debate but includes quantum mechanical phenomena regarding the electronic band structure of the nanoparticles [23] and highly uncoordinated atoms at corners and edges [24]. Surprisingly, for unsupported metallic nickel nanoparticles an optimum size of about 15 nm was observed. The data are presented in Fig. 10.3. To vary the nickel diameter, the spark-ablated particles were sintered at different temperatures before being employed in the methanation. As shown in Fig. 10.3, the specific BET surface area is reduced with increasing pretreatment temperature translating to an increasing primary particle diameter d_{pp}. However, the maximum catalytic activity, given by the TOR, is observed for d_{pp} ≈ 15 nm, when the available surface area has already dropped to about 25% of the initial value. Therefore, the size effect outweighs the surface reduction by far. The phenomenon seems to be related to the adsorption characteristics of the CO molecules on the nickel particles. The adsorption of the CO molecules can occur in bridged or linear position. Bhatia et al. [25] concluded that the methanation activity is increased by effects that enhance the formation of a weakly bound carbon monoxide species. They found a maximum activity for particles with a diameter of 14.4 nm. Larger crystallite sizes favor weakly bound adsorbed species, corresponding to the linear position, as observed by Eischens et al. [26] from infrared (IR) spectroscopy.

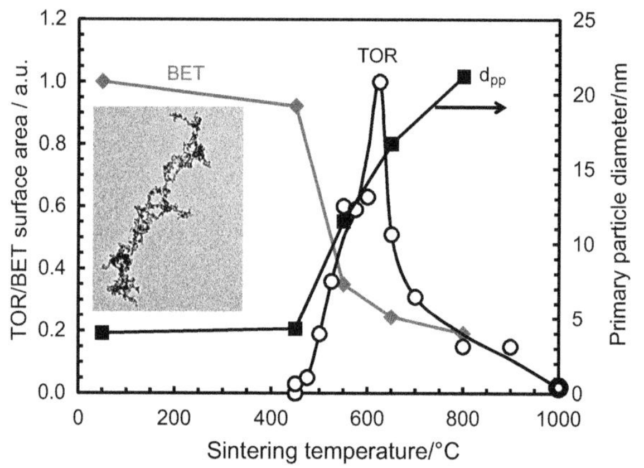

Figure 10.3 Influence of the presintering temperature of nickel nanoparticles on particle surface area (measured with BET and related to the initial value of 163 m^2/g), catalytic activity (TOR), and primary particle diameter d_{pp} (inset: TEM micrograph of Ni agglomerate).

The notion of an optimum enthalpy for the CO adsorption agrees with the principle of Sabatier, which infers that for different materials an optimal adsorption enthalpy for the dissociation of an adsorbed molecule exists. For lower adsorption enthalpies the interatomic bond is not weakened enough to be broken, and for higher adsorption enthalpies the molecule is bonded too strongly to the particle surface to desorb, occupying otherwise active sites. However, to understand the activity behavior found in Fig. 10.3, Sabatier's principle and size-dependent adsorption enthalpy have to be combined. Extended nickel surfaces exhibit adsorption enthalpies slightly on the left side of the volcano curve [27]. With decreasing crystallite size the adsorption enthalpy increases, reaches the peak of the volcano curve and drops again on the right side toward higher adsorption enthalpies, that is, smaller particles sizes. As will be shown later (cf. Section 10.4.4), the optimum size may also be influenced by the interaction with the support material.

10.4.3 Sintering and Poisoning Kinetics

As mentioned earlier, for reaction temperatures above 350°C poisoning effects were observed. At the same time, the Ni particles grow by coalescence. To investigate the kinetics of these two phenomena and their influence on the catalytic activity, time-resolved experiments were performed at 450°C. The Ni particles, which had been exposed to the methanation reaction for a defined time, were analyzed regarding their catalytic activity (FTIR measurement of CH_4 concentration) and their photoelectric activity (APES), and samples had been taken for off-line characterization of particle size (TEM) and deposited carbon (TGA). Catalytic and photoelectric activities decrease in a similar way with increasing carbon deposition (Fig. 10.4).

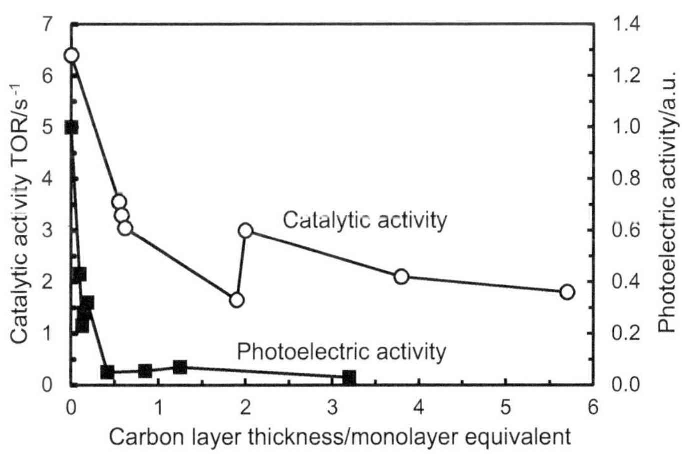

Figure 10.4 Influence of carbon deposits on catalytic and photoelectric activity of Ni nanoparticles. While the photoactivity is nearly completely suppressed at higher carbon deposits, the catalytic activity reaches a steady state.

Since no high-resolution TEM images were available, it was assumed that the amount of carbon on the Ni nanoparticles was homogeneously distributed over the nickel surface and therefore is given as monolayer equivalent. The time for one monolayer of carbon to build up was about 6 s. The observation that even for submonolayer coverage of the nickel surface with carbon a significant activity loss occurs indicates that more than one surface atom is needed to constitute an active site. In fact, Dalmon and Martin concluded that ensembles of about 13 Ni atoms are required for the methanation reaction to take place [28].

It is interesting to note that catalytic and photoelectric activity exhibit a similar behavior with respect to carbon deposition. Both processes depend on the availability of electrons: in catalysis the supplied electrons weaken bonds in the molecules so that they can dissociate [29], and in photoemission the density of electron states at the Fermi level determines the photoactivity [30]. However, due to basic differences the behavior is not identical.

In Fig. 10.5 the combined effects of poisoning and sintering are shown for two initial particle morphologies. In one case, the particles were sintered nearly to the optimum size for high-methanation activity. In the second case, the particles were taken as generated form the spark ablation, that is, highly ramified agglomerates of very small primary particles ($d_{pp} \approx 4$ nm). It can be nicely seen that the spherical particles of 19 nm do not undergo sintering but experience significant poisoning as reflected in the activity loss (TOR) and supported by additional TEM and TGA measurements [32].

In contrast, the Ni agglomerates undergo sintering and poisoning in parallel. First, the activity loss due to carbon deposition dominates until the particles have grown to the size where the optimum catalytic activity is reached (at about 15 nm). Here the TOR increases significantly (at residence times of about 1 s) before poisoning by carbon deposition takes over again for longer reaction times. Due to the aerosol state of the particles it was possible to observe these extremely fast processes without any transport limitations of the educts and without interference with the support that underlines the potential of aerosol catalysis for research studies.

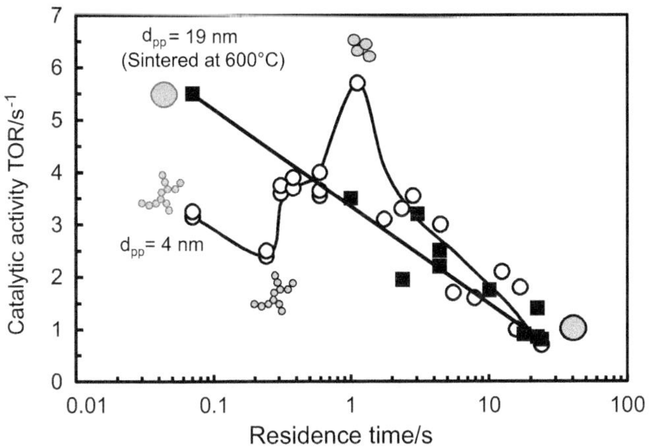

Figure 10.5 Evolution of the catalytic activity of two types of Ni nanoparticles at a reaction temperature of 450°C: (i) Ni agglomerates with primary particles of 4 nm (as produced with spark ablation) and (ii) Ni particles sintered at 600°C prior to methanation with primary particles of 19 nm. While for type (ii) particles, only poisoning by carbon deposits reduces the catalytic activity, for type (i) particles, sintering and size effects are superimposed to the poisoning. Reprinted by permission from Springer Nature Customer Service Centre GmbH: Springer Nature, *Journal of Nanoparticle Research*, Ref. [31], copyright (2003).

10.4.4 Comparison of Supported and Unsupported Methanation

Nanoparticles are usually employed in heterogeneous catalysis on support materials. To compare the results presented earlier for aerosol catalysis to supported systems, Ni nanoparticles were produced by spark ablation and conditioned, as outlined before. In addition to the experiments in the aerosol state, particles were immobilized on silica support materials by thermophoretic deposition and exposed to pulses of the educt gases at 450°C using a pulse delivery system. The pulsing process was necessary, since the hydrogen in the educt gases accelerates particle sintering, so that during the time to heat up the reactor to the required 450°C the Ni particles would have coalesced completely, making the study of size related effects impossible. To obtain the temporal evolution of the supported Ni particles, the duration of the single pulses were summed up.

First, the influence of the primary particle size on the catalytic activity of the Ni nanoparticles was studied. To exclude poisoning effects only very short contact times in the order of 100 ms were chosen. The results presented in Fig. 10.6 indicate that the optimum size for the aerosol particles is about 15 nm, while the highest activity for the supported Ni particles is observed for about 25 nm. As mentioned before, these sizes are too large for quantum confinement phenomena and may rather be related to adsorption of the CO molecules.

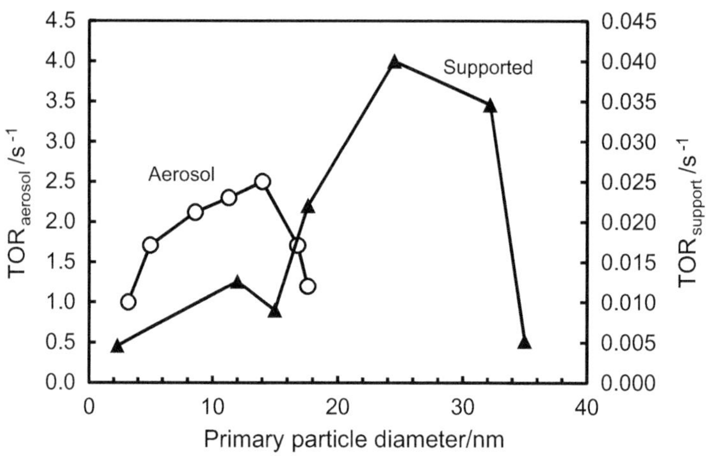

Figure 10.6 Size effects in the catalytic activity for aerosol and supported nickel nanoparticles for the methanation reaction at 450°C (please note the different scales for both systems). Reprinted by permission from Springer Nature Customer Service Centre GmbH: Springer Nature, *Journal of Nanoparticle Research*, Ref. [33], copyright (2006).

Following the discussion earlier regarding the interplay of size-dependent adsorption enthalpy and optimum adsorption enthalpy for the methanation (volcano curve), it seems that the influence of the support is to increase the adsorption enthalpy of the CO on the nickel particles. Therefore, the optimum adsorption enthalpy for supported particles is reached at a particle size larger than for the aerosol state.

Second, the temporal behavior of the catalytic activity starting with agglomerates of very small Ni nanoparticle (here about 2.2 nm) was investigated. For aerosol and supported particles a decrease of the activity is observed first due to poisoning (Fig. 10.7). Then the catalytic activity increases as a consequence of particle growth by sintering, reaching the optimum sizes given in Fig. 10.6. Finally, the poisoning takes over again, leading to a substantial loss of activity. However, sintering and poisoning are much faster in the aerosol state than on support. This time shift reflects the different transport kinetics that are much faster in the aerosol state than on support. While the poisoning depends on the formation rate of carbon, that is, the dissociation rate of CO, the sintering is driven by the availability of hydrogen that removes surface oxygen pining and promotes the coalescence between particles [34]. For both processes the aerosol reaction capitalizes on the fast transport, while the supported particles experience a transport limitation in the laminar boundary layer close to the support surface.

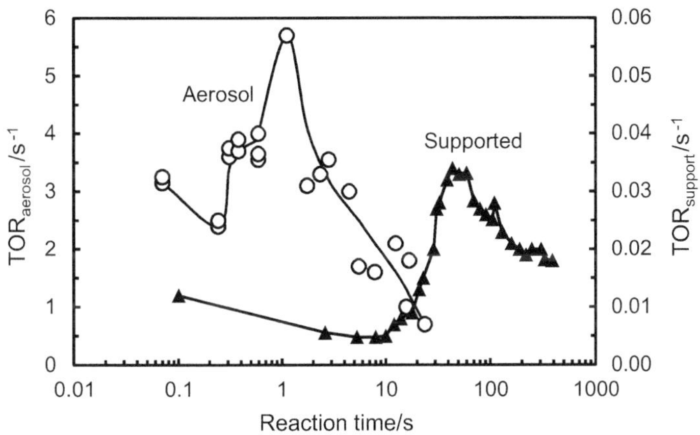

Figure 10.7 Temporal evolution of the catalytic activity for aerosol and supported nickel nanoparticles for the methanation reaction at 450°C (please note the different scales for both systems).

10.5 Platinum Nanoparticles for Hydrogen Oxidation

In contrast to the methanation discussed earlier the oxidation of hydrogen on platinum nanoparticles is a fast reaction that does not undergo poisoning. As in the case of methanation, the oxidation of hydrogen has no side reactions so that selectivity issues can be neglected. The overall reaction can be described by

$$2H_2 + O_2 \rightarrow 2H_2O. \quad (10.5)$$

Due to its highly exothermic character, the reaction already proceeds at ambient temperature. The liberation of the reaction enthalpy may increase locally the gas temperature, so the agglomerates can undergo restructuring and sintering. Therefore, it was necessary to perform the reaction at low educt concentrations of H_2 and O_2 in nitrogen (typically a few percentage points) in order to avoid thermal stress on the nanoparticles agglomerates. A typical TEM micrograph of a Pt agglomerate is shown in the inset of Fig. 10.8, showing the dendritic agglomerate structure that consists of single-digit nanoparticles.

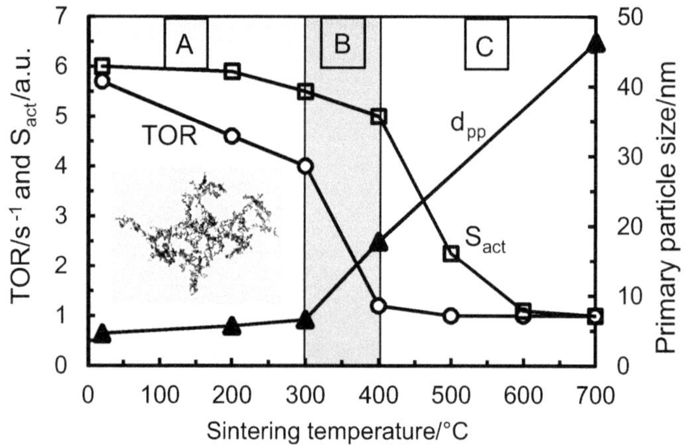

Figure 10.8 Catalytic activity related to the active surface area (given as TOR), amount of active surface area S_{act}, and primary particle size d_{pp} of Pt agglomerates as a function of the heating temperature prior to reaction. To discuss the results the figure is divided into three parts: A, B, and C (inset: TEM micrograph of Pt agglomerate).

10.5.1 Particle Morphology and Transport Kinetics

Due to extremely high number concentrations the Pt nanoparticles from spark ablation form agglomerates very quickly with dendritic structures, accommodating a large fraction of inner surface area. Considering the fast kinetics of the hydrogen oxidation, the question arises, which fraction of the total agglomerate surface contributes to the formation of water?

Since the hydrogen mobility far exceeds that of oxygen on the particle surface and in the gas phase, the inner surface becomes quickly saturated with hydrogen when adding both educt gases at the same time to the aerosol particles. Therefore, the surface of the Pt agglomerates was saturated with oxygen, that is, giving oxygen a head start, before adding the hydrogen. The reaction progress was monitored by measuring the water vapor content as products using an FTIR spectrometer with an aerosol cell. The results are shown in Fig. 10.9 as a function of the contact time that started when the hydrogen was added.

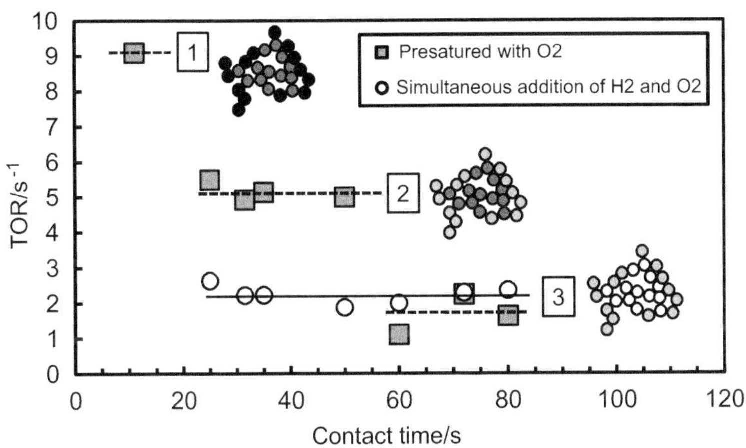

Figure 10.9 Water formation rate (given as TOR) as a function of contact time. Compared are the simultaneous addition of the educt gases and the presaturation of the Pt nanoparticles with oxygen prior to the addition of hydrogen. The time for the oxygen to saturate the particle surface was of the order of 10 s before the hydrogen was added. 1: oxygen saturated and easily accessible active surface; 2: active surface in the steady state, while the inner surface still contributes with high oxygen coverage; and 3: inner surface has been depleted for the presaturation case, and the water formation rate approaches the one of simultaneous addition.

While for the simultaneous addition of H_2 and O_2 the TOR remains constant for different contact times, the catalytic activity for the presaturated particles is initially about 4 times higher than at the end. As indicated in Fig. 10.9, in this case the exposed agglomerate surface is covered with oxygen and easily accessible for the hydrogen from the gas phase (No. 1 in Fig. 10.9). Once the exposed surface, which will be called active surface in what follows, is mostly depleted from oxygen, the reaction front continues to move into the inner part of the agglomerates by diffusion of hydrogen over the particle surface and in gas phase (No. 2 in Fig. 10.9). Finally, also the inner oxygen is consumed and only the active surface continues to produce water vapor (No. 3 in Fig 10.9). In this state the TOR is close to the one observed for the simultaneous addition of H_2 and O_2, where most of the surface area is covered with hydrogen. This is a consequence of the different adsorption behaviors of the educt gases. While oxygen needs two free neighboring surface sites for adsorption, the hydrogen can adsorb on only one free surface site. From the results in Fig. 10.9 it is obvious that about two-thirds of the total surface belongs to the inner fraction that does not contribute to the water formation for the simultaneous addition of the educt gases. As underlined by Monte Carlo simulations [35] it can be concluded that only the active surface of the Pt agglomerates takes part in the hydrogen oxidation. This indicates that catalytic activity, when related to the exposed surface area, may be used to reveal changes of the surface properties or size effects, for instance, when exposing the particles to higher temperatures, as treated in the next chapter.

10.5.2 Size Effects

The results of such tempering experiments, where the Pt agglomerates have been thermally treated before adding the educt gases, are presented in Fig. 10.8 and Fig. 10.10. To reveal the different effects, the results in Fig. 10.8 are divided into three temperature regions.

In region A, a small decrease in the catalytic activity can be related to a slight increase in the primary particle size, while the active surface remains nearly constant. However, in region B the activity drops by a factor of 4, and at the same time the primary particle size increases by a factor of 3. Again the active surface changes only slightly. Besides the indicated size dependence of the TOR, also a change of the surface properties could contribute to the activity reduction, since the agglomerate morphology was changed by tempering the particles. However, photoelectric emission measurements with APES indicated that in the temperature range between 300°C and 400°C (corresponding to d_{pp} values from 6 nm to 17 nm) work function and emission constant (cf. Eq. 10.1) remain virtually unchanged (Fig. 10.10).

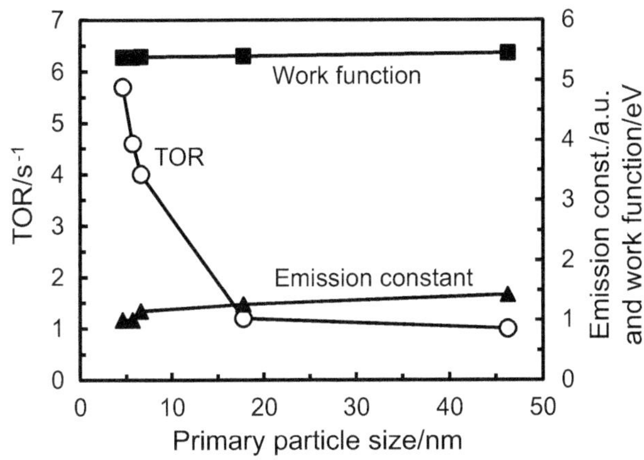

Figure 10.10 Size effects for the hydrogen oxidation on Pt nanoparticles and evolution of the photoelectric parameters (work function and emission constant) with particle size that were changed by tempering.

Therefore, no substantial changes of the particle surface of the Pt particles were introduced by tempering the agglomerates in region B. Finally, region C exhibits constant TOR although the active surface is greatly reduced and the primary particle size increases substantially. This confirms the importance of the active surface for the mass transport of the educt molecules to the agglomerates and their subsequent reactions.

Regarding the decrease of the catalytic activity with increasing temperature, a change of the active surface sites could be responsible due to either a change of the particle surface composition or a real particle size effect. On the one hand, the surface properties may change in consequence of adsorption of molecules or chemical reactions. However, the photoelectric emission measurements indicate rather minute changes of the particles surface properties with temperature (Fig. 10.10). The constancy of work function and emission constant are not a proof but at least a good indicator for stable surface conditions. On the other hand, with increasing particle size the proportion of highly oriented surface atoms increases and the fraction of exposed, low coordinated surface sites decreases. Especially the amount of low coordinated surface atoms may induce amazing catalytic activity effects [36]. Therefore, it is concluded that the observed size dependency of the catalytic activity of Pt nanoparticles in the hydrogen oxidation is a real quantum confinement effect in the size range typical for metallic nanoparticles [37].

10.6 Stabilization Strategies for Spark Ablation Catalyst Nanoparticles by Aerosol Coating

As outlined before, spark ablation leads to high-purity nanoparticles with diameters typically in the single-digit range. Even when agglomerated, these nanoparticles exhibit a high surface area with a corresponding high catalytic activity. However, at higher reaction temperatures the particles tend to reduce their surface area by coalescence. This effect is further aggravated by the melting point reduction induced by the high surface curvature of small particles [2]. Therefore, from a stability point of view additional measures have to be taken to preserve the particle properties with regard to sintering resistance, but also in some applications with respect to mechanical rigidity. While in the case of catalysts stabilization, many attempts have been undertaken to improve the high-temperature behavior, such as caging the active metal components using metal organic frameworks (MOFs) [38] or immobilization by employing support material with textured surfaces [39], interface passivation was used only sparsely [40]. Here, a new room-temperature aerosol method is introduced that can be applied to deposit tiny amounts of silica in the necks between nanoparticles. This technique is based on the combination of vapor condensation and plasma activation [41].

10.6.1 New Aerosol-Coating Technique

The schematic setup of this new coating technique is outlined in Fig. 10.11. The aerosol particles originating from spark ablation are mixed with tetraethyl orthosilicate (TEOS) vapor and with the plasma products that are produced by passing air through a dielectric barrier discharge (DBD). In the subsequent coating tube the TEOS precursor molecules react with the plasma species and form a condensable vapor, which forms a liquid film on the particles.

During the residence time in the order of minutes at room temperature this film solidifies, leading to a SiO_2 shell. Finally, the particles are analyzed with on-line methods such as condensation particle counter (CPC; number concentration) and SMPS (size distribution), and samples are taken for off-line characterization including TEM/energy-dispersive X-ray spectroscopy (EDX) (morphology and composition), FTIR (bonding types), and filtration (mass concentration).

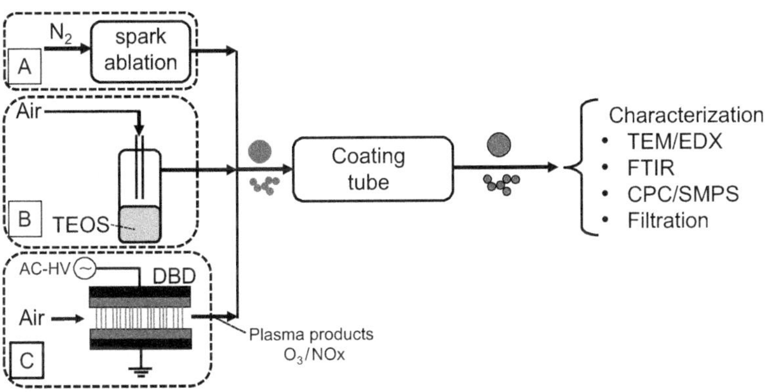

Figure 10.11 Schematic diagram of the experimental setup for coating of particles with a SiO_2 shell in an aerosol process at ambient temperature: (A) particle synthesis by spark ablation (possibly followed by sintering furnace), (B) saturation of air stream with the precursor vapor by passing a bubbler containing the liquid precursor (TEOS = tetraethyl orthosilicate), and (C) production of reactive species in a dielectric barrier discharge (DBD) plasma. After mixing the three gas streams, the particle coating proceeds in the subsequent chamber, where the reactive DBD species transform the TEOS vapor molecules into a condensable species, which condenses onto the particles as liquid layer and then solidifies in further chemical reactions.

To gain a better understanding of the coating process, aerosol Pt nanoparticles were sintered in a tube furnace and a monodisperse fraction was classified with a differential mobility analyzer (DMA), which was then exposed to the coating procedure described before. In this way, the uncoated particles were spherical and of the same size, which facilitated the evaluation of TEM micrographs regarding the thickness of the SiO_2 coating. The coated particles were also analyzed with SMPS. The determined coating volume as a function of the TEOS concentration is shown in Fig. 10.12.

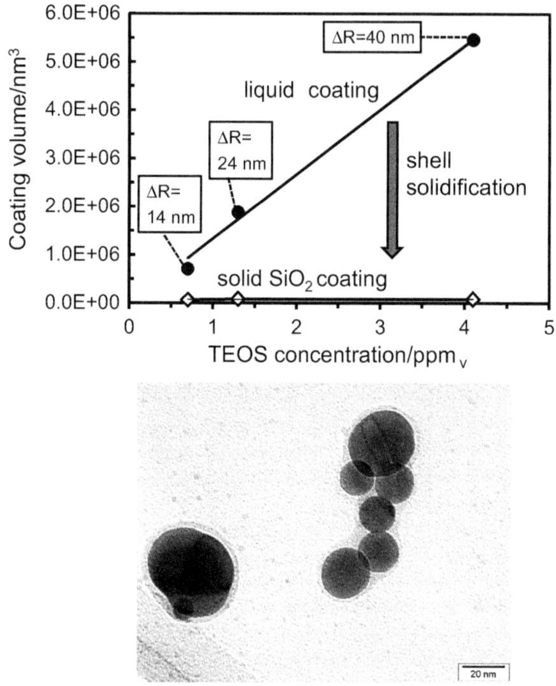

Figure 10.12 (Top) Volume of the coating shell as a liquid film (with a residence time of 83 s in the coating tube) and after solidification (showing thin remaining SiO_2 coatings of 2 to 3 nm thickness). (Bottom) TEM micrograph of coated Pt nanoparticles: spheres and agglomerate (scale bar = 20 nm).

Since the TEOS precursor transforms into the condensable species it may be expected that the volume of the condensed molecules increases with the TEOS concentration, as confirmed with the SMPS measurements (Fig. 10.12). However, the TEM analysis delivers much thinner coatings than obtained with SMPS, indicating that during the residence time only a fraction of the film has solidified. The rest, being still liquid, evaporates in the ultrahigh vacuum of the TEM. Although the reaction mechanisms during the coating process are not understood in detail, the TEM micrographs of agglomerates indicate that condensation followed by solidification plays an important role. The condensation allows in particular for the complete coating, even of agglomerates. However, at high TEOS concentrations also homogeneous nucleation takes place, leading to individual silica particles. On the other hand, at very low concentrations, condensation starts at the necks between the particles induced by capillary condensation.

10.6.2 Effect of Coating on Mechanical Stability

To evaluate the influence of the coating on mechanical stability, size-classified agglomerates were impacted onto a TEM grid in a low-pressure impactor and the projected agglomerate area was determined from micrograph analysis as a function of the impaction velocity (Fig. 10.13). For all coating degrees, the measured projection area corresponds well to the selected mobility, when particles were deposited by diffusion or at low impaction velocities.

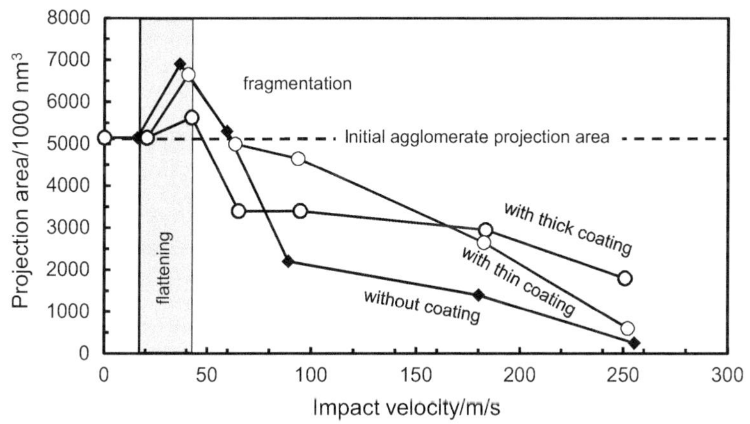

Figure 10.13 Influence of the impaction velocity on the projection area of Pt agglomerates for agglomerates of different coating thickness: after an initial flattening the agglomerates start to fragment (thin coating: Si/Pt = 0.39; thick coating: Si/Pt = 0.84). Reprinted from Ref. [42], Copyright (2018), with permission from Elsevier.

However, at higher impaction velocities the agglomerates exhibit flattening due to rearrangement of the dendritic branches that protrude from the TEM grid at lower velocities. At even higher velocities fragmentation sets in. The evolution of the projection area with velocity depends on the degree of coating. Between 16 and 40 m/s the increase of the projection area is more pronounced for the uncoated particles compared to thinly coated (Si/Pt = 0.39) and thickly coated agglomerates (Si/Pt = 0.84). This indicates that the thicker coating improved the rigidness of the protruding branches. Many of the branches seemed no longer flexible enough to flatten to the TEM grid and instead stood upright until the impaction velocity was high enough to fragment the modified structure. However, the fragmentation at higher velocities was reduced in comparison to the uncoated agglomerates for both coated agglomerates. While the projection area of the thinly coated agglomerates followed a linear downwards trend, the thicker coating resulted in a flatter progression after an initial decrease between 40 and 65 m/s, indicating strong cohesion between the remaining agglomerate parts. While the thicker coating increased the rigidness of the branches, as shown by the lower degree of flattening, the branches are presumably still the easiest parts to break. When the impaction velocity crosses an energy threshold, these parts might fragment directly instead of restructuring first, due to their coating-restrained flexibility. In contrast, the thinner coating (Si/Pt = 0.39) seems to slow down the fragmentation in comparison to the uncoated particles, but not prevent it completely. As conclusion, the deposition of SiO_2 on agglomerates may significantly improve their mechanical stability. It will be the content of the next chapter to show how this coating influences other agglomerate properties such as thermal stability and surface accessibility [42].

10.6.3 Effect of Coating on Photoelectric Activity

As shown earlier, photoemission is a very sensitive tool for monitoring changes of the particle surface. It was shown that in some cases, as for methanation, catalytic activity and photoactivity may be correlated (Fig. 10.14). Therefore, photoactivity was chosen here as an indicator for the effects of the coating on the surface accessibility and its sensitivity to elevated temperatures. In this case, not a hermetic coating was intended but mostly coating in the necks. Therefore, a low TEOS concentration of 0.3 ppmv and a residence time of 83 s were chosen. The results are shown in Fig. 10.14. The sintering of Pt agglomerates to spheres leads to a reduction of the photoactivity to 30% of the initial value. This reflects the loss of total particle surface, which can be attributed, on the one hand, to particle losses when passing the sintering furnace and, on the other hand, to the surface reduction of the individual particles due to coalescence. As shown in Fig. 10.14, even low TEOS concentrations result in a hermetic SiO_2 coating of the spheres, leading to a complete suppression of photoactivity. The situation is different when coating agglomerates.

For agglomerates, the silica coating reduces the photoactivity by about 5%. At elevated temperatures the photoactivity is reduced in both cases, however, to 70% at 500°C for the coated agglomerates, while the uncoated particles exhibit a remaining photoactivity of only 40% at 500°C. Up to 400°C the coated agglomerates retain even 90% of the initial photoactivity, clearly underlying the stabilization effect of the silica coating, which is mostly deposited in the necks, as observed with TEM micrographs.

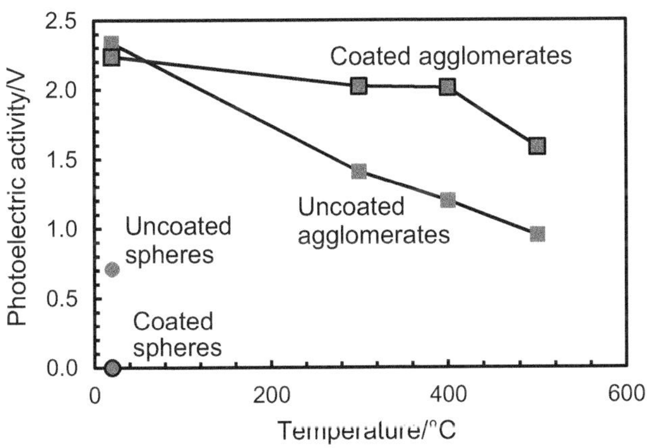

Figure 10.14 Reduction of the photoelectric activity with increasing temperature: comparison of coated and uncoated particles for agglomerates and spherical particles.

10.7 Spark Ablation Nanoparticles as Photoinitiator for Polymerization

The approach to tailor the surface of aerosol particles by combining condensation and subsequent reactions, as shown for silica coatings earlier, can also be extended to polymer coatings. In the following example a special property of titania nanoparticles is employed, that is, the creation of electron–hole pairs under UV irradiation, to start the polymerization on the particle surface. This feature is common to all semiconducting particles and may be extended to plasmonic nanoparticles, as mentioned at the end of this chapter.

10.7.1 Particle-Induced Aerosol Photopolymerization

The experimental setup for the polymer coating is shown in Fig. 10.15. Besides the particle generator, which was here spark ablation of titanium oxidized by trace amounts of oxygen in the nitrogen carrier gas, a condensable monomer vapor is produced by passing nitrogen through a bubbler containing the liquid monomer (here butyl acrylate, BuA).

By adjusting the temperature of the bubbler the saturation of the gas can be controlled. To avoid premature condensation, the line of the monomer gas stream is heated (typically to about 60°C) before mixing with the aerosol stream. In the mixer due to the temperature reduction the monomer vapor condenses onto the particles, for agglomerates beginning in the necks due to capillary condensation.

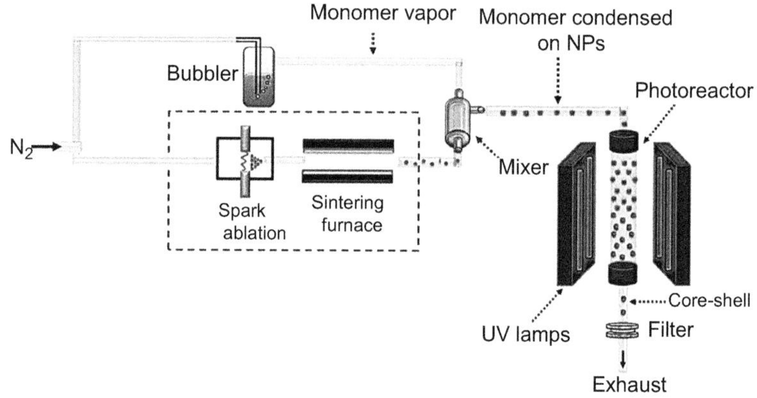

Figure 10.15 Schematic diagram of the experimental setup for coating nanoparticles with a hermetic polymer shell: A monomer vapor is mixed with the aerosol where the monomer condenses onto the particles, and in the subsequent UV chamber the polymerization on the particles is initiated by creating electron–hole pairs. From Ref. [43]. Reproduced by permission of The Royal Society of Chemistry.

To induce the polymerization step, the aerosol passes an irradiation chamber where the particles are exposed to UV light, which creates electron–hole pairs. Particularly the holes serve as starters for the polymerization reaction [44–46]. This aerosol method has two advantages: (i) the holes start the polymerization so that no additional species such as initiators have to be added, and (ii) since the polymerization starts all over the particle surface, a hermetic coating, tightly fitting the particle contour, is achieved, which is often not the case when adding starters from outside [47].

10.7.2 Improved Nanoparticle Dispersability

One problem for the integration of metal and metal oxide nanoparticles into organic matrices is their poor dispersability in organic solvents. Here the above-shown process can be used to coat TiO_2 nanoparticles with a hydrophobic polymer shell. To evaluate the success of this method, bare TiO_2 and TiO_2-PBuA nanoparticles were dispersed in cyclohexane as an organic solvent. The macroscopic observations of the hydrophobic shell effect are shown in Fig. 10.16.

The change in hydrophilicity of the core makes TiO_2-PBuA nanoparticles stably dispersed in cyclohexane, while bare TiO_2 nanoparticles precipitate in cyclohexane after a few minutes. This macroscopic result corroborates microscopic analyses of the shell properties by FTIR, TEM, and X-ray photoelectron spectroscopy (XPS), pointing to a different functionality of the coated nanoparticles [48].

Figure 10.16 TiO_2-PBuA nanohybrids and bare TiO_2 nanoparticle dispersions in cyclohexan (left) immediately after dispersing using an ultrasonic bath and (right) after a few minutes. From Ref. [43]. Reproduced by permission of The Royal Society of Chemistry.

10.7.3 Enhanced Photocatalytic Activity

The polymer coating on the TiO_2 nanoparticles may be taken a step further. By calcination of the coated powder in a temperature-programmed reaction the PBuA shell is transformed into graphite (first heated in air up to 240°C and then heated to the end temperature in nitrogen at a rate of 5°C/min). When comparing the end temperatures of 450°C, 500°C, and 600°C it was found that the highest degree of graphitization was achieved for 600°C. It was observed in the literature that a graphitic layer increases the absorption capacity and extends light absorption into the visible range [49]. Especially for TiO_2 nanoparticles it was found that the graphite shell also shifts the phase change from anatase to rutile to higher temperatures [50], similar to the observations made with SiO_2 coatings on titania [40].

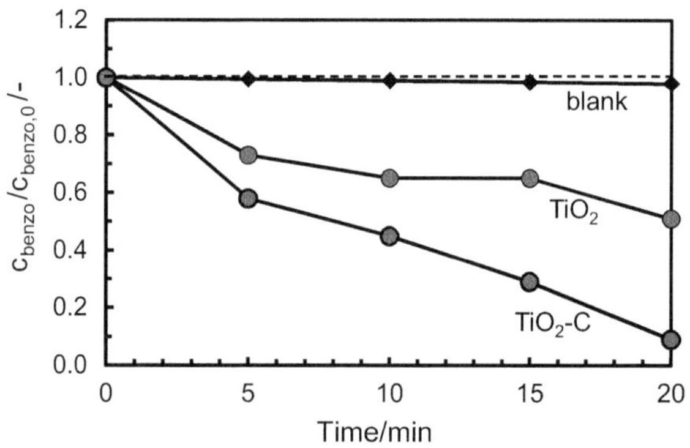

Figure 10.17 Photocatalytic degradation of benzoic acid without particles (blank), for bare titania nanoparticles (TiO_2) and for graphite-coated titania nanoparticles (TiO_2-C).

To evaluate the photoactivity of the graphite-coated titania nanoparticles (TiO_2-C) the photo-oxidative reduction of benzoic acid under UV irradiation was chosen. The decrease of the IR absorption of the C-H vibration was used as an indicator for the benzoic acid concentration. As shown in Fig. 10.17 after 20 min of irradiation benzoic acid concentration was reduced by 90% for TiO_2-C, while for bare TiO_2 only a reduction of 50% was achieved. The mechanisms of the enhanced photoactivity are based on the migration of the photoinduced electrons to the graphite/TiO_2 interface where they are absorbed, preventing recombination with holes [49]. Further investigations will be necessary to estimate the photoactivity enhancement due to extended absorption in the visible range [49].

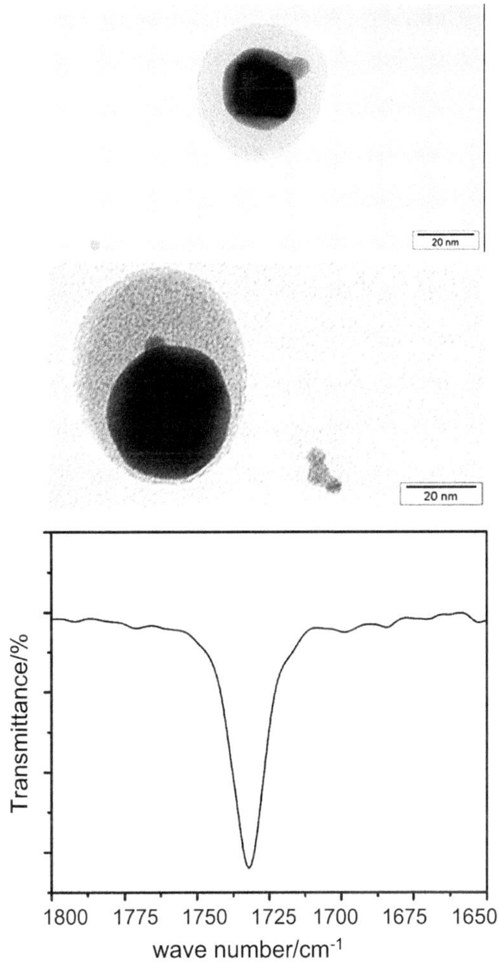

Figure 10.18 (Top) TEM micrographs of Au nanoparticles coated with PBuA. (Bottom) FTIR spectrum of the Au-PBuA nanoparticles, indicating the polymer shell formation.

As a final remark, polymerization cannot only be started with the creation of electron–hole pairs, as shown here for semiconducting nanoparticles, but also with metal nanoparticles using their plasmonic resonance. First experiments confirmed this concept. However, since the separation of the surface charge does not take place evenly over the complete particle surface, the coating may exhibit some asymmetry, as shown in TEM micrographs for Au nanoparticles produced by spark ablation (Fig. 10.18). Further studies will be required to elucidate the mechanisms of polymerization for nanoparticles with plasmonic resonance behavior.

10.8 Conclusions

It has been shown that spark ablation can be used as a versatile and solvent-free method to produce very well-defined metal and metal oxide nanoparticles with relatively narrow size distribution. Aerosol catalysis can be used to study basic properties and to optimize the particle size and composition of spark-produced particles. Agglomerates of these particles exhibit the desired large specific surface areas but require stabilization, and strategies to serve this purpose have been presented. These findings, as well as the results on initiating polymerization coating by particles from spark ablation, are meant to demonstrate the large potential of this nanoparticle production method. In view of applications, it is important to mention that spark ablation not only performs well in lab-scale experiments but also has potential with regard to scale-up [51, 52]. In addition, nanoparticles of different material combinations and shapes can be produced, such as metals, bimetals, alloys, and semiconducting particles. Concerning the ease of mixing materials, please refer to Chapter 5. A particularly interesting feature for supported catalysts is the strong adhesion of single-digit nanoparticles to substrates, as recently shown [53]. In summary, catalysts production by spark ablation is considered an emerging field with a high potential for fundamental studies and technical applications.

Acknowledgments

The valuable contributions of many colleagues are highly appreciated. Thanks go especially to PD Dr. habil. Martin Seipenbusch, Prof. Dr. Gerhard Kasper, M.Sc. Patrick Post, M.Sc. Malte Bierwirth, Dr. Jalal Poostforooshan, Dr. Masoom Shaban, and Dipl.-Min. Peggy Knospe.

The financial support of several projects by the German Research Foundation (Deutsche Forschungsgemeinschaft DFG) under grants WE 2331-10/12/15/18/19 is highly appreciated.

References

1. Fechete, I., Wang, Y. and Védrine, J. C. (2012). The past, present and future of heterogeneous catalysis, *Catal. Today*, **189**, pp. 2–27.
2. Buffat, Ph. and Borel, J.-P. (1976). Size effect on the melting temperature of gold particles, *Phys. Rev. A*, **13**, pp. 2287–2298.
3. Chen, L., Li, Y. Zhang, X., Zhang, Q., Wang, T. and Ma, L. (2014). Mechanistic insights into the effects of support on the reaction pathway for aqueous-phase hydrogenation of carboxylic acid over the supported Ru catalysts, *Appl. Catal. A*, **478**, pp. 117–128.
4. Weber, A. P., Seipenbusch, M., Thanner, Ch. and Kasper, G. (1999). Aerosol catalysis on nickel nanoparticles, *J. Nanopart. Res.*, **1**, pp. 253–265.
5. Tabrizi, N. S., Ullmann, M., Vons, V. A., Lafont, U. and Schmidt-Ott, A. (2009). Generation of nanoparticles by spark discharge, *J. Nanopart. Res.*, **11**, pp. 315–332.
6. Feng, J., Biskos, G. and Schmidt-Ott, A. (2015). Towards industrial scale synthesis of ultrapure singlet nanoparticles with controllable sizes in a continuous gas-phase process, *Sci. Rep.*, **5**, p. 15788, doi:10.1038/srep15788.
7. Vons, V. A. (2010). Spark discharge generated nanoparticles for hydrogen storage applications, PhD Thesis, TU Delft.
8. Seipenbusch, M., Weber, A. P., Schiel, A. and Kasper, G. (2003). Influence of the carrier gas atmosphere on restructuring and sintering kinetics of nickel and platinum aerosol nanoparticle agglomerates, *J. Aerosol Sci.*, **34**, pp. 1699–1709.
9. Feng, J. (2016). Scalable spark ablation synthesis of nanoparticles, PhD Thesis, TU Delft, ISBN: 9789461867346.
10. Seipenbusch, M., Toneva, P., Peukert, W. and Weber, A. P. (2007). Impact fragmentation metal nanoparticle agglomerates, *Part. Part. Syst. Char.*, **24**, pp. 193–200.
11. Burtscher, H., Scherrer, L., Siegmann, H.-C., Schmidt-Ott, A. and Federer, B. (1982). Probing aerosols by photoelectric charging, *J. Appl. Phys.*, **53**, pp. 3787–3791.

12. Leonardi, A. (1991). Feinste Schwebeteilchen aus Dieselmotoren, PhD thesis, ETH Zurich No. 9515, Switzerland.

13. Kelley, R. D. and Goodman, D. W. (1982). Catalytic methanation over single crystal nickel and ruthenium: reaction kinetics on different crystal planes and the correlation of surface carbide concentration with reaction rate, *Surf. Sci.*, **123**, pp. L743–L749.

14. Balandin, A. A. (1969). Modern state of the multiplet theory of heterogeneous catalysis, *Adv. Catal. Rel. Sub.*, **19**, pp. 1–210.

15. Dalmon, J. A. and Martin, G. A. (1983). The kinetics and mechanism of carbon monoxide methanation over silica-supported nickel catalysts, *J. Catal.*, **84**, pp. 45–54.

16. Tahri, A., Amariglio, A. and Amariglio, H. (1994). Thermodesorption et thermoreactions avec l'hydrogene du monoxyde de carbone adsorbe sur un catalyseur Au nickel, *J. Chim. Phys.*, **158**, pp. 1490–1506.

17. Alstrup, I. (1995). On the kinetics of CO methanation on nickel surfaces, *J. Catal.*, **151**, pp. 216–225.

18. Klose, J. and Baerns, M. (1984). Kinetics of the methanation of carbon monoxide on an alumina-supported nickel catalyst, *J. Catal.*, **85**, p. 105

19. Vannice, M. A. (1976). The catalytic synthesis of hydrocarbons from carbon monoxide and hydrogen, *Catal. Rev. Sci. Eng.*, **14**, pp. 153–191.

20. Goodman, D. W., Kelley, R. D., Madey, T. E. and Yates, Jr., J. T. (1980). Kinetics of the hydrogenation of CO over a single crystal nickel catalyst, *J. Catal.*, **63**, pp. 226–234.

21. Polizzotti, R. S. and Schwarz, J. A. (1982). Hydrogenation of CO to methane: kinetic studies on polycrystalline nickel foils, *J. Catal.*, **77**, pp. 1–15.

22. Che, M. and Bennett, C. O. (1989). The influence of particle size on catalytic properties of supported metals, *Adv. Catal.*, **36**, pp. 55–172.

23. Valden, M., Pak, S., Lai, X. and Goodman, D. W. (1998). Structure sensitivity of CO oxidation over model Au/TiO_2 catalysts, *Catal. Lett.*, **56**, pp. 7–10.

24. Lemire, C., Meyer, R., Shaikhutdinov, S. K. and Freund, H.-J. (2004). CO adsorption on oxide supported gold: from small clusters to monolayer islands and three-dimensional nanoparticles, *Surf. Sci.*, **552**, pp. 27–34.

25. Bhatia, S., Bakhshi, N. N. and Mathews, J. F. (1978). Characterization and methanation activity of supported nickel catalysts, *Can. J. Chem. Eng.*, **56**, pp. 575–581.

26. Eischens, R. P., Francis, S. A. and Pliskin, W. A. (1956). The effect of surface coverage on the spectra of chemisorbed CO, *Adv. Catal.*, **60**, pp. 194–201.

27. Medford, A. J., Vojvodic, A., Hummelshoj, J. S., Voss, J., Abild-Petersen, F. Studt. F., Bligaard, T., Nilsson, A. and Norskov, J. K. (2015). From the Sabatier principle to a predictive theory of transition-metal heterogeneous catalysis, *J. Catal.*, **328**, pp. 36–42.

28. Dalmon, J. A. and Martin, G. A. (1981). *Proc. 7th Int. Congr. Catal.*, Tokyo, Japan, p. 402.

29. Thomas, J. M. and Thomas, W. J. (1997). *Principles and Practice of Heterogeneous Catalysis*, VCH Verlagsgesellschaft mbH, Weinheim.

30. Müller, U., Ammann, M., Burtscher, H. and Schmidt-Ott, A. (1991). Photoemission from clean and oxygen-covered ultrafine nickel particles, *Phys. Rev. B*, **44**, pp. 8284–8287.

31. Weber, A. P., Seipenbusch, M. and Kasper, G. (2003). Size effects in the catalytic activity of unsupported metallic nanoparticles, *J. Nanoparticle Res.*, **5**, pp. 293–298.

32. Seipenbusch, M. (2003). Katalytische und photoelektrische Aktivität gasgetragener Nanopartikel-Agglomerate, PhD thesis, University of Karlsruhe (KIT).

33. Weber, A. P., Davoodi, P., Seipenbusch, M. and Kasper, G. (2006). Catalytic behavior of nickel nanoparticles: gasborne vs. supported state, *J. Nanoparticle Res.*, **8**, pp. 445–453.

34. Wu, N. L. (2001). Sol-gel synthesis and characterization of metal oxide nanocrystallites, *Proc. of the sixth WCCE*, Melbourne.

35. Seipenbusch, M., Binnig, J., Heim, M., Weber, A. P. and Kasper, G. (2002). Aerosol catalysis: the influence of particle structure on the catalytic activity of platinum nanoparticle agglomerates, *Helv. Chim. Acta*, **84**, pp. 3686–3701.

36. Lemire, C., Meyer, R., Shaikhutdinov, S. K. and Freund, H.-J. (2004). Do quantum size effects control CO adsorption on gold nanoparticles? *Angew. Chem. Int. Ed.*, **43**, pp. 118–121.

37. El-Sayed, M. A. (2001). Some interesting properties of metals confined in time and nanometer space of different shapes, *Acc. Chem. Res.*, **34**, pp. 257–264.

38. Dhakshinamoorthy, A. and Garcia, H. (2012). Catalysis by metal nanoparticles embedded on metla-organic frameworks, *Chem. Sco. Rev.*, **41**, pp. 5262–5284.

39. Stevenson, S. A., Dumesic, J. A., Baker, R. T. and Ruckenstein, E. (1987). *Metal-Support Interactions in Catalysis, Sintering and Redispersion*, Van Nostrand Reinhold, New York.

40. Qi, F., Moiseev, A., Deubener, J. and Weber, A. (2011). Thermostable photocatalytically active TiO_2 anatase nanoparticles, *J. Nanopart. Res.*, **13**, pp. 1325–1334.

41. Post, P., Jidenko, N., Weber, A. P. and Borra, J.-P. (2016). Post-plasma SiOx coatings of metal and meto oxide nanoparticles for enhanced thermal stability and tunable photoactivity applications, *Nanomaterials*, **6**(5), p. 91.

42. Post, P., Bierwirth, M. and Weber, A. P. (2018). Mechanical stability measurements of surface modified nanoparticle agglomerates, *J. Aerosol Sci.*, **126**, pp. 33–46.

43. Shaban, M., Poostforooshan, J. and Weber, A. P. (2017). Surface-initiated polymerization on unmodified inorganic semiconductor nanoparticles via surfactant-free aerosol-based synthesis toward core-shell nanohybrids with tunable shell thickness, *J. Mater. Chem. A*, **5**, pp. 18651–18663.

44. Stroyuk, A. L., Sobran, I. V. and Kuchmiy, S. Y. (2007). Photoinitiation of acrylamide polymerization by Fe_2O_3 nanoparticles, *J. Photochem. Photobiol., A*, **192**, pp. 98–104.

45. Dadashi-Silab, S., Asiri, A. M., Khan, S. B., Alamry, K. A. and Yagci, Y. (2014). Semiconductor nanoparticles for photoinitiation of Free radical polymerization in aqueous and organic media, *J. Polym. Sci., Part A: Polym. Chem.*, **52**, pp. 1500–1507.

46. Dong, C. and Ni, X. (2004). The photopolymerization and characterization of methyl methacrylate initiated by nanosized titanium dioxide, *J. Macromol. Sci. A*, **41**, pp. 547–563.

47. Poostforooshan, J., Rennecke, S., Gensch, M., Beuermann, S., Brunotte, G.-P., Ziegmann, G. and Weber, A. P. (2014). Aerosol process fort he in siut coating of nanoparticles with a polymer shell, *Aerosol Sci. Technol.*, **48**, pp. 1111–1122.

48. Shaban, M., Poostforooshan, J. and Weber, A. P. (2017). Surface-initiated polymerization on unmodified inorganic semiconductor nanoparticles via surfactant-free aerosol-based synthesis toward core-shell nanohybrids with tunable shell thickness, *J. Mater. Chem. A*, **5**, pp. 18651–18663.

49. Zhang, L., Fu, H. and Zhu, Y. (2008). Efficient TiO_2 photocatalysts from surface hybridization of TiO_2 particles with graphite-like carbon, *Adv. Funct. Mater.*, **18**, pp. 2180–2209.

50. Inagaki, M., Kojin, F., Tryba, B. and Toyoda, M. (2005). Carbon-coated anatase: the role of the carbon layer for photocatalytic performance, *Carbon*, **43**, pp. 1652–1659.

51. Biskos, G., Vons, V. A., Yurteri, C. U. and Schmidt-Ott, A. (2008). Generation and sizing of particles for aerosol-based nanotechnology, *KONA Powder Part. J.*, **26**, pp. 13–35.

52. Pfeiffer, T. V., Feng, J. and Schmidt-Ott, A. (2014). New developments in spark production of nanoparticles, *Adv. Powder Technol.*, **25**, pp. 56–70.

53. Feng, J., Hontanon, E., Blanes, M., Meyer, J., Guo, X., Santos, L., Paltrinieri, L., Ramlawi, N., de Smet, L. C. P. M., Nischl, H., Kruis, F. E., Schmidt-Ott, A. and Biskos, G. (2016). Scalable and environmental benign process for smart textile nanofinishing, *ACS Appl. Mater. Interface*, **8**, pp. 14756–14765.

Chapter 11

Nanoparticles and Nanoparticle-Based Materials Produced by Spark Ablation for Environmental Gas Sensors

George Biskos
Energy Environment and Water Research Centre, The Cyprus Institute, Nicosia 2121, Cyprus
Faculty of Civil Engineering and Geosciences, Delft University of Technology, Delft 2628CN, The Netherlands
g.biskos@cyi.ac.cy; g.biskos@tudelft.nl

This chapter describes how chemiresistive gas sensors can benefit from aerosol-based techniques—and in particular from nanoparticle synthesis by spark ablation—for synthesizing their sensing materials. First it gives a brief description and highlights the unique features of using spark ablation for synthesizing building-block nanoparticles. Then it discusses the operating principles and several characteristic features of chemiresistive gas sensors for use in air quality monitoring, and how these can be attributed by using spark ablation in combination with a number of aerosol nanoparticle deposition methods to fabricate their sensing elements (i.e., the

Spark Ablation: Building Blocks for Nanotechnology
Edited by Andreas Schmidt-Ott

ISBN 978-981-4800-82-2 (Hardcover), 978-0-367-81709-1 (eBook)
www.jennystanford.com

chemiresistors). Distinction is made between metal oxide and metal nanoparticle–based gas sensors, as the role of nanoparticles in each category differs substantially. The chapter closes by highlighting the most important conclusions and the advantages of using the techniques described for synthesizing materials for environmental gas sensors.

11.1 Introduction

Gas sensors are widely used in our everyday lives as over the years they have become standard practices for many industrial processes [1], the automotive and aerospace sector [2], the agriculture and food industry [3], as well as in medicine and healthcare [4]. This huge variety of applications supports a ca. $2 billion market (estimation for 2017; [5]), which is continuously growing as new technologies emerge to meet increasing demands.

At the same time, air pollution has become a global threat for human health and the environment because it can increase the risk for respiratory and cardiovascular diseases and take part in processes leading to acidic rain and photochemical smog [6–9]. Although gas sensors that could, in principle, be used to monitor atmospheric gaseous pollutants (e.g., CO, SO_2, and NOx) are already on the market, their high limits of detection (which are typically in the ppm range) prohibit direct use in air quality monitoring. What makes application of these sensors for environmental monitoring even more challenging is that myriad compounds are present in the atmospheric environment at the ppb range, which warrants for sensors of high selectivity and low cross-sensitivity that go beyond existing standards. As a result, the development of gas sensors for environmental monitoring still remains a challenge, and an area where additional research efforts are required for delivering systems that can meet requirements. As argued in the rest of this chapter, these efforts can benefit from bottom-up synthesis methods used in aerosol-based nanotechnology, and particularly from nanoparticle synthesis by spark ablation.

Gas sensors can be classified based on their operating principles as acoustic wave devices [10], quartz microbalances [11], calorimetric sensors (also referred to as anemometers or pellistors)

[12], sensors using electrochemical cells [13], sensors based on field-effect transistors [14], and chemiresistive sensors [15]. Among those, the latter are most commonly used mainly due to their simple way of manufacturing and associated low cost.

Chemiresistive sensors rely on changes of the resistance—or the conductivity—of their sensing material (i.e., the chemoresistor) upon exposure to a target gas. The effect was first reported by Bratain and Bardeen [16] in the 1950s for germanium, before it was observed to take places also with a number of metal oxides [17–19]. Soon after these reports, chemiresistors were patented by Taguchi [20] and brought to the market by Figaro Engineering, which is currently one of the major manufacturers of gas sensors.

Significant research efforts for improving the sensitivity, selectivity, stability, as well as the response and recovery time of chemiresistive sensors have been made since they were first introduced on the market. Despite the progress made thus far, a fundamental understanding of the mechanisms that alter the resistance of chemiresistors upon exposure to specific gases still has some important gaps [21]. At the same time, new fabrication methods that can attribute unique features to these sensors and help to understand the underlying physical/chemical mechanisms describing their performance still need to be developed. In this respect, novel methods of nanotechnology, and particularly bottom-up synthesis approaches, are expected to have a great impact in fabricating versatile nanomaterials of well-defined composition and structure that can be used as chemiresistors for the next generation of sensitive and selective sensors.

Chemiresistors consists of a thin layer deposited over an insulating substrate, with electrodes attached to it for measuring its resistance and, for some types, heating them. The sensitivity of the sensors is proportional to their surface-to-volume ratio, so the thinner the films (and thus the higher their surface-to-volume ratio), the higher their sensitivity. An elegant way of increasing the surface-to-volume ratio even further is by making the films porous and nanostructured. Bottom-up synthesis methods are much more efficient for achieving such structures compared to top-down techniques. For example, nanostructured materials composed of very small particles (i.e., particles having sizes smaller than ca. 20 nm) lead to higher sensitivities as a result of their high surface-to-volume ratio. Producing first the nanoparticle building blocks for these materials in a controlled, yet flexible, way regarding their size and chemical composition and using a subsequent step to deposit them on insulating surfaces to form nanostructured layers are therefore elements of fabrication techniques that are highly required in this field.

This chapter provides a brief overview of nanoparticle synthesis by spark ablation, which, combined with a number aerosol-based deposition methods, provide processes that are surprisingly versatile and powerful for designing nanostructured chemiresistors with tunable properties. Emphasis is given to the unique features of these methods and how they can address existing challenges both toward the fundamental understanding of the processes defining their performance, as well as for the fabrication of sensors at industrial scales.

11.2 Nanoparticle/Nanomaterial Synthesis by Spark Ablation and Characteristic Features

This section provides an overview of nanomaterial synthesis using nanoparticle building blocks produced by spark ablation. First, it describes the operating principle of spark discharge nanoparticle generators and the unique features of the technique for building chemiresistors. Second, it gives a brief overview of the methods for aerosol nanoparticle deposition and nanomaterial fabrication that can be coupled with spark ablation to produce well-defined nanostructured materials. The overview here is kept short as more details on the spark ablation process and the nanoparticle deposition principles are provided in Chapter 2.

Electric discharges have been used to produce gas-suspended nanoparticles since the 1980s [22–24]. An electric discharge can be either continuous or pulsed, forming, respectively, an arc or repeated sparks between two electrodes. The main difference between the two techniques is the throughput and the final size of the nanoparticles [25], as well as the option of producing alloyed particles, which can be achieved only by spark ablation (cf. Chapter 5). Arc discharges exhibit higher throughputs and produce agglomerates consisting of larger primary particles (typically larger than 20 nm). Spark discharges, on the other hand, have lower throughputs but also produce smaller primary particles (typically smaller than 20 nm) when operated at atmospheric conditions. In both cases, the resulting agglomerated particles can be restructured to spherical particles by sintering in the gas phase [26]. Recent efforts have shown that the technique can be used to produce singlet (i.e., non-agglomerated) nanoparticles, without any additional heating stage, and clusters that can have diameters down to the atomic size [27, 28]. Nanoparticles in this size range (i.e., sub–20 nm) are particularly interesting for sensors as they can be used to synthesize sensing materials of high surface-to-volume ratio, a property that defines the lowest detection level and the sensitivity of the sensors.

Chapter 2 gives a detailed description of the spark discharge generator. In brief, electric discharges created between two electrodes ablate electrode material to produce vapor clouds. A gas flow passing through the gap between the electrodes serves for quenching the resulting vapor clouds, which subsequently condense to form atomic clusters and nanoparticles while being carried out of the discharge chamber. A great advantage of the spark discharge nanoparticle generator is that one can easily tune the size of the resulting nanoparticles. Upon formation of the first atomic clusters from the vapor cloud at the vicinity of the discharge, subsequent condensation yields nanoparticles that grow until they reach a certain size just before those are deposited—a process that depends on the operating parameters of the generator (i.e., gas flow rate, discharge energy, and repetition frequency). As described in Chapter 2 and in more detail by Feng et al. [28], the mean size of the particles from the moment they are formed and until they are deposited is controlled and can be fully predicted by the operating parameters of the spark discharge generator. The size distributions of the resulting nanoparticles are log-normal with a geometric standard deviation of ca. 1.4 [29, 30].

It should be noted here that the size distributions of nanoparticles produced by spark discharge generators are typically wider compared to those produced by wet-chemistry techniques. This, however, is not a problem for chemical sensors at the production stage. It can be a limitation for conducting basic research to understand the influence of nanoparticle size on the sensor characteristics, where nanoparticles with narrower size distributions would certainly be preferable. Nearly monodisperse nanoparticles can, in fact, be produced by coupling a spark discharge generator with a differential mobility analyzer (DMA) [35] (such an integrated system is available commercially; VSParicle B.V., Delft, the Netherlands) or a particle mass analyzer (PMA) [36–39]. Both a DMA and a PMA can be used for selecting nanoparticles within a narrow size range based on the electrical mobility or the mass-to-charge ratio of the particles, respectively.

Another great advantage of spark ablation is the ease with which a wide variety of mixed nanoparticles having sizes from the atomic (alloys) to the nanoscale, including core-shell structures (cf. Chapter 2, Chapter 5, and Refs. [31–33]), can be produced. Recently, it was demonstrated that ternary mixtures can also be produced by spark ablation (just as, in principle, mixtures with any number of components) [34], expanding the capabilities of the technique. Because conductive electrodes are needed to form a spark, metal and metal oxide (semiconducting) particles are the primary products. Metal oxide nanoparticles can be synthesized by introducing small amounts of gaseous reactants into the gas stream (either upstream or downstream the spark; see Chapter 2).

Because optimized chemiresistors can hardly be expected to consist of pure (one element/compound) metal or metal oxide nanoparticle building blocks, the mixing feature of spark ablation holds great potential of revolutionizing the field of gas sensors. Utilizing mixed nanoparticles of well-defined composition and size can help, on the one hand, to fully understand the operating mechanisms of chemiresistors and, on the other hand, in fabricating chemiresistive sensors that will exhibit sensitivities and selectivities that go beyond the current state of the art and thus meet the requirements for use in air quality monitoring.

Application of spark ablation for nanomaterial synthesis at industrial scales, in general, has been limited by its relatively low throughput. This, however, has been overcome in the recent years by modulating the sparking frequency, as briefly discussed before, thereby increasing the nanoparticle production rates to a few g/h (depending on the material) from a single unit [29]. This increase can now enable use of the technique in manufacturing lines, especially when one considers using multiple systems. For the particular case of synthesizing nanomaterials for gas sensors, where only a few tens/hundreds of nanograms of nanoparticles are required for each sensor, the technique can prove highly useful and thus be employed to fabricate thousands of sensors per day, even with one generator.

One of the main prerequisites in the fabrication of chemiresistors (and in general of nanomaterials produced at industrial scales) is the high robustness and repeatability of the production process. Being a continuous gas-phase technique, nanoparticle synthesis by spark ablation can benefit from state-of-the-art online tools for characterizing aerosol nanoparticles. For example, using a DMA, coupled with a particle counter, one can monitor the size distributions of the nanoparticles during synthesis. This provides a unique advantage compared to liquid-based processes that produce nanoparticles in batches, and thus their characterization can take place only after the process is completed. In contrast to classical methods, spark ablation offers also an environmentally friendly way of producing nanoparticle building blocks for the synthesis of novel nanomaterials as it involves a very limited number of preparation steps, it avoids chemical precursors, and it produces minimal wastes.

Another advantage of synthesizing nanomaterials using nanoparticle building blocks suspended in a gaseous medium is the possibility of using an array of deposition techniques to immobilize them. Inertial impaction, electrostatic deposition, and filtration (cf. Fig. 2.10) are three standard ways of depositing the nanoparticles, and in principle, all can be used for producing materials for chemiresistive sensors. Inertial impaction can also be used in conjunction with aerodynamic focusing, enabling production of versatile nanostructured chemiresistors.

In summary, spark ablation provides a versatile way of producing nanoparticles of well-defined size and composition that are suspended in gas media. As a result it can be used in combination with standard nanoparticle classification (based on their mobility or mass) and deposition techniques for further refining the end products, depending on the application.

11.3 Nanomaterials for Chemiresistive Gas Sensors

As stated in the introduction, synthesizing porous nanostructured materials using bottom-up approaches holds great promise for producing highly sensitive and selective chemiresistors for gas sensors. This section provides an overview of the operating principles of chemiresistors that employ metal oxide or metallic nanoparticles and highlights the unique features of spark ablation that can attribute desired properties to the building-block nanoparticles and the end sensing nanomaterials.

11.3.1 Chemiresistors Based on Metal Oxide Nanoparticles

Metal oxide chemiresistive gas sensors based on metal oxide films rely on changes of the resistance—or conductivity—of their sensing materials (i.e., the chemoresistors) upon exposure to a target gas, with *n*-type materials being more sensitive compared to *p*-type materials [40]. These films are assembled using metal oxide nanoparticle building blocks in ways that attribute a high surface-to-volume ratio and porosity to the material and thus higher sensitivity compared to uniform films. Figure 11.1 provides a schematic layout of the entire sensor and the operating principle. Upon exposure to air, oxygen adsorbed on the grains captures electrons from the conduction band, thereby inducing an electron-depleted layer that increases the surface potential and the work function of the material. If the nanoparticle building blocks only gently touch one another, a potential barrier, often called the Schottky barrier, is formed, limiting the migration of electrons across each nanoparticle boundary and through the entire chemiresistor.

The Schottky barrier plays a dominant role in determining the resistance of the chemiresistor. When the target gas is reducing, it consumes the adsorbed oxygen, reducing this boundary potential barrier and consequently decreasing the resistance of the chemiresistor. In contrast, when the target gas is oxidizing, it is adsorbed competitively with oxygen on the surface of the nanoparticle building blocks. In that case, if the target has a higher adsorption affinity to the surface of the chemiresistors compared to oxygen, the surface potential barrier increases and thus the resistance of the chemiresistive material is increased.

Evidently, the sensitivity of chemiresistors strongly depends on the structure, which is defined by the size of the nanoparticle building blocks. Reducing the diameter of the nanoparticles to levels that are comparable to the thickness of the electron-depleted layers shortens the distances between Schottky barriers. As a result, the depleted zones start to overlap and consequently the electrical properties of the chemiresistors are predominantly defined by surface states. At the same time, reducing the particle size increases the surface-to-volume-ratio of the chemiresistors, providing more space for the surface reactions (adsorption of oxygen and the target gas) to take place, thereby increasing the sensitivity of the chemiresistors. This has been demonstrated experimentally by a number of studies that have shown a substantial increase in the sensitivity of chemiresistors with decreasing size of the nanoparticle building blocks below ca. 20 nm [41–43], and corroborating theoretical calculations [44]. Apart from the size of the nanoparticles, the sensitivity of the chemiresistors is defined by the ability of the target gas to efficiently percolate through the entire material, which is, in turn, defined by the porosity and thickness of the nanostructured film. Low-thickness and high-porosity chemiresistors exhibit high sensitivity and fast response time [45, 46], two elements that are highly required for environmental gas sensors.

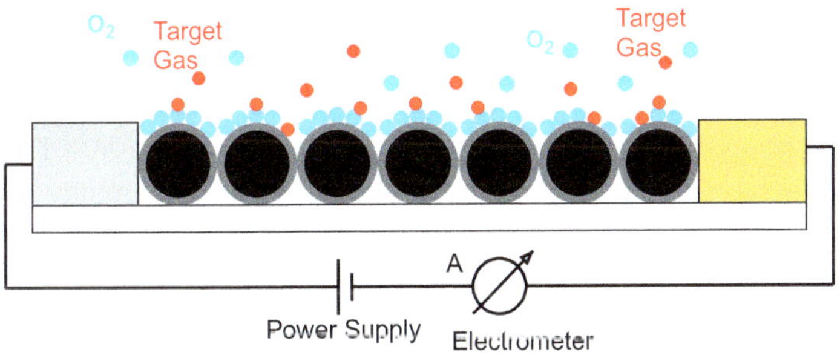

Figure 11.1 Schematic diagram showing the operating principle of a gas sensor using a chemiresistor based on metal oxide nanoparticle building blocks (gray/black spheres). Apart from the nanostructured chemiresistor, the sensor consists of a compact electrometer and a power supply. Adsorbed oxygen and target gas molecules onto the surface of the nanoparticle building blocks create an electron-depleted layer (depicted as the gray part of the nanoparticles), thereby altering the resistance of the chemiresistor. When the concentration of the target gas in the overlaying air varies, the number of associated adsorbed molecules on the surface of the chemiresistors varies proportionally, yielding variations to the number of respective captured electrons and thus the resistance of the film.

Nanoparticle synthesis by spark ablation, coupled with state-of-the-art deposition methods offered by aerosol-based techniques, provides high flexibility in synthesizing nanostructured chemiresistors of well-defined properties. First, the ability of spark generators to produce metal oxide nanoparticles of different sizes offers one very vital parameter for tuning the sensitivity of the chemiresistors. Second, by controlling the amount of nanoparticles deposited to build the chemiresistive nanomaterial (mainly by adjusting the deposition time provided that the concentration of the nanoparticles in the resulting aerosol is measured), and the operating parameters of the selected deposition process, one can in principle affect the thickness and porosity of these materials, thereby providing an additional tool for tuning their sensitivity.

Spark ablation has only recently been used to synthesize nanostructured films of oxidized tungsten for sensing NO_2 [47]. In that work, polydisperse agglomerated particles of oxidized tungsten with a mean diameter of 100 nm were produced by a spark discharge nanoparticle generator and deposited on interdigitated Al_2O_3 substrates by inertial deposition. Figure 11.2 shows increases in the resistance of the material upon repeated exposure to 10 ppm of NO_2. The sensitivity of these materials at the optimum temperatures (which was determined to be around 200°C, as shown in the inset of Fig. 11.2) was high enough also at 500 ppb NO_2. Those results should be considered as preliminary, and certainly only partly demonstrate the capabilities of the technique to produce chemiresistors having low levels of detection (in the few-ppb range) that can meet requirements in air quality monitoring. Further to this work, investigations on how the size and degree of oxidation of the nanoparticles produced by spark ablation, as well as how different deposition strategies can affect the sensitivity of the sensors, need to be carried out.

Another very important parameter of metal oxide chemiresistive gas sensors is their selectivity, that is, their ability to alter their resistance due to the adsorption of one or a very small family of target gas species. It should be noted here that although one can, in principle, use different metal oxides to measure different target gases, doing so is not enough to ensure good sensor selectivity. One strategy of making chemiresistors selective is by doping them with metals or other metal oxide species. It should be noted here that in the case of the nanoparticle-based metal oxide chemiresistors, doping refers to the addition of catalytically active sites to the surface of the base material. Apart from the selectivity, doping improves overall sensor performance by increasing the sensitivity, decreasing the response and recovery times, reducing the working temperature, as well as enhancing the thermal and long-term stability of the chemiresistors.

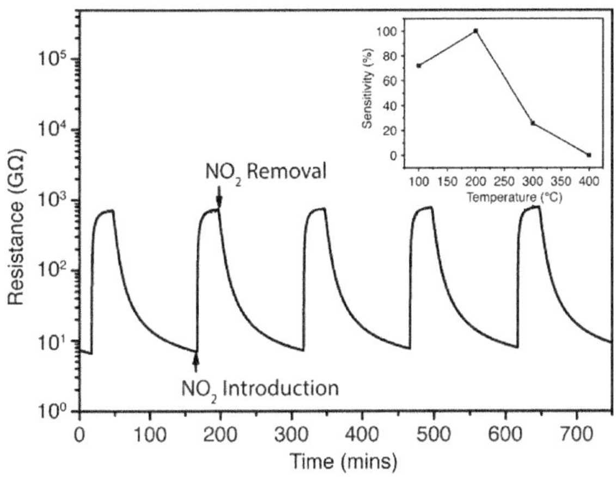

Figure 11.2 Changes in the resistance of nanoparticulate tungsten oxide films deposited on interdigitated electrodes when exposed alternatively to 0 and 10 ppm of nitrogen dioxide in air at 200°C. Inset: Normalized sensitivity of the nanoparticulate films as a function of temperature. Adapted with permission from Ref. [47]. Copyright (2016) American Chemical Society.

Doping chemiresistive materials induces two sensitization mechanisms, namely electronic and chemical sensitization [48]. In the first case, decorating the surface of the base semiconductor with smaller metal nanoparticles that oxidize upon exposure to air induces an electron-depleted space-charge layer near the interface. Upon reaction with a reducing target gas, these additive nanoparticles are reduced and thus release electrons back to the base semiconductor. In the second case, chemical sensitization results from catalytic surface reactions on the decorating nanoparticles. Those nanoparticles provide selective activation sites for the target gas molecules, spilling over activated fragments onto the semiconductor to react with the adsorbed oxygen. As a result, the surface coverage of the base semiconductor with oxygen changes, altering the resistance of the chemiresistors, providing a material that is more selective compared to the pure metal oxide semiconductors described before.

In reality, the performance of semiconductors is far more complicated than the picture drawn up here. As a result, searching for the material synthesis process that will yield highly sensitive and selective sensors for applications in air quality—and beyond that—still requires purely empirical approaches. In fact, the field of sensor development still requires advanced experimental techniques that will allow good control of a number of parameters that can be altered independently of each other. In this respect, nanoparticle synthesis by spark ablation can provide an excellent tool for synthesizing nanoparticles for both the base semiconductor of the chemiresistors as well as the decorating nanoparticles that can be used to enhance their performance. Of particular interest is the ability of spark ablation to produce mixed nanoparticles, both at the atomic and at the nanometer scales, as described in Chapters 2 and 5. Identifying the composition of the decorating nanoparticles, which in combination with the base semiconductor nanoparticle building blocks could enhance selectivity of the reactions taking place on the surface of the chemiresistors, should be a matter of extensive experimentation. Nevertheless, literature reports thus far already give good hints on which combination of materials could be the most promising, thereby providing good starting points.

11.3.2 Chemiresistors Based on Metal Nanoparticles

Another type of chemiresistive sensors relies on polymers whose conductivity can be manipulated by the addition of conductive metal nanoparticles. Early attempts in this direction were made in the 1980s, when polymeric materials containing conductive fillers (e.g., vanadium or carbon black particles) were first tested as chemiresistive sensors [49, 50]. This principle was refined by Wohltjen and Snow, who synthesized 2 nm gold nanoparticles coated with a octanethiol film [51]. To fabricate them, the metal nanoparticles were encapsulated by polymer monolayers in a controlled way and deposited on an interdigitated electrode. The resulting chemiresistive sensor was shown to measure concentrations of toluene, tetrachloroethene, 1-propanol, and water vapor with an extremely high sensitivity.

Figure 11.3 provides a schematic layout of nanoparticlc-bascd polymeric sensors. The operating principle relies on changes of the electronic interaction (electron hopping and tunneling) between the nanoparticles that is defined by the level of penetration of target gas molecules into the organic matrix. Such polymeric semiconductors offer great selectivity toward specific target gases through the selection of an appropriate polymer for coating the nanoparticles, and thus the design of those sensors has to be made at the molecular level [52].

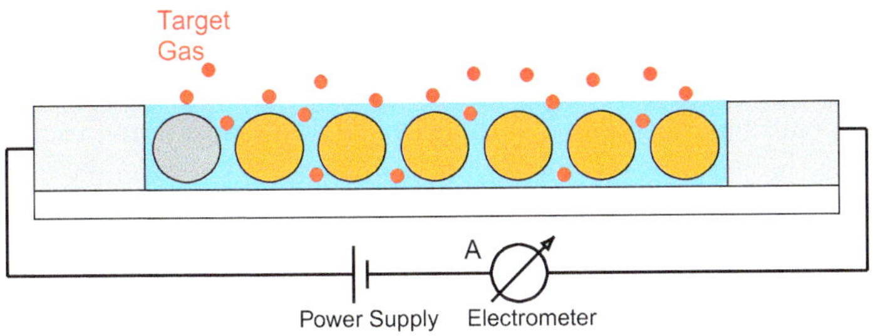

Figure 11.3 Schematic diagram showing the operating principle of a gas sensor using a polymeric chemiresistor containing metal nanoparticles. Apart from the polymer-nanoparticle-based chemiresistor, the sensor consists of a compact electrometer and a power supply. When the concentration of the target gas in the overlaying air varies, the number of molecules adsorbed by the chemiresistor alters proportionally, yielding measurable changes to its resistance.

Given the high selectivity of polymer-nanoparticle-based chemiresistors, a number of attempts have been made to use them as environmental gas sensors. For example, Joseph et al. [53] made a chemiresistor by cross-linking Pt and Au nanoparticles with nonanedithiol for measuring concentrations of CO and NH_3 at environmentally relevant concentrations (i.e., down to the ppb range). In contrast to earlier attempts, the nanoparticles were interconnected via covalent bonds, while a good fraction of the linker groups that were not bound to the nanoparticle cores provided binding sites for the target gas molecules.

Despite that the mechanism of how the different target gas molecules induce changes in the resistance on the polymer–nanoparticle matrix is not fully understood yet, experiments reported in the literature thus far show that such hybrid systems can provide chemiresistors for highly selective gas sensors. Compared to the metal oxide materials, another advantage of the chemiresistors discussed in this section is that they operate at room temperature. This, however, comes at the expense of a low thermal stability, which is one of the main bottlenecks for using them for air quality monitoring, especially if the measurements have to take place outdoors, under variable temperatures.

Preparation of the nanoparticle–polymer chemiresistors is typically performed by wet-chemistry techniques, mainly because preparation of both the nanoparticles and the polymer films can take place in the liquid phase. Gas-phase techniques, and particularly spark ablation coupled with deposition methods described in Section 1.2.2, could provide important advantages and additional tools for tuning the performance of such chemiresistors. For example, spark ablation, coupled with size selection by a DMA, can be employed to produce monodisperse nanoparticles of different sizes before those are coated with a polymer (either in the gas phase by evaporation/condensation or in the liquid phase). This will provide a way of decoupling nanoparticle synthesis from the formation of the polymer layer (which is difficult to achieve with wet-chemistry techniques), thereby allowing systematic investigations of how nanoparticle size can affect the overall sensor performance, something that currently is not fully understood. Another advantage of this decoupling is that it can enable investigations of how the use of nanoparticles of different composition can affect the performance of those sensors, given that altering nanoparticle compositions (and also making alloy/mixed nanoparticles) can be achieved relatively easily by spark ablation.

11.4 Conclusions

Spark ablation is an excellent tool for synthesizing nanoparticle building blocks that can subsequently be deposited (using the deposition techniques described briefly in Section 11.2 and in Chapter 2 in more detail) to form nanostructured materials. The high versatility of the technique in synthesizing nanoparticles of well-defined size and composition makes it an excellent research tool for studying the fundamental mechanisms that define the performances of these sensors. Considering also that recent developments have yielded commercially available spark discharge generators that can now reach production rates of a few g/h (depending on the material) of singlet nanoparticles from a single unit, the technique can also be used in production lines. This is particularly favorable for the fabrication of chemiresistors as the amounts of materials required are in the order of a few tens to a few hundreds of nanograms.

References

1. Yang, J., Zhou, J., Lv, Z., Wei, W. and Song, H. (2015). A real-time monitoring system of industry carbon monoxide based on wireless sensor networks, *Sensors*, **15**(11), pp. 29535–29546.
2. Bhatt, G., Manoharan, K., Chauhan, P. S. and Bhattacharya, S. (2019). MEMS sensors for automotive applications: a review, in Bhattacharya, S., Agarwal, A. K., Prakash, O. and Singh, S. (eds.) *Sensors for Automotive and Aerospace Applications*, Springer, Singapore, pp. 223–239.
3. Li, S., Simonian, A. and Chin, B. A. (2010). Sensors for agriculture and the food industry, *Electrochem. Soc. Interface*, **19**(4), pp. 41–46.
4. Oberg, P. A., Togawa, T. and Spelman, F. A. (2006). *Sensors in Medicine and Health Care*, John Wiley & Sons.
5. Gas Sensor Market Size & Share, Industry Analysis Report, 2018–2025, [Online]. Available at: https://www.grandviewresearch.com/industry-analysis/gas-sensors-market [Accessed: 26-Feb-2019].
6. Lelieveld, J., Evans, J. S., Fnais, M., Giannadaki, D. and Pozzer, A. (2015). The contribution of outdoor air pollution sources to premature mortality on a global scale, *Nature*, **525**(7569), pp. 367–371.
7. Organization, W. H. (2016). Ambient air pollution: a global assessment of exposure and burden of disease, World Health Organization.
8. Menz, F. C. and Seip, H. M. (2004). Acid rain in Europe and the United States: an update, *Environ. Sci. Policy*, **7**(4), pp. 253–265.
9. Hallquist, M., et al. (2016). Photochemical smog in China: scientific challenges and implications for air-quality policies, *Natl. Sci. Rev.*, **3**(4), pp. 401–403.
10. Jakubik, W. P. (2011). Surface acoustic wave-based gas sensors, *Thin Solid Films*, **520**(3), pp. 986–993.
11. Vashist, S. K. and Vashist, P. (2011). Recent advances in quartz crystal microbalance-based sensors, *J. Sens.*, [Online]. Available at: https://www.hindawi.com/journals/js/2011/571405/abs/ [Accessed: 26-Feb-2019].

12. Vereshchagina, E., Tiggelaar, R. M., Sanders, R. G. P., Wolters, R. A. M. and Gardeniers, J. G. E. (2015). Low power micro-calorimetric sensors for analysis of gaseous samples, *Sens. Actuators, B*, **206**, pp. 772–787.
13. Schwandt, C., Kumar, R. V. and Hills, M. P. (2018). Solid state electrochemical gas sensor for the quantitative determination of carbon dioxide, *Sens. Actuators, B*, **265**, pp. 27–34.
14. Lv, A., Pan, Y. and Chi, L. (2017). Gas sensors based on polymer field-effect transistors, *Sensors*, **17**(1), p. 213.
15. Metal and metal oxide nanoparticles in chemiresistors: does the nanoscale matter? - Franke - 2006 - Small - Wiley Online Library, [Online]. Available at: https://onlinelibrary.wiley.com/doi/full/10.1002/smll.200500261 [Accessed: 26-Feb-2019].
16. Brattain, W. H. and Bardeen, J. (1953). Surface properties of germanium, *Bell Syst. Tech. J.*, **32**(1), pp. 1–41.
17. Heiland, G. (1957). Zum Einfluß von Wasserstoff auf die elektrische Leitfähigkeit an der Oberfläche von Zinkoxydkristallen, *Z. Für Phys.*, **148**(1), pp. 15–27.
18. Bielański, A., Dereń, J. and Haber, J. (1957). Electric conductivity and catalytic activity of semiconducting oxide catalysts, *Nature*, **179**(4561), p. 668.
19. Seiyama, T., Kato, A., Fujiishi, K. and Nagatani, M. (1962). A new detector for gaseous components using semiconductive thin films, *Anal. Chem.*, **34**(11), pp. 1502–1503.
20. Taguchi, N. (1962). Patent 45–38200.
21. Barsan, N. and Weimar, U. (2001). Conduction model of metal oxide gas sensors, *J. Electroceram.*, **7**(3), pp. 143–167.
22. Altgeld, H., Schnettler, A. and Stehmeier, D. (1980). Spark discharge particle generator for laser Doppler anemometry, *J. Phys. E*, **13**(4), pp. 437–441.
23. Ando, Y. and Ohkohchi, M. (1982). Production of ultrafine powder of β-Sic by arc discharge, *J. Cryst. Growth*, **60**(1), pp. 147–149.
24. Schwyn, S., Garwin, E. and Schmidt-Ott, A. (1988). Aerosol generation by spark discharge, *J. Aerosol Sci.*, **19**(5), pp. 639–642.
25. Hontañón, E., et al. (2013). The transition from spark to arc discharge and its implications with respect to nanoparticle production, *J. Nanopart. Res.*, **15**(9), p. 1957.
26. Kruis, F. E., Fissan, H. and Rellinghaus, B. (2000). Sintering and evaporation characteristics of gas-phase synthesis of size-selected PbS nanoparticles, *Mater. Sci. Eng. B*, **69–70**, pp. 329–334.

27. Maisser, A., Barmpounis, K., Attoui, M. B., Biskos, G. and Schmidt-Ott, A. (2015). Atomic cluster generation with an atmospheric pressure spark discharge generator, *Aerosol Sci. Technol.*, **49**(10), pp. 886–894.

28. Feng, J., Biskos, G. and Schmidt-Ott, A. (2015). Toward industrial scale synthesis of ultrapure singlet nanoparticles with controllable sizes in a continuous gas-phase process, *Sci. Rep.*, **5**, p. 15788.

29. Feng, J., et al. (2016). Green manufacturing of metallic nanoparticles: a facile and universal approach to scaling up, *J. Mater. Chem. A*, **4**(29), pp. 11222–11227.

30. Wang, C. S. and Friedlander, S. K. (1967). The self-preserving particle size distribution for coagulation by Brownian motion: II. Small particle slip correction and simultaneous shear flow, *J. Colloid Interface Sci.*, **24**(2), pp. 170–179.

31. Tabrizi, N. S., Xu, Q., van der Pers, N. M., Lafont, U. and Schmidt-Ott, A. (2008). Synthesis of mixed metallic nanoparticles by spark discharge, *J. Nanopart. Res.*, **11**(5), p. 1209.

32. Lafont, U., Simonin, L., Tabrizi, N. S., Schmidt-Ott, A. and Kelder, E. M. (2009). Synthesis of nanoparticles of Cu, Sb, Sn, SnSb and Cu2Sb by densification and atomization process, [Online]. Available at: https://www.ingentaconnect.com/content/asp/jnn/2009/00000009/00000004/art00045 [Accessed: 28-Feb-2019].

33. Tabrizi, N. S., Xu, Q., van der Pers, N. M. and Schmidt-Ott, A. (2010). Generation of mixed metallic nanoparticles from immiscible metals by spark discharge, *J. Nanopart. Res.*, **12**(1), pp. 247–259.

34. Feng, J., et al. (2018). Magnetic phase transition in spark-produced ternary LaFeSi nanoalloys, *ACS Appl. Mater. Interfaces*, **10**(7), pp. 6073–6078.

35. Knutson, E. O. and Whitby, K. T. (1975). Aerosol classification by electric mobility: apparatus, theory, and applications, *J. Aerosol Sci.*, **6**(6), pp. 443–451.

36. Fernández de la Mora, J. and Kozlowski, J. (2013). Hand-held differential mobility analyzers of high resolution for 1–30 nm particles: design and fabrication considerations, *J. Aerosol Sci.*, **57**, pp. 45–53.

37. Ehara, K., Hagwood, C. and Coakley, K. J. (1996). Novel method to classify aerosol particles according to their mass-to-charge ratio—aerosol particle mass analyser, *J. Aerosol Sci.*, **27**(2), pp. 217–234.

38. Tajima, N., Fukushima, N., Ehara, K. and Sakurai, H. (2011). Mass range and optimized operation of the aerosol particle mass analyzer, *Aerosol Sci. Technol.*, **45**(2), pp. 196–214.

39. Broßell, D., Valenti, M., Bezantakos, S., Schmidt-Ott, A. and Biskos, G. (2015). The nano-particle mass classifier (nano-PMC): development, characterization, and application for determining the mass, apparent density, and shape of particles with masses down to the zeptogram range, *Aerosol Sci. Technol.*, **49**(7), pp. 495–507.

40. Bârsan, N., Huebner, M. and Weimar, U. (2013). Conduction mechanism in semiconducting metal oxide sensing films: impact on transduction, in Jaaniso, R. and Tan, O. K. (eds.) *Semiconductor Gas Sensors*, Woodhead Publishing, pp. 35–63.

41. Lu, F., Liu, Y., Dong, M. and Wang, X. (2000). Nanosized tin oxide as the novel material with simultaneous detection towards CO, H2 and CH4, *Sens. Actuators, B*, **66**(1), pp. 225–227.

42. Korotcenkov, G., et al. (2004). In2O3 films deposited by spray pyrolysis: gas response to reducing (CO, H2) gases, *Sens. Actuators, B*, **98**(2), pp. 122–129.

43. Korotcenkov, G., et al. (2004). In2O3 films deposited by spray pyrolysis as a material for ozone gas sensors, *Sens. Actuators, B*, **99**(2), pp. 297–303.

44. Rothschild, A. and Komem, Y. (2004). The effect of grain size on the sensitivity of nanocrystalline metal-oxide gas sensors, *J. Appl. Phys.*, **95**(11), pp. 6374–6380.

45. Matsunaga, N., Sakai, G., Shimanoe, K. and Yamazoe, N. (2003). Formulation of gas diffusion dynamics for thin film semiconductor gas sensor based on simple reaction–diffusion equation, *Sens. Actuators, B*, **96**(1), pp. 226–233.

46. Vuong, D. D., Sakai, G., Shimanoe, K. and Yamazoe, N. (2005). Hydrogen sulfide gas sensing properties of thin films derived from SnO2 sols different in grain size, *Sens. Actuators, B*, **105**(2), pp. 437–442.

47. Isaac, N. A., Valenti, M., Schmidt-Ott, A. and Biskos, G. (2016). Characterization of tungsten oxide thin films produced by spark ablation for NO_2 gas sensing, *Appl. Mater.*, **8**(6), pp. 3933–3939.

48. Yamazoe, N., Sakai, G. and Shimanoe, K. (2003). Oxide semiconductor gas sensors, *Catal. Surv. Asia*, **7**(1), pp. 63–75.

49. Lundberg, B. and Sundqvist, B. (1986). Resistivity of a composite conducting polymer as a function of temperature, pressure, and environment: Applications as a pressure and gas concentration transducer, *J. Appl. Phys.*, **60**(3), pp. 1074–1079.

50. Ruschau, G. R., Newnham, R. E., Runt, J. and Smith, B. E. (1989). 0–3 ceramic/polymer composite chemical sensors, *Sens. Actuators*, **20**(3), pp. 269–275.

51. Wohltjen, H. and Snow, A. W. (1998). Colloidal metal–insulator–metal ensemble chemiresistor sensor, *Anal. Chem.*, **70**(14), pp. 2856–2859.

52. Evans, S. D., Johnson, S. R., Cheng, Y. L. and Shen, T. (2000). Vapour sensing using hybrid organic–inorganic nanostructured materials, *J. Mater. Chem.*, **10**(1), pp. 183–188.

53. Joseph, Y., Guse, B., Yasuda, A. and Vossmeyer, T. (2004). Chemiresistor coatings from Pt- and Au-nanoparticle/nonanedithiol films: sensitivity to gases and solvent vapors, *Sens. Actuators, B*, **98**(2), pp. 188–195.

Chapter 12

Alloy Plasmonic Nanoparticles and Their Applications

Marco Valenti
Nanophotonics, AMOLF, Science Park 104 Amsterdam, 1098 XG, The Netherlands
valenti@amolf.nl

This chapter highlights the ability of the spark discharge generator (SDG) to produce well-defined spherical metallic nanoparticles, which have a strong interaction with visible light due to the localized surface plasmon resonance phenomenon. Hereby, an overview of the metallic nanoparticles' optical properties and their applications is given. An emphasis is put on the ability of the metallic nanoparticles to concentrate the electric field, generate hot electrons, and heat up on the nanoscale, upon illumination. Throughout the chapter the advantages of the SDG to control the alloy composition of metallic nanoparticles are explained through specific applications.

Spark Ablation: Building Blocks for Nanotechnology
Edited by Andreas Schmidt-Ott

ISBN 978-981-4800-82-2 (Hardcover), 978-0-367-81709-1 (eBook)
www.jennystanford.com

12.1 Introduction

The spark discharge generator (SDG) can be used to produce nanoparticles of a wide range of sizes (from a couple of nanometers to a couple of microns) [1]. Nonspherical agglomerate production can be avoided by raising the temperature in the particle growth phase [2]. Alternatively, coalescence of agglomerates that have formed can be induced for forming spherical metallic nanoparticles by adding a tube oven downstream of the SDG [3]. These spherical metallic nanoparticles exhibit very interesting optical properties, having a large absorption and scattering cross section in the visible part of the spectrum (Fig. 12.1). The scattering, extinction, and absorption cross sections (σ_{sca}, σ_{ext}, and σ_{abs}, respectively) can be estimated with the following relations [4]:

$$\sigma_{sca} = \frac{k^4}{6\pi}|\alpha|^2 \tag{12.1}$$

$$\sigma_{ext} = k\mathrm{Im}(\alpha) \tag{12.2}$$

$$\sigma_{abs} = \sigma_{ext} - \sigma_{sca}, \tag{12.3}$$

where k is the wavenumber and α is the nanoparticle polarizability, which in the quasistatic approximation (valid when $ak < 1$, where a is the nanoparticle radius) is

$$\alpha = 3V\varepsilon_m \frac{\varepsilon(\omega) - \varepsilon_m}{\varepsilon(\omega) + 2\varepsilon_m}, \tag{12.4}$$

where V is the volume of the plasmonic nanoparticle (PNP), $\varepsilon(\omega)$ is the complex electrical permittivity of the metal, and ε_m is the electrical permittivity of the surrounding medium.

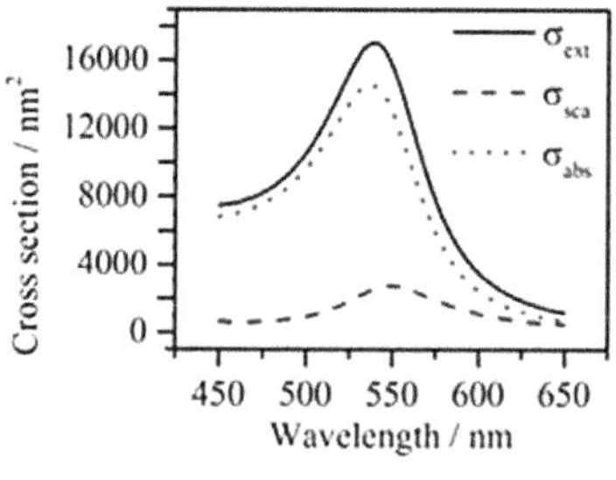

Figure 12.1 Extinction, scattering, and absorption spectra of a gold nanoparticle with a radius of 35 nm.

The large optical response is due to the collective oscillation of free electrons induced by the incoming light at specific frequencies (resonance frequencies). This phenomenon is called localized surface plasmon resonance (LSPR) (see also Chapter 13). The resonance frequencies highly depend on the shape, size, and material of the metallic nanoparticles. For spherical PNPs, the resonance arises at the angular frequency (or wavelength) where the real part of the dielectric permittivity of the metal $\varepsilon(\omega)$ equals $-2\varepsilon_m$ and the polarizability of the nanoparticle reaches a maximum. The LSPR wavelengths (λ_{res}) at standardized conditions (e.g., spherical geometry) [6] for a variety of plasmonic materials are given in Table 12.1. Alloyed PNPs allow the ability to tune their absorption spectra by varying their composition [5]. The LSPR mode of alloy PNPs is located between the LSPR modes of the pure components [5]. Figure 12.2 shows the absorption spectrum of composition variant Au-Ag alloy PNPs. Clearly, the LSPR mode red-shifts with increasing Au composition.

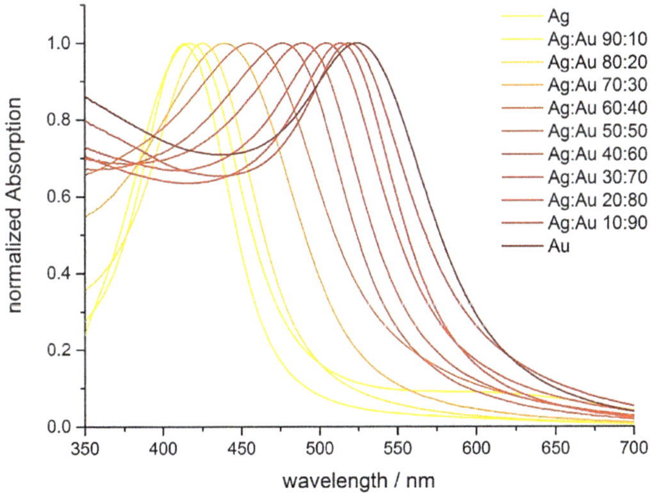

Figure 12.2 Plasmon resonance absorption spectra of Ag:Au nanoalloys with different compositions. The color of each curve corresponds to the color of the corresponding particle dispersion (silver: yellow; gold: red). From Ref. [5]. Reproduced by permission of The Royal Society of Chemistry.

Table 12.1 Figures of merit for the ability of materials to enhance the near-field (Fa^0) and generate heat (Jo^0), together with the corresponding resonance wavelengths and melting temperatures [6]

	λ_{res} (nm)	Fa^0	Jo^0	$T_{melting}$ (°C)
Au	528	19.6	6.32	1064
Ag	357	118	52	961
Cu	585	15.2	2.65	1085
Ni	218	11.6	21.4	1455
Pt	323	10.5	12	1768
W	159	14.2	25.6	3422

The ability to tune the PNPs' absorption spectrum (through their size, shape, and composition) [7] gives them a great advantage over semiconductor materials for solar harvesting devices, since unlike the absorption of semiconductors, PNPs are not limited by a bandgap. This is where the flexibility of spark ablation, at least regarding size and composition, comes into play. If the absorbed energy could be converted into electrical or chemical energy, PNP arrays could be designed to exploit the entire solar spectrum [8]. However, solar cell devices are by no means the only application envisioned for PNPs (see, for example, Chapter 13 for medical applications). One of the most outstanding features of PNPs is that they act as optical antennas, confining the incoming light at/near their metallic surfaces (near field) [7], which means that PNPs can bring visible light to the nanoscale. The high intensity of the near field brings about a variety of applications [9–11]. One important application is surface-enhanced Raman spectroscopy (SERS) [11], where the Raman signal is enhanced by several orders of magnitude, as originally found by Jeanmaire and van Duyne in 1977 [12], allowing us to identify even single-molecule vibrations. Another feature of PNPs is that when the LSPR decays, the energy can dissipate as heat, increasing the temperature of the particle and its surroundings. The ability to heat up on the nanoscale has a broad range of applications and gave rise to the field of thermoplasmonics [13].

The different applications for PNPs are categorized here into (i) near-field, (ii) hot-electron, and (iii) thermoplasmonic applications. This categorization arises naturally when looking into the decaying processes of the LSPR (Fig. 12.3). Initially, light induces the LSPR that exhibits a large-field confinement near the PNP surface (Fig. 12.3a), which is exploited in SERS. The LSPR decays after a few femtoseconds by emitting light (light scattering) or by creating hot-electron carriers (light absorption, Fig. 12.3b). These hot carriers can be extracted for solar cell applications and sensors [14]. If the hot electrons are not extracted, their energy is reduced through electron–electron scattering (Fig. 12.3c), generating a distribution of electrons with a higher electronic temperature in ~1 ps. Finally, the electrons couple with phonons releasing the plasmonic energy as heat (Fig. 12.3d).

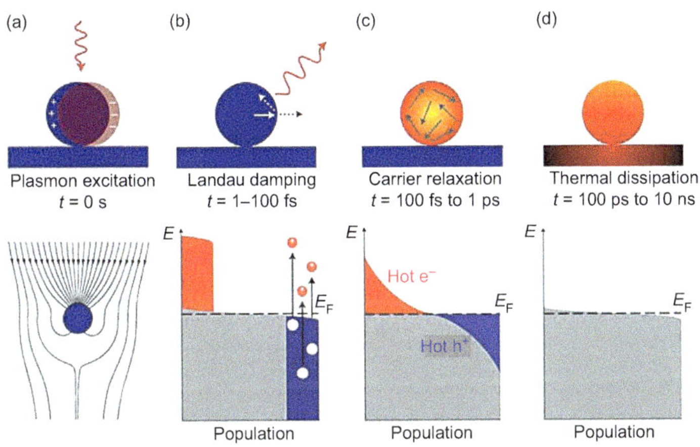

Figure 12.3 (a–d) Photoexcitation and subsequent relaxation processes following the illumination of a metal nanoparticle with a laser pulse, and characteristic timescales. (a) First, the excitation of a localized surface plasmon redirects the flow of light (Poynting vector) toward and into the nanoparticle. (b–d) Schematic representations of the population of the electronic states (gray) following plasmon excitation: hot electrons are represented by the red areas above the Fermi energy E_F, and hot-hole distributions are represented by the blue area below E_F. (b) In the first 1–100 fs following Landau damping, the athermal distribution of electron–hole pairs decays either through re-emission of photons or through carrier multiplication caused by electron–electron interactions. During this very short time interval τ_{nth}, the hot-carrier distribution is highly nonthermal. (c) The hot carriers will redistribute their energy by electron–electron scattering processes on a timescale τ_{el} ranging from 100 fs to 1 ps. (d) Finally, heat is transferred to the surroundings of the metallic structure on a longer timescale τ_{ph} ranging from 100 ps to 10 ns, via thermal conduction. Reprinted by permission from Springer Nature Customer Service Centre GmbH: Springer Nature, *Nature Nanotechnology*, Ref. [15], copyright (2015).

In this chapter the main applications of spherical PNPs are reviewed, with emphasis on SERS, hot-electron applications, and thermoplasmonics. Even though, PNP rods, cubes, and other PNPs with sharp edges are of great interest in plasmonics, this chapter focuses on spherical PNPs since a sphere is the most natural PNP geometry that can be generated with the SDG, a tube oven, and a differential mobility analyzer (DMA) with absolute control over the size and composition. A typical setup that allows the SDG to easily produce monodispersed aerosols of spherical PNPs is shown in Fig. 12.4. This ability to control the size and composition (including alloys) makes the SDG an excellent tool to study the size and material dependency of plasmonic properties. Especially, the flexibility of the SDG to produce a broad range of alloy PNPs allows us to fine-tune the absorption spectra of the PNPs by varying their composition. The SDG can produce alloy PNPs, either by using already alloyed SDG electrodes (cathode and anode of the same alloy composition, Fig. 12.5A) or by using two different electrode materials (Fig. 12.5B) that allows us to modify the resulting alloy composition by tuning the spark parameters (see Chapter 5). The main aim of the present chapter is to identify the plasmonic properties that can be studied or obtained with the SDG for specific applications.

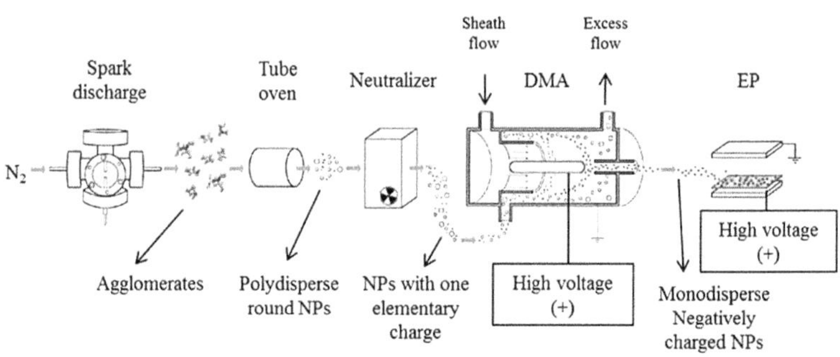

Figure 12.4 Setup to produce monodispersed PNPs with control over size and composition. Initially the SDG generates primary nanoparticles of ~5 nm, if the initial agglomeration takes place at room temperature in a N_2 flow. The particles diffuse and collide with one another, forming agglomerates. The agglomerates are then sintered in a tube oven into spherical PNPs, which are subsequently size-selected in a differential mobility analyzer (DMA). The PNPs coming out of the DMA are singly charged and can be deposited electrostatically on a conducting substrate. Reprinted with permission from Ref. [3]. Copyright (2015) American Chemical Society.

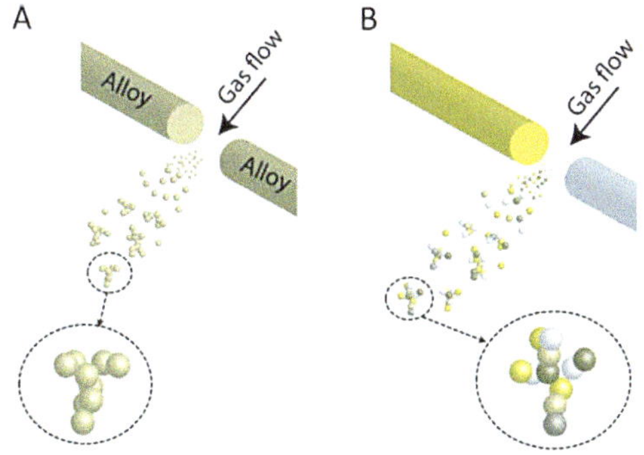

Figure 12.5 Illustration of the agglomerate's composition produced by the SDG for the case of room temperature in the growth zone, when two alloy electrodes are used (A) and when two electrodes of two different materials are used (B). In both cases, alloy PNPs are produced when the agglomerates are sintered in a tube oven.

12.2 Surface-Enhanced Raman Spectroscopy

Raman scattering of a molecule is an inelastic light scattering process in which the molecule can lose or gain energy that is specific to their vibrational mode transitions. A molecule Raman spectrum is a fingerprint of the molecule, but a large number of molecules is needed in order to obtain a measurable Raman signal. However, it was discovered that when molecules are near a PNP or on a rough metallic surface, the Raman signal can be amplified by a few orders of magnitude, allowing us to identify single molecules. This technique is called SERS (see also Chapter 13). Identification of molecules in dilute samples is of great interest for biological sensor applications and in catalysis. Below, the principles of Raman spectroscopy and SERS are introduced and the opportunities for carrying out fundamental studies with SDG-made PNPs are discussed.

12.2.1 Raman Spectroscopy

In the light-scattering process of a molecule, light is initially absorbed, bringing the molecule from an electronic state to an excited nonstationary state (usually called virtual state). In Rayleigh scattering, the nonstationary state instantly relaxes back into the initial electronic state (Fig. 12.6), emitting light of the same energy (i.e., frequency) as the excitation light. However, the nonstationary state can also relax into excited vibrational states, emitting light with lower energies than the excitation light. This scattering process is called Stokes Raman scattering (Fig. 12.6). A second Raman scattering process is the anti-Stokes Raman scattering, where the incoming light is absorbed by bringing an already excited state to a virtual state, which subsequently relaxes into the ground state (Fig. 12.6). In the anti-Stokes process the emitted light has more energy than the irradiating light.

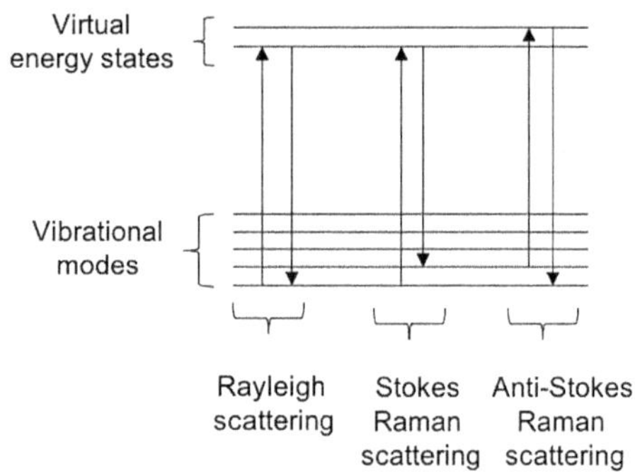

Figure 12.6 Illustration of the energy transitions in Rayleigh scattering, Stokes Raman scattering, and anti-Stokes Raman scattering.

12.2.2 Surface-Enhanced Raman Spectroscopy

When the PNP's near field overlaps with the irradiated molecule, the Raman signal is significantly enhanced. The enhancement of the signal is proportional to the intensity of the electric field, as suggested by the following equation[4]:

$$P_s(\omega_s) = N\,\sigma_{SERS} L(\omega_i)^2 L(\omega_s)^2 I(\omega_i), \tag{12.5}$$

where, $L(\omega_i)$ is the plasmonic field enhancement at the wavelength of the irradiating light, $L(\omega_s)$ is the plasmonic field enhancement at the wavelength of the Raman scattering (Stokes or anti-Stokes), N is the number of molecules, σ_{SERS} is the Raman scattering cross section, and $I(\omega_i)$ is the irradiating laser intensity. From Eq. 12.5 it is clear that the maximum Raman enhancement is obtained when the

PNP resonance wavelength (where the field enhancement is largest) coincides with the Raman scattering wavelength (Stokes or anti-Stokes). Therefore, the PNP should be designed such that its LSPR frequency overlaps with the wavelength of the Raman scattering signal of the molecule of interest. The matching of the spectrum can be obtained by fine-tuning the composition and size of the PNP with the SDG.

12.2.2.1 PNP composition

Due to the flexibility of the SDG in tuning the composition of the PNPs, the LSPR frequency can be tuned with bimetallic (e.g., alloy) PNP to match the Raman scattering wavelength of interest. However, the electric field enhancement is significantly affected by the composition of the PNP. The maximum electric field amplitude on the PNP surface can be approximated with the following relation [13]:

$$E_{\max} = \left(1 + 2\frac{\varepsilon(\omega) - \varepsilon_{\mathrm{m}}}{\varepsilon(\omega) + 2\varepsilon_{\mathrm{m}}}\right) E_0 \tag{12.6}$$

Clearly the enhancement highly depends on the complex electrical permittivity of the material, and generally a "good" plasmonic material has a large real part of the electrical permittivity (ε') and a small imaginary part of the electrical permittivity (ε''). Specifically, a figure of merit for the ability of a plasmonic material to enhance the electric field is the Faradaic number (Fa^0)[6],

$$\mathrm{Fa}^0 = \left|\frac{\text{Maximum electric field over space}}{\text{Incoming electric field}}\right|^2, \tag{12.7}$$

which for small spherical PNPs at their LSPR frequency is [6]

$$\mathrm{Fa}^0 = 9\left(1 + \left|\frac{\varepsilon'}{\varepsilon''}\right|^2\right). \tag{12.8}$$

Fa^0 values for common plasmonic materials are shown in Table 12.1, with silver being the best material to enhance the field. However, due to oxidation layers on the metallic PNPs, the theoretical performance of the Ag PNPs [6] (e.g., Fa^0 derived with bulk electrical permittivity) deviates significantly from the one obtained experimentally [16]. The great flexibility of the SDG to change the material not only allows us to study the ability of bimetallic PNPs to enhance the field but also allows for the fabrication of alloy PNPs that are stable in the environment of interest and that have a matching LSPR with the Raman signal of interest.

Spherical PNPs made with the SDG (i.e., setup shown in Fig. 12.4) are ideal for fundamental studies since analytical solutions are available for the near field, extinction cross sections, etc. However, the SDG can also produce small plasmonic aggregates with geometries that can surpass the field enhancement of the spherical PNPs. An experimental setup used by Abd El-Aal et al. [17] to produce plasmonic aggregates for SERS is shown in Fig. 12.7, where PNPs aggregates are directly deposited by diffusion within the spark chamber. By optimizing the operation parameters of the SDG, Abd El-Aal et al. obtained arrays of silver PNP aggregates that can detect the molecule Rh B to the single molecular level [17].

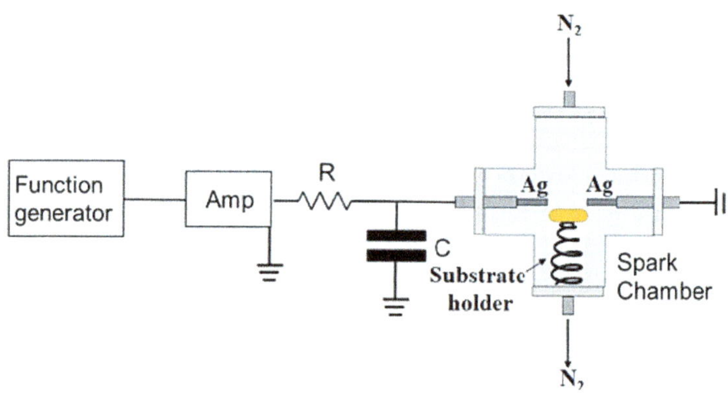

Figure 12.7 SDG setup to deposit an array of plasmonic aggregates for SERS. Reprinted from Ref. [17], Copyright (2018), with permission from Elsevier.

12.3 Hot-Electron Applications

LSPR decays by generating hot carriers that gain the plasmon energy [18]. The lifetime of hot electrons is in the order of femtoseconds, since they can readily lose their energy by electron–electron scattering [19]. In the simplest case, the particle is not in contact with any surface, and the emission of an electron from the particle can be observed for high enough $h\nu$ (photoelectron emission). Back in the 1980s, a strong enhancement of the electron emission yield from small particles was already reported by Schmidt-Ott et al. [20], even if $h\nu$ is shifted with respect to the LSPR resonance frequency, and surface plasmon excitation was identified as a relevant factor explaining this phenomenon. In contact with a semiconductor (rather than emission into vacuum), a much lower photon energy is required for transfer of the electron, and for all metals there is a much better match of this photon energy with the LSPR energy. If the hot carriers can be removed before electron scattering occurs, they can be extracted for solar cell applications or to drive chemical reactions in an electrochemical cell [18]. Contacting the PNP with an n-type semiconductor like TiO_2 generates a Schottky barrier where hot electrons can be extracted with the following efficiency [21]:

$$IE \approx C_{\mathrm{F}} \frac{(h\nu - q\phi_{\mathrm{B}})^2}{h\nu} S(\nu), \tag{12.9}$$

where C_F is the Fowler emission coefficient, $h\nu$ is the energy of the hot electron, $q\ \phi_B$ is the energy of the barrier at the metal–semiconductor interface (or the work function for photoelectron emission), and $S(\nu)$ is the PNP absorption spectrum.

The above relation (i.e., Fowler relation) shows that the efficiency of extraction increases with the hot-electron energy, which was assumed to be $h\nu$ in Eq. 12.9. Even though hot electrons can have energies up to $h\nu$, their energy highly depends on the nature of the excitation (i.e., intraband or interband excitations). The valence band of transition metals consists of the *s* and *d* bands, as shown in Fig. 12.8. For Au and Ag, the *d* band is located ~2.3 and 3.7 eV below the Fermi level, respectively. When illuminated with visible light (~1.8 eV to ~3.1 eV) the LSPR energy of Au PNPs is sufficient to excite an electron from either the s band (intraband) or the d band (interband). On the other hand, the visible-light-induced LSPR in Ag can only decay by intraband excitations. It is clear from the figure below that intraband excitations can generate hot electrons with energies up to $h\nu$ if the electron is excited near the Fermi level. On the other hand, if electrons are excited from the d band, most of the energy stays in the hot hole (at least the interband energy). The role of the electronic structure on the hot-electron extraction has been well studied theoretically [22] and experimentally by using spark discharge [23].

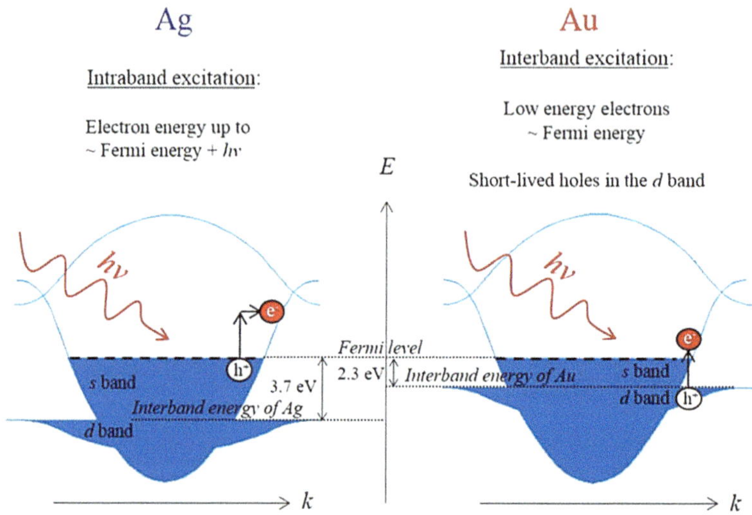

Figure 12.8 Illustration of the intraband electronic transitions in Ag PNPs (A) and interband electronic transition in Au PNPs (B) when illuminated with visible light. Specifically, interband transitions in Au are caused by violet, blue, and green light in the visible spectrum, which have higher energies than the interband energy threshold of Au. Reprinted with permission from Ref. [23]. Copyright (2017) American Chemical Society. https://pubs.acs.org/doi/10.1021/acsphotonics.6b01048. Further permissions related to the material excerpted should be directed to the ACS.

Compositionally differing (Ag, Au, and Au-Ag alloy) spherical PNPs of the same size were produced with the SDG (using the setup illustrated in Fig. 12.4) in order to elucidate the role of the composition on the hot-electron extraction efficiency. For the alloy PNP synthesis, 50/50 (atomic) Au-Ag alloy electrodes were used in the SDG. The size-selected and electrostatically deposited PNPs have ~50/50 (atomic) composition (same as the electrodes used), as measured by X-ray photoelectron spectroscopy (XPS), which demonstrates the ability of the SDG to make alloy PNPs of a predefined composition and absorption. Figure 12.9 shows that the absorption spectrum of the synthesized alloy PNPs exhibits an LSPR mode in between the absorption modes of the pure PNPs (synthesized with the same method), which is in good agreement with the one predicted with Eqs. 12.1–12.4.

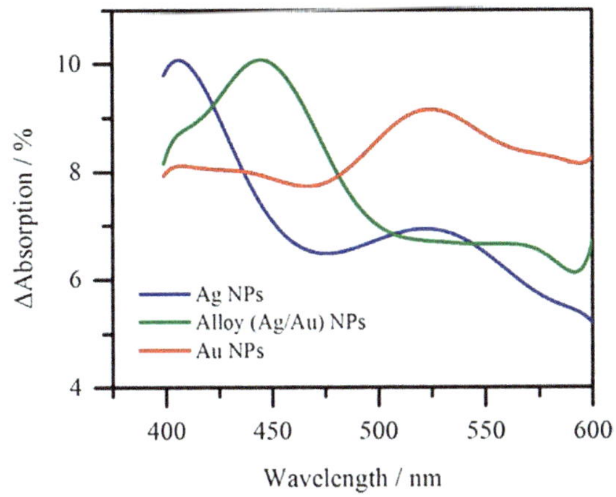

Figure 12.9 Absorption spectra of 15 nm Ag, Ag-Au alloy, and Au PNPs. Reprinted with permission from Ref. [23]. Copyright (2017) American Chemical Society. https://pubs.acs.org/doi/10.1021/acsphotonics.6b01048. Further permissions related to the material excerpted should be directed to the ACS.

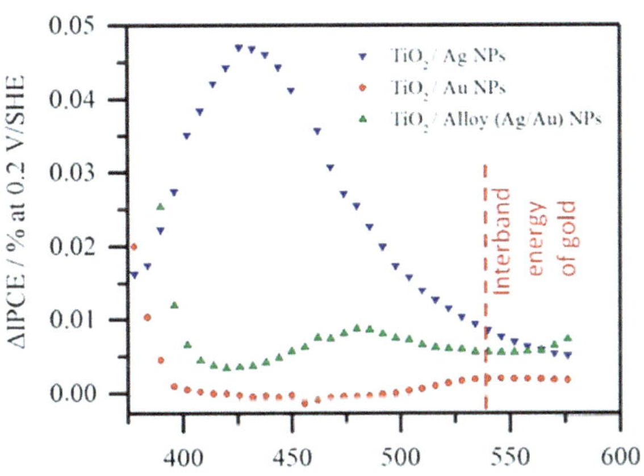

Figure 12.10 IPCE enhancement of TiO_2 films when decorated with 15 nm Ag, alloy (Ag/Au), and Au PNPs at 0.2 V vs. a standard hydrogen electrode (SHE). Reprinted with permission from Ref. [23]. Copyright (2017) American Chemical Society. https://pubs.acs.org/doi/10.1021/acsphotonics.6b01048. Further permissions related to the material excerpted should be directed to the ACS.

Figure 12.10 shows the corresponding hot-electron injection efficiency of the PNPs in terms of the incoming photon-to-current efficiency (IPCE). Clearly, the Ag PNPs exhibit the largest efficiency.

The alloy PNPs exhibit a lower hot-electron extraction efficiency and the Au PNP's hot-electron extraction is not detected. An explanation for the decrease in hot-electron extraction with increasing Au composition is given in Fig. 12.11. The increase in interband excitation with increasing Au portion results in a decrease of the hot-electron energies and therefore a decrease in the hot-electron extraction.

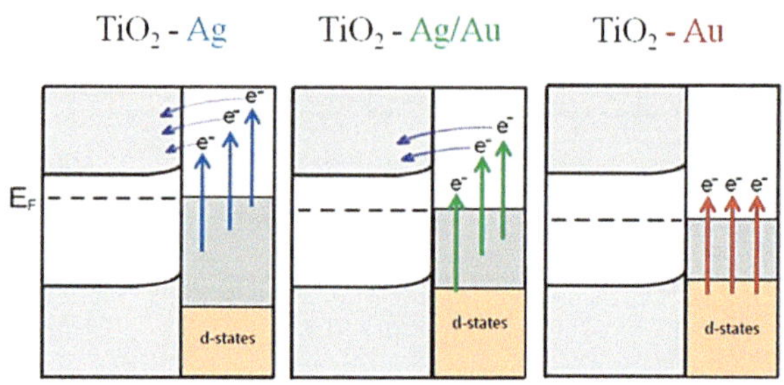

Figure 12.11 Illustration of the relative Fermi equilibration energies (E_F) and hot-electron energy (above the Fermi-level equilibration) of Ag, alloy (Ag/Au), and Au PNPs when illuminated with visible light. When Ag is illuminated, only intraband transitions occur, since the energy difference between the position of the d states and the Fermi level is larger than that of the irradiating visible light. On the other hand, Au allows for prominent interband excitations when illuminated with violet, blue, and green light, due to its relatively shorter distance between the d states and the Fermi level. When Ag/Au is illuminated, it undergoes less interband excitations than Au, since its d states are shifted further away from the Femi level. Interband excitations exhibit low hot-electron energies (~Fermi energy), limiting the hot-electron injection process. Reprinted with permission from Ref. [23]. Copyright (2017) American Chemical Society. https://pubs.acs.org/doi/10.1021/acsphotonics.6b01048. Further permissions related to the material excerpted should be directed to the ACS.

12.4 Thermoplasmonics

When plasmon-induced hot electrons are not extracted, all the absorbed energy is dissipated as heat. There are a variety of applications that take advantage of the PNP's ability to heat up with nanoscale resolution [13]. One of the most promising applications is plasmonic photothermal therapy (PPTT; see also Chapter 13) [24]. In PPTT, PNPs are delivered to cancer cells in the human body. If the PNPs absorb light in a region of the spectrum that is highly transparent in human tissue, a laser at the LSPR mode can be used to heat up the PNPs and induce cancer cell death. PPTT is a growing field of research that has already reached clinical trials. Other thermoplasmonics applications can be found in chemistry, where heating up PNPs with a laser can locally increase the rate of a reaction on or near the PNPs [25].

The temperature increase in a spherical PNP when illuminated with irradiance I (power per unit surface) is [26]

$$\delta T(r) = \frac{\sigma_{\text{abs}} I}{4\pi\kappa a}, \tag{12.10}$$

where κ is the thermal conductivity of the surrounding material.

For any thermoplasmonic applications, a key feature of the PNPs is their stability during heating in the different environments. For example, Ga and In cannot be heated significantly since they melt at relatively low temperatures, changing their plasmonic properties. Even Au exhibits a low melting point for important plasmonic applications, such as heat-assisted magnetic recording [27]. Plasmonic materials that can withstand high temperatures without losing their plasmonic properties are called refractory metals and are being increasingly investigated [27]. Some examples of refractory materials are W, Ta, and metal nitrides.

As for the electric field enhancement, for plasmonic heating there is also a useful figure of merit called the Joule number (J_0) that attempts to quantify the ability of the plasmonic materials to generate heat [6]:

$$J_0 = 9e\frac{|\varepsilon'|^{3/2}}{\varepsilon''} \tag{12.11}$$

The melting temperatures and the Joule numbers of several plasmonic materials are given in Table 12.1. The SDG can play an important role in developing alloy PNPs that fulfil the melting temperature, plasmonic heating, and resonance frequency needed for specific thermoplasmonic applications.

12.5 Conclusions

The SDG can be used to produce well-defined spherical PNPs with a predefined composition to obtain the plasmonic properties needed for a specific application. Three plasmonic properties were highlighted in this chapter: (i) field enhancement, (ii) hot-electron generation, and (iii) heat generation. The material figure of merit for (i) and (iii) can be estimated with Eqs. 12.8 and 12.11, while the ability to generate high-energy hot electrons is given in terms of the electronic structure (interband energy). The figures of merit of several plasmonic materials are given in Table 2.1, along with their resonant wavelengths and melting temperatures. Alloying these plasmonic materials with the SDG can optimize the absorption spectra and physical-chemical stability of the PNPs for the manufacture of plasmonic devices. Moreover, the flexibility of the SDG also allows us to study the ability of alloy PNPs to (i) enhance the field, (ii) generate hot electrons, and (iii) generate heat on the nanoscale. As for other areas covered in the present book, the SDG is an excellent research tool for exploiting the huge potential of PNPs.

References

1. Pfeiffer, T. V., Feng, J. and Schmidt-Ott, A. (2014). New developments in spark production of nanoparticles, *Adv. Powder Technol.*, **25**(1), pp. 56–70.
2. Feng, J., Biskos, G. and Schmidt-Ott, A. (2015). Toward industrial scale synthesis of ultrapure singlet nanoparticles with controllable sizes in a continuous gas-phase process, *Sci. Rep.*, **5**, p. 15788.
3. Valenti, M., Dolat, D., Biskos, G., Schmidt-Ott, A. and Smith, W. A. (2015). Enhancement of the photoelectrochemical performance of CuWO4 thin films for solar water splitting by plasmonic nanoparticle functionalization, *J. Phys. Chem. C*, **119**(4), pp. 2096–2104.
4. Maier, S. A. (2007). Surface plasmon polaritons at metal/insulator interfaces, in Maier, S. A. (ed.) *Plasmonics: Fundamentals and Applications*, Springer, New York, NY, pp. 21–37.
5. Ristig, S., Prymak, O., Loza, K., Gocyla, M., Meyer-Zaika, W., Heggen, M., Raabe, D. and Epple, M. (2015). Nanostructure of wet-chemically prepared, polymer-stabilized silver–gold nanoalloys (6 nm) over the entire composition range, *J. Mater. Chem. B*, **3**(23), pp. 4654–4662.
6. Lalisse, A., Tessier, G., Plain, J. and Baffou, G. (2015). Quantifying the efficiency of plasmonic materials for near-field enhancement and photothermal conversion, *J. Phys. Chem. C*, **119**(45), pp. 25518–25528.

7. Kelly, K. L., Coronado, E., Zhao, L. L. and Schatz, G. C. (2003). The optical properties of metal nanoparticles: the influence of size, shape, and dielectric environment, *J. Phys. Chem. B*, **107**(3), pp. 668–677.

8. Mubeen, S., Lee, J., Liu, D., Stucky, G. D. and Moskovits, M. (2015). Panchromatic photoproduction of H2 with surface plasmons, *Nano Lett.*, **15**(3), pp. 2132–2136.

9. Lozano, G., Rodriguez, S. R. K., Verschuuren, M. A. and Gómez Rivas, J. (2016). Metallic nanostructures for efficient LED lighting, *Light Sci. Appl.*, **5**, p. e16080.

10. Atwater, H. A. and Polman, A. (2010). Plasmonics for improved photovoltaic devices, *Nat. Mater.*, **9**, p. 205.

11. Sharma, B., Frontiera, R. R., Henry, A.-I., Ringe, E. and Van Duyne, R. P. (2012). SERS: materials, applications, and the future, *Mater. Today*, **15**(1), pp. 16–25.

12. Jeanmaire, D. L. and Van Duyne, R. P. (1977). Surface Raman spectroelectrochemistry: Part I. Heterocyclic, aromatic, and aliphatic amines adsorbed on the anodized silver electrode, *J. Electroanal. Chem. Interfacial Electrochem.*, **84**(1), pp. 1–20.

13. Baffou, G. (2017). *Thermoplasmonics: Heating Metal Nanoparticles Using Light*, Cambridge University Press, Cambridge.

14. Knight, M. W., Sobhani, H., Nordlander, P. and Halas, N. J. (2011). Photodetection with active optical antennas, *Science (80-)*, **332**(6030), p. 702 LP-704.

15. Brongersma, M. L., Halas, N. J. and Nordlander, P. (2015). Plasmon-induced hot carrier science and technology, *Nat. Nanotechnol.*, **10**, p. 25.

16. Mcmahon, M. D., Lopez, R., Meyer, H. M., Feldman, L. C. and Haglund, R. F. (2005). Rapid tarnishing of silver nanoparticles in ambient laboratory air, *Appl. Phys. B*, **80**(7), pp. 915–921.

17. El-Aal, M. A., Seto, T., Kumita, M., Abdelaziz, A. A. and Otani, Y. (2018). Synthesis of silver nanoparticles film by spark discharge deposition for surface-enhanced Raman scattering, *Opt. Mater. (Amst)*, **83**, pp. 263–271.

18. Clavero, C. (2014). Plasmon-induced hot-electron generation at nanoparticle/metal-oxide interfaces for photovoltaic and photocatalytic devices, *Nat. Photonics*, **8**, p. 95.

19. Sundararaman, R., Narang, P., Jermyn, A. S., Goddard III, W. A. and Atwater, H. A. (2014). Theoretical predictions for hot-carrier generation from surface plasmon decay, *Nat. Commun.*, **5**, p. 5788.

20. Schmidt-Ott, A., Schurtenberger, P. and Siegmann, H. C. (1980). Enormous yield of photoelectrons from small particles, *Phys. Rev. Lett.*, **45**(15), pp. 1284–1287.

21. Fowler, R. H. (1931). The analysis of photoelectric sensitivity curves for clean metals at various temperatures, *Phys. Rev.*, **38**(1), pp. 45–56.

22. Brown, A. M., Sundararaman, R., Narang, P., Goddard, W. A. and Atwater, H. A. (2016). Nonradiative plasmon decay and hot carrier dynamics: effects of phonons, surfaces, and geometry, *ACS Nano*, **10**(1), pp. 957–966.

23. Valenti, M., Venugopal, A., Tordera, D., Jonsson, M. P., Biskos, G., Schmidt-Ott, A. and Smith, W. A. (2017). Hot carrier generation and extraction of plasmonic alloy nanoparticles, *ACS Photonics*, **4**(5), pp. 1146–1152.

24. Huang, X., Jain, P. K., El-Sayed, I. H. and El-Sayed, M. A. (2007). Plasmonic photothermal therapy (PPTT) using gold nanoparticles, *Lasers Med. Sci.*, **23**(3), p. 217.

25. Baffou, G. and Quidant, R. (2014). Nanoplasmonics for chemistry, *Chem. Soc. Rev.*, **43**(11), pp. 3898–3907.

26. Baffou, G. and Quidant, R. (2012). Thermo-plasmonics: using metallic nanostructures as nano-sources of heat, *Laser Photon. Rev.*, **7**(2), pp. 171–187.

27. O'Connor, D. and Zayats, A. V. (2010). The third plasmonic revolution, *Nat. Nanotechnol.*, **5**, p. 482.

Chapter 13

Spark Ablation for Biomedical Application

Bijay Kumar Poudel,[a] Jungho Hwang,[a]
Jae Hong Park,[b] and Jeong Hoon Byeon[c]
[a]*School of Mechanical Engineering, Yonsei University,*
Seoul 03722, Republic of Korea
[b]*School of Health Sciences, Purdue University,*
IN 47907, USA
[c]*School of Mechanical Engineering, Yeungnam University,*
Gyeongsan 38541, Republic of Korea
postjb@yu.ac.kr

This chapter examines bioapplications of spark ablation (SA)-derived nanoparticles (NPs) and general principles behind these applications. As the bioapplication of SA-derived NPs is still in its infancy, many examples are drawn from NPs prepared from other synthetic routes, but could also be prepared by the continuous and versatile method of SA, and the general principles should also apply to designing SA-derived NPs in the future.

Spark Ablation: Building Blocks for Nanotechnology
Edited by Andreas Schmidt-Ott

ISBN 978-981-4800-82-2 (Hardcover), 978-0-367-81709-1 (eBook)
www.jennystanford.com

13.1 Introduction

Unlike their bulk counterparts, nanoparticles (NPs) possess a large surface-to-volume ratio, which endows them with unique physicochemical properties that can be exploited for biomedical applications. These remarkable physical, chemical, thermal, electrical, and optical properties of NPs are related to quantum size or surface effects. Some properties have more significance in biomedical applications, while other properties have relevance in industrial and environmental applications. Hence, it is imperative to design NPs with desirable size, shape, and surface properties for use in a diverse range of diagnostic and therapeutic interventions.

Spark ablation (SA), introduced by Schwyn et al. in 1988 [1], offers an interesting alternative to conventional wet-chemistry methods for the synthesis of metal NPs. The principle of SA is described in Chapter 2. Contrary to colloidal sol-gel methods of NP production, this gas-phase process is continuous and energy efficient and has the potential of controllable scalability capable of achieving ca. 100 kg/day production rates by using multiple electrodes (BUONAPART-E project) [2]. In addition, SA has the advantage of high-purity NP yields owing to the use of ultrapure carrier gas and electrode material without the contaminants associated with liquid solvents and surfactants.

Spark discharge has been used for the production of a variety of NPs by using electrodes of pure metals, alloyed metals, and some doped semiconductors like Si, Sb, and Ge (Fig. 13.1). The method is simple, and the only pertinent requirement for the starting material is that the electrodes should have suitable conductivity ($\rho < 0.2$ Ω.cm) [3–5]. And if the high temperatures reached during the spark decompose the electrodes, the compound is not necessarily formed again in condensation of the vapor (e.g., palladium hydride). Bimetallic or mixed metallic NPs can be made in a single spark by using two different electrode materials (cf. Chapter 5), electrodes consisting of an alloy, or electrodes made by magnetic compaction of different micrometer-size powders. Tabrizi [3] proved by using SA with copper and tungsten electrodes that even materials that are considered immiscible can be mixed on "a scale larger than the lattice constants but smaller than the particle diameter." Production of NPs from semiconductor materials is possible, especially if doped electrodes are used [4]. By using oxygen in the carrier gas, generation of metal oxide NPs such as CdO and Fe_2O_3 is also possible [6] (cf. Chapter 2). Other materials synthesized by SA include atomic clusters, carbon NPs, and carbon composites with metals [7]. More compounds like halogenides or sulfides could be prepared by adding respective gaseous reactants either upstream or downstream of the spark, although the formation of such compounds from SA has, to our knowledge, not been reported so far. A tube reactor can be added downstream, where an elevated temperature is necessary to induce the reaction (cf. Chapter 2). We expect that engineering of such solutions will be driven, among others, by attractive applications in the biomedical field.

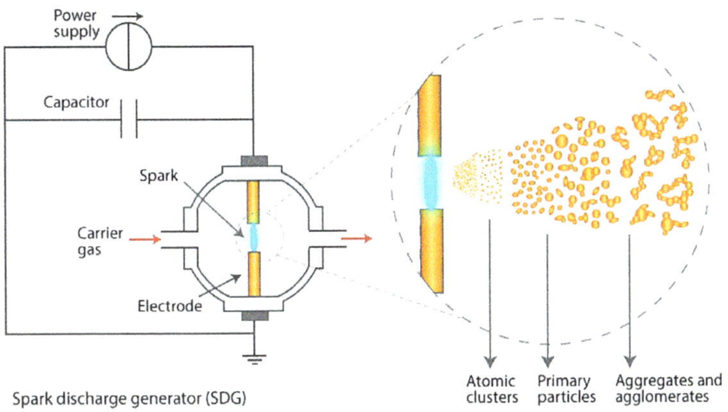

Figure 13.1 Spark discharge generator.

It should be noted that as-synthesized NPs from SA can rarely be used directly for clinical applications. Usually, inorganic NPs prepared by different synthetic routes are integrated or functionalized with biomaterials to enhance the biocompatibility for in vivo applications. NPs are loaded with drugs and/or imaging molecules and conjugated with targeting moieties for targeted and controlled delivery to the diseased site in colloidally stable form. These reactions are usually performed by conventional wet-chemical methods. Owing to high temperature, SA cannot be used directly to produce functionalized NPs, as organic/polymeric ligands decompose at temperatures >300°C [8]. One roundabout but tedious way can be collection of spark-generated NPs in a suitable liquid and use of conventional methods for conjugating organic functional groups. Chapter 2 (Section 2.4.1) refers to methods of bringing particles from SA into liquid suspension. Byeon and Roberts (2012) [9] performed the step of combining the inorganic and organic species in the aerosol phase by using a collison atomizer downstream of SA to directly combine NPs with an atomized carrier particle suspension containing also the functional material of interest, followed by heat treatment in a tubular heater to evaporate the solvent. They successfully utilized this continuous aerosol-based strategy to fabricate nanomagnetosols (comprising Fe NPs, the drug ketoprofen, and Eudragit shells), hybrid nanogels (comprising Fe NPs and *N*-isopropylacrylamide [NIPAM] gel or Fe-Au NPs and NIPAM gel), and nanoinorganics (gold-immobilized silica). These nanocomposites showed cell viability of >80% in human embryonic kidney cells, indicating excellent biocompatibility [9]. Byeon (2016) used a similar strategy for preparing multibiofunctional nanoplatforms consisting of a hard NP (silica, Au-decorated graphene oxide, or thiol-capped Au NP) as the core and a modified-polymer (cholesterol-chitosan linked with polyethyleneimine) as the shell for gene delivery applications [10].

13.2 Biomedical Applications

13.2.1 Metal-Derived NPs

13.2.1.1 Antimicrobial agents

Microbial infectious diseases are one of the main causes of global morbidity and mortality. In recent years, the emergence and spread of multidrug-resistant (MDR) bacterial strains has become increasingly widespread and has posed a major threat to public health worldwide. Metallic NPs such as Ag, Cu, Ti, Mg, Zn, and Au and their oxides display strong bactericidal potential and are now considered viable antimicrobial agents. Among these, Ag NPs have proven to be the most effective broad-spectrum antibacterial agents against over 650 microorganisms, including bacteria (both gram-negative and gram-positive), viruses, and pathogenic eukaryotes [11, 12]. Unlike conventionally used chemical antibiotics, these NPs do not pose direct and acute adverse effects, although chronic toxicological analysis is mandatory prior to clinical use. These NPs used as antimicrobial agents offer many distinctive advantages over conventional antibiotics, such as cost-effective synthesis, sterilization and purification, enhanced stability even during harsh environmental fluctuations, biocompatibility and nontoxicity, targetability to diseased site or cells via targeting ligands, lower potential of development of microbial resistance, controllable release, and enhanced microbial internalization [13].

The elucidation of the molecular mechanism of antimicrobial (bactericidal or bacteriostatic) activities of metallic and metallic oxide NPs has garnered considerable interest in recent years, especially with the emergence of drug-resistant pathogenic microorganisms. While the principal mechanism has been shown to be the generation of intracellular reactive oxygen species (ROS) and affinity of ions to biomolecular thiol groups, the overall activity is more likely to be the combination of multiple mechanisms that depend on the NP characteristics. Simultaneous action via multiple antimicrobial pathways significantly decreases the probability of concurrent acquisition of multiple resistance genes by microbes [14]. Colloidal metallic NPs continuously release ions due to the electrochemical potential in solution. These ions are believed to be the major contributor to the antibacterial mechanism of NPs. However, the antibacterial mechanism of these NPs has only been partially understood.

The bacterial cell wall carries patches of negatively charged domains possibly due to the presence of carboxylic acid groups of the membrane proteins, phospholipids, and lipopolysaccharides [15]. Cationic metal NPs or ions released from the NPs can electrostatically adhere and accumulate on the cell wall. NPs have a large specific surface-to-volume ratio, thus enabling increased contact and attachment with bacteria. This might result in the alteration of membrane integrity and permeability (further enabling uncontrollable transport of NP and ions into cells), pit formation and subsequent leakage of cellular contents, and disruption of chemiosmotic homeostasis of bacteria. For MgO NPs, electrostatic aggregation on cell walls without internalization is the major mechanism of antibacterial activity [16]. The destruction of the cell wall from physical contact of NPs is more pronounced in gram-negative bacteria, as they lack the thick protective peptidoglycan layer (ca. 80 nm) in contrast to gram-positive bacteria [17].

These NPs induce endogenous ROS generation (metabolic intermediates with highly positive redox potential, such as the superoxide anion, hydroxyl radicals, organic hydroperoxides, and hydrogen peroxides) [18]. ROS are possibly generated due to the reaction of metallic ions with thiol-containing amino acids of respiratory enzymes, resulting in the inhibition of the mitochondrial respiratory chain [19]. ROS can react and cause irreparable damage to almost all structural, metabolic, and enzymatic biomolecules [20–22].

Very small metal NPs (<12 nm) can directly penetrate the bacterial cell membrane and bind to different proteins and enzymes [17]. Their production by SA is extremely simple. Larger NPs can release metal ions that react with –SH groups of amino acids side chains within the proteins, leading to protein and enzyme dysfunction [23–27]. Ag^+ ions produced from Ag NPs also can bind directly to the DNA, leading to cell cycle arrest [25].

Photocatalytic antibacterial activity using NPs involves photoirradiation of specific wavelengths to generate ROS. Coatings of TiO_2 and ZnO have been studied for their virucidal, fungicidal, and bactericidal activity in applications like disinfection of medical devices, water filtration, and food hygiene [28–31]. Joost et al. (2014) used nanostructured films prepared from TiO_2 NPs to demonstrate excellent antibacterial activity against *E. coli* under UV-A light (λ = 315–400 nm, 22 W m^{-2}) of intensity comparable to that of the sunlight [32]. Upon illumination, electrons from the valence band absorb high-energy photons (λ < 400 nm) and are excited to the conduction band. The photoholes in the valence band they leave behind oxidize surface hydroxyl and titanol groups and absorb water into highly reactive hydroxyl radicals ($\cdot$OH) or hydrogen peroxide. The photoexcited electrons react with atmospheric oxygen (O_2) to form superoxide radicals ($\cdot O^{2-}$). The $\cdot$OH radicals can diffuse over short distances and are considered primary ROS, responsible for photocatalytic and antibacterial activity of TiO_2. ZnO NPs show similar energetic mechanism of ROS generation [32].

In a study conducted by Wypij et al. (2018), biosynthesized Ag NPs (ca. 5–20 nm) showed synergistic activity with commercial antibacterials (ampicillin, kanamycin, and tetracycline) against *E. coli*,

Klebsiella pneumoniae, *Staphylococcus aureus*, and *Bacillus subtilis*, and antifungals (amphotericin B, fluconazole, and ketoconazole) against *Candida albicans*. The fractional inhibitory concentration (FIC) index was found to be 0.12, which means microbial growth was inhibited by 1/16th of the minimum inhibitory concentration (MIC) of both Ag NPs and antimicrobial agents compared to when antimicrobials were used alone [33]. The same group also reported synergism of actinobacteria-synthesized Ag NPs (4–36 nm) with ketoconazole against pathogenic fungi (*Trichophyton rubrum*, *C. albicans*, *C. tropicalis*, and *Malassezia furfur*) responsible for superficial mycosis [34]. Similar reports have been published previously [35–37]. ZnO NPs (ca. 20 nm) prepared by the gel combustion method showed excellent antimicrobial activity against both gram-negative (*E. coli* and *P. aeruginosa*) and gram-positive (*B. subtilis* and *S. aureus*) bacteria [38].

Recently, multimetal oxide NPs were investigated to combine different properties of individual metals or oxides. Alzahrani et al. (2017) synthesized CuZnFe oxide NPs to combine antibacterial activity of ZnO and CuO with the biocompatibility of Fe_2O_3 NPs and found them to have higher antibacterial activity against *E. coli* compared to individual CuO and ZnO NPs [39]. The metal(s) or metal oxide components can be chosen judiciously to exhibit other advantageous properties, such as antibacterial selectivity (e.g., by MgO of ZnMgO NPs), colloidal stability (e.g., by Fe_3O_4 of Zn/Fe_3O_4 NPs), and photocatalytic antibacterial activity (by Au of Au/ZnO NPs) [40]. These are examples for the benefit of using mixed materials, which could be done with great ease by SA (see Chapter 5), and we expect that this feature will be of great value for biomedical applications.

The smaller size of NPs ensures a higher degree of interaction with bacterial cells due to the larger surface area-to-volume ratio, and thus higher antimicrobial activity. Nanosized Cu and CuO NPs showed higher toxicity against *E. coli* compared to micron-size Cu and CuO particles at the same Cu concentration, and interestingly, they exhibited a distinctly different mode of toxicity [41]. Similarly, the antimicrobial activity of Ag NPs has been shown to be size dependent [42, 43]. The shape dependency of the antimicrobial potential of Ag NPs was studied by Pal et al. (2007), who reported that triangular NPs were most active, while spherical NPs were more active than rod-shaped NPs against *E. coli* [44]. Generally, deformed NPs with many corners, edges (small radii of curvature), and defects exhibit increased antimicrobial activity [17]. Furthermore, the capping agent that is used to increase the colloidal stability and aqueous dispersibility of NPs can affect the antibacterial activity.

SA-derived metal NPs typically (though not necessarily; cf. Chapter 2) consist of agglomerates of very small primary particles that exhibit extremely small radii of curvature. These NPs or composites have been used for their antimicrobial properties. Byeon (2016) produced chitosan Cu-Ag nanobunches by injecting spark-produced Cu NPs into an ultrasound probe connected cell containing a solution of $AgNO_3$ and chitosan (used both as a reducing and a stabilizing agent). The resulting Cu-Ag nanobunches were electrosprayed (16 μL min^{-1}) over a glass plate or carbon fiber contacted with a plate heater. The coating was evaluated for its antimicrobial activity. The coating significantly inhibited the growth of *E. coli* and *S. aureus* in a concentration-dependent manner. The chitosan coating also increased the biocompatibility of Cu-Ag nanobunches in human embryonic kidney cells [45].

In another example, a thermoresponsive antimicrobial coating was prepared by SA. Si NPs and carbon nanotubes (CNTs) were synthesized as support by using Si-Si and Ag-graphitic carbon electrodes. The synthesized Si NPs and CNTs were separately injected into cells containing polydimethylsiloxane (PDMS), poly(*N*-isopropylacrylamide) (PNIPAM) (thermoresponsive polymer), and $AgNO_3$ under ultrasound irradiation. This resulted in the reduction of $AgNO_3$ to Ag NPs that attached to Si NPs and CNTs, which were then electrosprayed onto a touch-screen panel protection film. The nanocomposite and coating showed enhanced antimicrobial activity against *E. coli* and *S. epidermidis* [46].

13.2.1.2 Diagnostics, biosensing, and bioimaging

The recent development in nanotechnology has enabled the development of a wide range of biosensing platforms for fast, precise, affordable, and accessible diagnosis, prediction of relapse, and response to therapy, which ultimately translates to better prognosis. For example, many diseases/physiological conditions can be diagnosed quickly using point-of-care testing (POCT) and in vitro diagnosis (IVD) instead of traditional centralized lab-based testing.

Most biosensing diagnostic platforms use immunoassay modality, which involves antibodies to recognize and specifically bind to a targeted biomarker/analyte from the body fluid/medium (e.g., serum, urine). This antigen–antibody interaction provides high specificity to the detection of targeted low-level biomarkers from the complex media, while the diagnostic platforms contribute to high-sensitivity readouts. Typically, immunoassay detection is performed in "sandwich" format that involves multiple (mostly two) antigen/antibody bindings (Fig. 13.2). In the first step, the analyte (acts as antigen) in the test medium is bound by the antibody attached to a substrate. The substrate can be an inert glass platform or NPs. The second antibody that attaches to the captured analyte carries a label or tag capable of producing a concentration-dependent, measurable readout, such as fluorescence, chemiluminescence, or enzymatic action.

Biomarkers are mostly proteins, carbohydrates, or nucleic acids associated with a disease and detectable in blood or body fluid (saliva, urine, feces). Some of the Food and Drug Administration (FDA)-approved/cleared biomarkers are mentioned in Ref. [47].

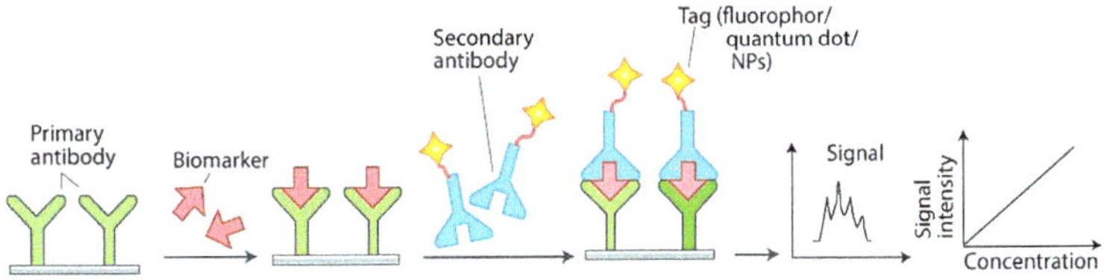

Figure 13.2 Schematics of the "sandwich" biosensor. Primary antibodies are immobilized on a substrate. After incubation with biomarkers, tagged secondary antibodies are used for signal generation.

13.2.1.2.1 *Localized surface plasmon resonance biosensing*

Diagnostic tests based on Au NPs and novel metal NPs are possible because of the NPs' distinct and uniquely tunable photophysical properties, including localized surface plasmon resonance (LSPR; see also Chapter 12, dedicated to this effect). They can be described as collective oscillations of the conductive electrons of NPs upon interaction with incident electromagnetic (EM) radiation (e.g., a laser beam) of resonant frequency (Fig. 13.3A). The NPs exhibit maximum optical extinction at this resonant frequency, which can be easily measured using ultraviolet-visible (UV-Vis) spectroscopy. The wavelength (λ_{max}) of LSPR is dependent on the geometry (shape, size, interparticle distance), composition, dielectric constant of the medium, and refractive environment, as described by Mie theory [48].

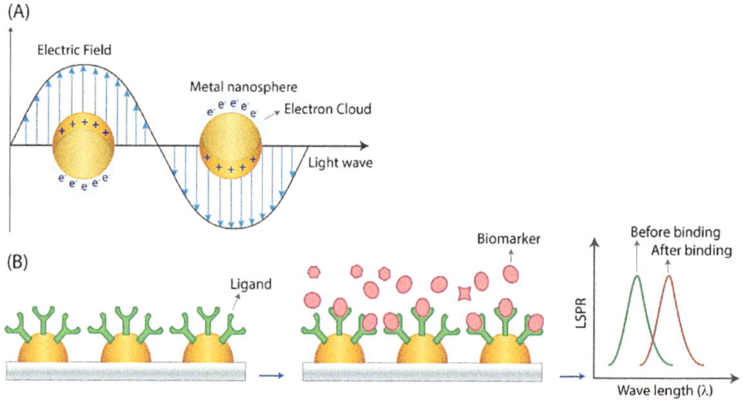

Figure 13.3 (A) LSPR in a metal nanosphere and (B) principle of an LSPR biosensor. Ligands on metal nanospheres bind with specific biomarkers, which increases the local refractive index and causes a shift in the LSPR position.

Apart from inherent characteristics of NPs, the dielectric constant of the surrounding medium has a large effect on the extinction of plasmonic NPs, which forms the basis of LSPR biosensing. In LSPR biosensors, a biological analyte binds or adsorbs on the plasmonic NPs, thereby changing the local refractive index, which, in turn, shifts the LSPR position (Fig. 13.3B). The LSPR shift upon analyte binding primarily depends on three variables: the change in the refractive index, the size of the analyte molecules, and the shape of NPs [49, 50]. The resonant wavelength λ_{max} of spherical noble metal NPs (gold, silver, copper) occurs in the visible range of the spectrum, while that of semiconductor NPs occurs in the near- or mid-infrared range [51, 52]. Spherical Au NPs have a deep-red color. As the size increases, the LSPR shifts, changing the color of the colloidal solution into blue or purple. The LSPR also shifts toward a higher wavelength when highly dispersed NPs aggregate, which might happen by design due to the presence of an analyte. This forms the basis of the colorimetric LSPR biosensor, which can be qualitative or quantitative. Hone et al. (2003) used mannopyranoside-functionalized Au NPs that bind to lectin molecules, which have multiple ligands and thus undergo aggregation, as multiple Au NPs can bind to a single lectin molecule. As the concentration of lectin increases, this induces Au NP growth, which is spectroscopically detectable by the resulting LSPR shift to longer wavelengths [53].

Ag NPs possess the sharpest peak among all metallic NPs, thus having the highest sensitivity to changes in the local refractive index [49]. However, Au NPs are preferred to Ag NPs because of their chemical inertness, biocompatibility, and surface modifiability. Also, Au nanostructures are considered well suited for human use because their surfaces are believed to be oxide free, unlike silver and copper nanostructures. However, for external plasmonic applications, silver nanostructures have been reported to perform better than gold nanostructures. Zhao et al. (2014) developed LSPR biosensors based on substrate-immobilized trianagular Ag NPs functionalized with monoclonal anti–squamous cell carcinoma antigen (anti-SCCa) antibodies that can directly detect the SCCa in blood samples via LSPR changes [54].

13.2.1.2.2 *Surface-enhanced Raman spectroscopy*

Surface-enhanced Raman spectroscopy (SERS) provides rapid, highly sensitive and label-free detection/quantification of small target analytes (e.g., specific biomarkers) without the need for expensive reagents. It also finds application in environmental monitoring, forensics, and analytical studies. Unlike fluorescent signals, SERS signals are narrower, so the label overlap is minimized and multiplex detection is possible. Also, SERS labels are not susceptible to photobleaching.

When photons are incident on an analyte molecule, the molecule scatters photons either of same frequency, that is, the same energy as incident photons (Rayleigh scattering) or photons of shifted frequency (different energy). The latter phenomenon is called Raman scattering. Raman scattering is an inelastic process in which scattered photons gain or lose energy to vibrational and rotational motion of analyte molecules. This produces bands corresponding to vibrational or rotational transitions specific to the molecular structure that serves as a chemical "fingerprint" to that molecule and provides information regarding the structure, interaction, or environment of the analyte [55, 56]. However, the probability of Raman scattering is extremely low as only about 1 in 10^6–10^{10} photons are scattered inelastically, which hinders its application as a routine platform for trace detection and diagnosis. In the 1970s, it was found that the amplification of Raman signal by factors from 10^6 to 10^{11} was possible when the targeted analyte was placed on or near (up to ca. 3 nm) NPs or nanostructured or roughened surfaces of plasmonic metals [57, 58]. This drastic amplification in Raman scattering signals is known as SERS. It allows the development of sensitive procedures for the characterization of target molecules. This signal amplification is attributed to the combination of EM and chemical enhancement mechanisms. See also Chapter 12 on SERS.

The EM enhancement is due to LSPR of nanostructured surfaces that is capable of increasing the incident electric field in their vicinity by 10^2–10^4, which amplifies Raman signals by factors of 10^6–10^8 [56]. The chemical enhancement of Raman signals is via chemisorption of the analytes on the SERS substrates, allowing electronic interactions between the analyte and the metal surface. Chemical enhancement factors as large as 10^5 have been reported [58].

Other than Au or Ag, metals such as Al, In, Cu, Pd, Pt, and Ga can also support plasmon resonances in the ultraviolet-visible-near-infrared (UV-Vis-NIR) range. Although abundant and cost effective, the rapid formation of their oxides has proven to be challenging in using Al and Cu for SERS, as oxides dampen the LSPR [56].

SERS has been used to detect exogeneous biomarkers, such as calcium dipicolinate, a biomarker for anthrax spores, using silver films over nanospheres (AgFON) as substrates with a limit of detection (LOD) of 10^{-14}–10^{-12} M [59] and endogenous biomarkers, such as glutathione with an LOD of 50 nM using colloidal Ag NPs as substrates [60].

SERS-based immunoassays have been used to detect and quantify proteins [61], immunoglobulins [62, 63], viruses [64, 65], bacteria [66], etc. The "sandwich" format (Fig. 13.4) is judiciously designed to read out highly amplified SERS signals. For example, it can involve a neutral substrate (e.g., glass, microtiter plate) coated with a primary antibody to the analyte under investigation. After binding with the targeted analyte from the medium, the substrate is exposed to Raman-labeled NPs, which are basically NPs conjugated with secondary antibodies, and the SERS signals are read. In another case, the substrate can be roughened metal or an NP-assembled surface, which acts as a SERS-active substrate, on which primary antibodies are immobilized. When exposed to the test medium, the analytes are captured by the primary antibodies. Raman-labeled NPs (metallic NPs conjugated with secondary antibodies) are then "sandwiched" with the SERS-active substrate. A strong SERS spectrum can be obtained under laser irradiation [67]. Depending upon the proximity, use of SERS-active substrates provides plasmonic coupling with Raman-labeled NPs, generating EM "hot spots" that further amplify the SERS signal. Thus, SERS signals can be modulated by tuning the gap between the "sandwich" and the LSPR of Raman-labeled tags [68]. In many studies, metallic NPs with monolayers of Raman-active molecules with large Raman cross sections, such as Cy3, 6G, Cy5, and rhodamine, have been used for in vivo imaging applications [67].

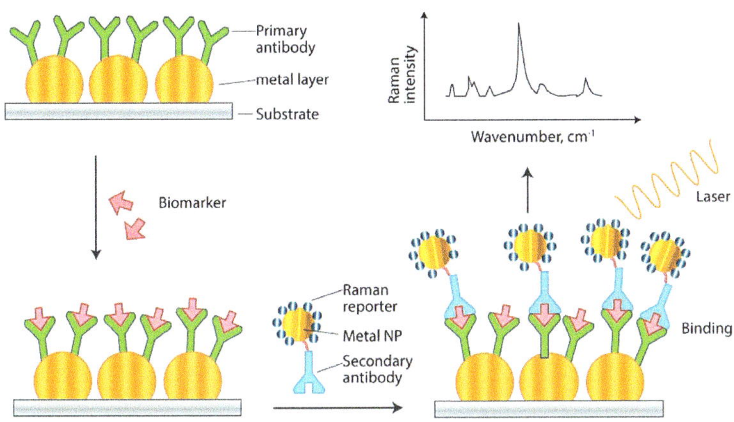

Figure 13.4 Principle of a SERS biosensor. Primary antibodies are immobilized on the metal layer that binds to specific biomarkers. Metal NPs conjugated with a Raman label and a secondary antibody are used to generate a strong SERS spectrum under laser irradiation.

SA has been used to fabricate SERS substrates. Byeon et al. (2015) fabricated a Ag dot and a ring-patterned silicon substrate for SERS. The spark-produced Pd particles were patterned on a silicon wafer through a hole of a shadow mask by thermophoretic deposition. The Pd-patterned wafer was immersed in a $AgNO_3$ solution for further growth of Ag on the Pd particles via electroless deposition. It showed a SERS enhancement by a factor of 1.8×10^5 [69].

Jung et al. (2014) fabricated a multipetal flower-shaped 3D metallic assembly composed of Cu NPs as a SERS substrate using pin-to-plate type SA and ion-induced electrostatic focusing. They demonstrated the control of the location and intensity of plasmonic hot spots as a function of the petal number. The SERS enhancement was reported to be as high as 1.91×10^7 [70]. The same group used SA and repeated ion-assisted aerosol lithography to vertically stack flower-shaped assemblies of Cu NPs to further increase SERS signals [71]. As reported in Chapter 12, the plasma resonance can easily be tuned in SA via the composition and particle size. This feature gives SA additional potential in SERS biomolecule sensing.

13.2.1.2.3 *Fluorescence modulation*

Fluorescence is an optical phenomenon where the absorption of a photon is followed by the emission of a photon of longer wavelength (less energy). The energy difference is the result of the vibrational relaxation and other photophysical events [47].

Fluorescence labeling and spectroscopy has become a dominant research tool in biological science, medical research, microscopy imaging, and optical devices because of its high sensitivity, low cost, ease of use, and versatility. It has been widely used in diagnostics procedures, such as in real-time quantitative polymerase chain reaction (qPCR) [72], DNA sequencing [73], flow cytometry [74], fluorescence microscopy [75], and in vivo imaging [76]. Fluorescent probes (fluorophores) have been engineered to possess higher emission intensity than background autofluorescence noise of the medium for use as an imaging tool in clinical diagnosis and for studying biological systems.

In recent years, much interest has been invested into improving the performance of existing fluorophores through incorporation with metallic NPs. When fluorophores are placed near metallic NPs, the emission intensity can be drastically enhanced, accompanied by a reduction of emission life time. Metal-enhanced fluorescence (MEF) increases the quantum yield and increases the photostability by reducing the excited-state lifetime and self-quenching.

Upon irradiation, the local EM field in the vicinity of metallic NPs increases strongly due to LSPR. A fluorophore placed in optimal proximity experiences (a) an increased excitation rate, (b) EM/plasmonic coupling, and (c) accelerated nonradiative decay, leading to increased quantum yield and decreased fluorescence emission life time. LSPR (SPR in the case of 2D metals)-induced changes in the excitation rate can be described by the Maxwell equation [77–79]. If the fluorophore is placed very near or onto the metal surface (<5 nm), the fluorescence is quenched through Förster energy transfer between them [80]. However, MEF does not occur if the fluorophore is placed too far from the NPs/metal surface, as the electric field enhancement due to LSPR decreases almost exponentially with distance from the metallic surface. Therefore, metallic NPs are able to induce both suppression (quenching) and enhancement of the fluorescence, depending on the proximity and other factors, and this requires fine-tuning to obtain the desired effect. Both effects have been used for the development of biosensors. Enhanced emission due to MEF has been used for higher signal-to-noise ratios (SNRs) during the detection of analytes, while the quenching effect has been used for the development of "on/off" sensors.

Various metal NP materials have been examined as MEF agents, including Au [81, 82], Ag [83, 84], Cu [85, 86], Al [87], Zn [88], and Rh [89] with various shapes and sizes. However, Au and Ag are typically preferred, as their LSPRs are in the visible or infrared region and they are chemically inert. The fluorophores are also selected on the basis of overlapping emission and absorption spectra. Fluorescent dyes emitting in the NIR region (700–900 nm) are preferred for bioapplication as tissue/water penetration is maximum with minimal background autofluorescence. MEF phenomena have been used for highly sensitive bioanalytical applications like DNA detection [90–92], RNA sensing [93], imaging [94], and immunoassay [95].

Many other factors play a significant role in fluorescent properties. According to Mie theory, small particles quench fluorescence by increasing absorption, while large particles enhance fluorescence by increasing scattering [96]. Metallic NP characteristics that modulate LSPR (as described before) are also factors in MEF, that is, stronger fluorescence enhancement was found when anisotropic NPs or surfaces with irregular topography, edges, and tips were used. Design factors such as the metal–fluorophor spacing, fluorophore density, and interparticle distance (for a multi-NP system) should be judiciously decided [97, 98]. The metal–fluorophor spacing and their spatial positioning are particularly important for colloidal NP-based MEF systems, and a variety of materials such as silica, polymers, DNA, and proteins have been used as spacers to separate fluorophores from NPs to prevent quenching [97].

As mentioned earlier, metallic NPs have also been used as fluorescence quenchers (on/off approach) in which fluorescence emission is dependent on the presence/absence of the biomarker. For example:

- The NP is functionalized with fluorophor-labeled single-stranded DNA (ssDNA) strands that form a hairpin structure, such that its stem loops to bring the fluorophore close to the NP, thereby quenching the fluorescence by the mechanism called nanoparticle surface energy transfer (NSET). In the presence of a target biomarker, the hairpin structure is disrupted by hybridization, leading to fluorescence recovery (Fig. 13.5A) [99, 100].
- NPs and fluorophores are functionalized with a reporter probe and ac apture probe, respectively. In the presence of a biomarker that cohybridizes with capture and reporter probes, the fluorophores are brought into close proximity, causing fluorescence quenching (Fig. 13.5B) [101].

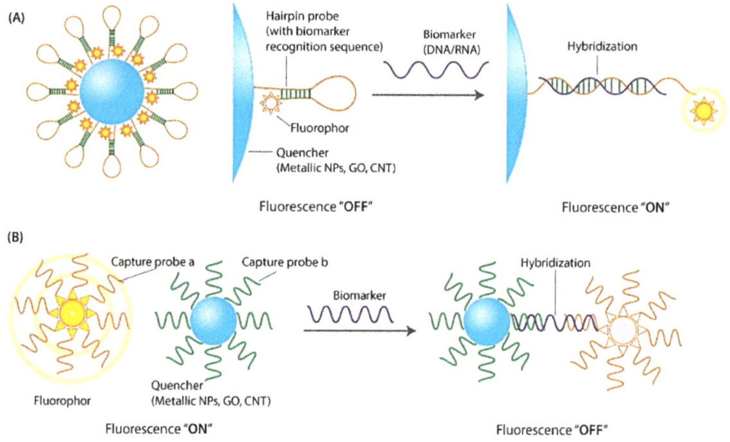

Figure 13.5 (A) The metal NP is functionalized with fluorophor-labeled ssDNA with a hairpin structure, such that its stem loops to bring the fluorophore close to the NP, thereby quenching the fluorescence. Binding with target biomarkers disrupts the hair-pin structure, causing the recovery of fluorescence. (B) NPs and fluorophores are functionalized with a reporter probe and a capture probe, respectively. The biomarker can cohybridize with both probes, leading to close proximity between NPs and the fluorophore, causing fluorescence quenching.

It would be quite straightforward to apply SA to produce multi-NP systems in fluorescence modulation. As with the other LSPR-related methods described before, the freedom in tuning the plasma resonance that SA provides would be beneficial.

13.2.1.2.4 *Photoacoustic imaging*

In photoacoustic (PA; optoacoustic) imaging, the tissue is irradiated with a nanosecond-pulse laser (5–10 ns; 25–30 mJ/cm^3) in the NIR region, which results in localized increase in temperature (ΔT) and a small transient thermoelastic expansion of tissue, causing a mechanical pressure rise (ρ_0) [102] according to

$$\rho_0 = \frac{\beta \cdot \Delta T}{\kappa}, \tag{13.1}$$

where, β is the thermal coefficient of volume expansion, which is ca. 4×10^{-4} K^{-1} for water or soft tissue, and κ is the isothermal compressibility, which is ca. 5×10^{-10} Pa^{-1} for muscle tissue.

Thus for soft tissue, each millikelvin temperature rise generates a pressure rise of 800 Pa. This pressure rise generates acoustic waves in the ultrasonic frequency range that is detected by a pressure transducer [103]. PA imaging is particularly useful because of its low cost, good depth of penetration (up to 6 cm), and high spatial contrast (as low as 6 μm) due to low ultrasonic scattering and a good SNR without using ionizing radiation [104].

The PA contrast is endogenously derived from variations in infrared absorption by intrinsic tissue chromophores such as oxygenated and deoxygenated hemoglobin, melanin, and lipid [105]. However, they lack the required specificity for diagnosis disease or tracking biological changes. Thus, exogenous contrast agents (CAs) are being used. Organic NIR dyes such as indocyanine green, rhodamine, and methylene blue have been traditionally used, but they are susceptible to photobleaching and are rapidly cleared from the blood.

Plasmonic metal NPs, especially Au NPs, are used in PA imaging as CAs due to their strong and tunable optical absorption resulting from the LSPR effect. The shape, size, and composition of these NPs can be changed to tune LSPR in the NIR range (λ = 680–900 nm) called "biological window" within which hemoglobin, water, and tissue components have the lowest absorption coefficient. For example, several configurations of Au NPs have been applied for PA imaging, including nanocages [106], nanorods [107, 108], hollow nanospheres [107], nanotubes [109], and nanoclusters [110]. The covalent surface modifiability of Au NPs via the strong gold–thiolate bond enables ligand conjugation for optimizing biocompatibility, for example, poly(ethyleneglycol) (PEG) functionalization and active targeting.

Carbon particles, including the allotropes of nanotubes, graphene, and nanodiamonds, can be synthesized to exhibit NIR absorption and offer high photostability [111–114]. Though carbonaceous nanostructures possess lower molar extinction coefficients and broad absorption spectra in the NIR region, they can be easily functionalized with dyes, drugs, polymers, and targeting moieties [115]. Another class of inorganic nanomaterials that have been investigated for excellent PA imaging are transition metal dichalcogenides, including TiS_2 [116], MoS_2 [117], Cu_2S [118], Co_9Se_8 [119], MoS_2/Bi_2S_3 nanocomposites [120], and Cu_2–xS [121].

It should be noted that similar to PA imaging in which photothermal heating is converted to acoustic waves, photothermal therapy (PTT) involves NIR light absorption to kill cancer cells. Hence, with judicious functionalization, PA agents might also be used in PTT for therapeutic and/or diagostic application. However, unlike PTT agents, which also work under irradiation with a continuous laser, PA agents can only work under pulsed laser irradiation.

Among the materials mentioned in this section, SA has shown to be well suited for production of round and fractal-like agglomerated metal and carbon NPs. The carbon NPs produced by SA have been well characterized and exhibit strong optical absorption [122]. Sulfides would require reactive modification of a metal, and selenides could be produced as alloys (see Chapters 2 and 5).

13.2.1.2.5 *MRI contrast agents*

Magnetic resonance imaging (MRI) has become one of the most promising imaging tools for clinical diagnosis of a wide range of diseases, such as cancer or neurological and cardiovascular diseases. This non-invasive technique provides high spatial and temporal resolution of soft tissues with good penetration depth without utilizing ionizing radiation (e.g., X-rays) and radiotracers (e.g., computed tomography [CT] scanning) [123].

MRI employs a strong static magnetic field (B_0) that aligns the magnetic moments of protons (^{1}H from water) in a sample in parallel (low-energy) or antiparallel (high-energy) configurations. Slightly more protons align parallel to B_0, that is, to the z axis or longitudinal axis, and the system gains "net magnetization" along the longitudinal axis. As the proton has angular momentum, it will rotate (precess) around the B_0 axis at a velocity called Larmor frequency.

When a second radiofrequency (RF) magnetic field $\mathbf{B_1}$ with frequency resonant to the Larmor frequency is applied perpendicularly to B_0 in short pulse, the proton absorbs energy and transfers from higher to lower energy levels and vice versa. At the same time, protons precess in phase. The net magnetization vector experiences torque and rotates away from the longitudinal direction at an angle called flip angle. The excited protons gradually relax back into their lower energy states [124]. The relaxation times depend on the body fluid or tissue, and space-resolved determination of the relaxation times composes the image. Where the difference between the relaxation times is not sufficient to provide a contrast, exogenous CAs absorbed by specific tissues are used modify the relaxation time here and therewith to increase the contrast with the neighboring tissue and thus to enhance the sensitivity of MRI. The major relaxation centers used for this purpose are transition metal ions (Fe^{2+}/Fe^{3+}, Mn^{2+}, Co^{2+}, and Cu^{2+}) and lanthanide metal ions (Gd^{3+}, Eu^{3+}, Ho^{3+}, and Dy^{3+}) [125, 126]. The most commonly used CAs involve gadolinium (III) ions chelates. However, Gd^{3+} ions have a high affinity to serum proteins, leading to very short circulation time and fast renal clearance, and can cause nephrogenic systemic toxicity [127, 128]. Recently, nanoparticulate CAs are being developed for their size-dependent magnetic properties, surface modifiability with functional moieties, biocompatibility, and shape regulation. This is where SA can play an important role in the future.

Gd^{3+}-based NPs such as $NaGdF_4$, [129] Gd_2O_3 [130], and GdF_3 [131] NPs have been developed as promising CAs. These NPs were prepared with precise size control, as their effect depends on their magnetic moment, and their average size is therefore important. They should be in a size range (typically smaller than 20 nm), where they are superparamagnetic at body temperature. Similarly, Mn-based NPs, such as MnO [132] and Mn_3O_4 [133], are another group of promising CAs. The surface Mn^{2+} ions have 5 unpaired electrons, which interact with proton spins, and this alters their relaxation properties. Fe-based NPs, especially superparamagnetic iron oxide nanoparticles (SPIONs), of size <20 nm form another class of highly investigated CAs [134, 135]. They are a nontoxic alternative to Gd^{3+} chelate–based CAs. Recently, ultrasmall (<5 nm) iron oxide NPs are being investigated as potential T1 CAs. Kim et al. (2011) prepared ultrasmall iron oxide NPs (1.5–12 nm) and investigated the size-dependent relaxation properties [136].

To date, there are no reports of MRI CA synthesis via SA. However, given that the synthesis of iron oxide [6] (see also Section 1.4.15, Chapter 1) and iron composite [137, 138] NPs has been reported using SA, the feasibility of MRI CA synthesis is to be expected. The versatility of SA regarding average size and composition (see Chapters 2 and 5) gives it great potential also for this application, especially if SA is followed by a reactor for chemical conversion by gas reaction.

13.2.2 Carbon Nanostructures

The discovery of buckminsterfullerenes (nano-C_{60}, buckyballs) by Kroto et al. in 1985 [139], CNTs by Iijima in 1991 [140], and exfoliated graphene by Novoselov and Geim in 2004 [141] spurred a rapid surge in the study of carbon nanostructures and allotropes in the field of basic science (photonics, electronics, and plasmonics) and applied research in energy storage, catalysis, sensor, bioimaging, and drug delivery.

13.2.2.1 Graphene

Graphene has an sp^2-hybridized planar (2D) honeycomb crystal lattice structure and offers a large surface area (>2600 $m^2\ g^{-1}$) and superior optical transmittance (97.7%), which is advantageous for the delivery of therapeutic agents and phototherapy [142]. Graphene or graphite can be chemically oxidized by a mixture of strong oxidants and concentrated acids to form graphene oxide (GO), which introduces various oxygen-containing functional groups (hydroxyl, carbonyl, lactols, epoxyl, carboxylic) at the basal plane or edge of the graphene sheet [143, 144]. The presence of oxygen functional groups imparts useful properties to GO, including amphiphilicity, aqueous dispersibility and stability, and facile surface modifiability. This has led to its use in drug delivery and biosensor platforms [145].

GO can adsorb many hydrophobic drugs with an aromatic structure, such as doxorubicin (DOX), via π–π stacking, hydrophobic interaction, hydrogen bonding, and electrostatic coupling. Thus, GO has been explored as a nanocarrier platform for drug delivery [146]. GO can strongly absorb NIR radiation and generate sufficient heat for killing cancer cells [158]. Several studies have proved that GO can protect noncovalently bound oligonucleotides from nuclease cleavage during inter- or intracellular delivery [147, 148].

Graphene acts as a fluorescence-quenching platform, similar to metallic NPs. Fluorescence resonance energy transfer (FRET) biosensors are based on the quenching or enhancement of fluorescence by the target biomarker as with Au NPs. Graphene-related materials have also shown superquenching capability toward various organic dyes and quantum dots (QDs) due to the long-range nanoscale resonance energy transfer. The large planar surface of GO allows physisorption of multiple DNA probes labeled with different fluorophores for the simultaneous detection of multiple targets in the same solution [149]. It shows higher binding affinity to ssDNA than to double-stranded DNA (dsDNA) and can be adsorbed on both sides by π–π stacking interaction between the ring structures in the nucleic acid bases and aromatic cells of the graphene [150]. This interaction is prevented in dsDNA as nucleic acid bases are shielded within the negatively charged phosphate backbone [151]. The adsorption of ssDNA increases the hole density in the graphene layer in the order of 5.61×10^{12} cm^{-2}. This ssDNA–graphene biointerface has been used in a field-effect transistor (FET) for the label free, reversible detection of complementary ssDNA [152].

GO also exhibits strong antimicrobial properties against a broad spectrum of pathogenic microorganisms. The antimicrobial property is due to physical and chemical interaction that causes membrane damage, probably by atomically thin sharp edges [153].

SA has been used to synthesize carbon NPs that were used as precursor for graphene synthesis. Byeon and Kim (2013) produced reduced graphene oxide (rGO) from spark-generated graphite NPs by reacting in simplified Hummer's solution (40 mL H_2SO_4 and 1.8 g $KMnO_4$) under ultrasound, followed by photoreduction as a continuous process [154]. In another study, GO particles of lateral dimension ca. 35 nm were produced from spark-generated carbon NPs, followed by Hummer's reaction. This GO was loaded with DOX and encapsulated within liposomes composed of 1,2-dipalmitoyl-sn-glycero-3-phosphocholine (DPPC), cholesterol (CHO), 1,2-distearoyl-snglycero-3-phosphoethanolamine-*N*-[methoxy(polyethylene glycol)-2000] (DSPE-PEG), and rapamycin. The lipid-coated GO containing DOX and rapamycin presented synergistic anticancer effects in MDA-MB-231, MCF-7, and BT474 cells, which was further enhanced by the NIR laser irradiation–induced photothermal effects of GO [166]. The same group layered spark-produced GO with chitosan-PEG and loaded DOX for synergistic chemotherapy and PTT in PC3 tumor-bearing Balb/c nude mice [155].

13.2.2.2 Carbon nanotubes

CNTs can be visualized as graphene rolled up as a seamless hollow cylinder with diameter of ~1 nm and length from 1 to 100 μm. Pristine CNTs have a very hydrophobic surface that imposes great limitations on their biomedical applications. Thus, CNTs are functionalized to increase their dispersibility and to conjugate moieties by covalent or noncovalent functionalization. CNTs are oxidized by refluxing with concentrated HNO_3, which adds defect sites and carboxylic groups at the tip, which are used for covalent functionalization. Thus similar to GO, oxidized CNTs are preferred for bioapplication as they present with functional groups and contain less metallic impurities. Oxidation of CNTs also breaks or shortens longer CNTs, which is desirable for transmembrane entry. Also, similar to GO, CNTs can form π–π interactions with drugs with an aromatic structure. However, oxidized CNTs form weaker π–π coupling [156]. Small molecular drugs like cisplatin can undergo endohedral filling in the inner empty cavity of CNTs [157]. To increase the aqueous dispersibility, CNTs are coated with amphiphilic surfactants or molecules. The large surface-to-volume ratio and hollow tubular structure contribute to higher density biomolecular conjugation [158].

CNTs show an extremely broad optical absorbance spectrum, covering both NIR I and II windows, that is, the biological transparency window that enables its use in in vivo photothermal ablation of cancer [159]. Drug-loaded CNTs are used in cancer therapy for combined chemotherapy and PTT. CNTs can bind DNA via π-electrons, and bound DNA is protected from enzymatic cleavage and interference from nucleic acid–binding proteins [160]. CNTs can efficiently cross the cellular barrier, which might be because of a passive "needle"-like structure.

Nanocomposite hydrogels were functionalized with GO to increase the chemical and physical stability and construct elastomeric scaffolds for tissue engineering [161]. Multiwalled COOH-functionalized CNTs were used to improve the conductivity of hydrogels, especially for scaffolds for cardiomyocytes [162, 163]. Multiwalled CNTs (MWCNTs) have shown great promise in orthopedic scaffolding and monitoring bone growth via electrochemical sensing techniques [164].

CNTs have been extensively studied for electrochemical glucose biosensing because they can increase electron transfer kinetics and sensitivity. Other analytes detected by CNT-based sensors include dopamine, insulin, streptavidin, immunoglobulin G (IgG), α-fetoproteins, C-reactive protein, human immunodeficiency virus 1 (HIV-1) protease, and prostate-specific antigen [155]. CNTs can be used as antimicrobial agents. They kill cells by inducing oxidative stress, cell membrane injury, and genotoxicity (by penetrating the nuclear membrane and binding to the G-C-rich region in the DNA sequence) [165, 166].

SA has been used for synthesizing CNTs and carbon-Ag nanocomposites. Byeon and Kim (2010) produced MWCNTs (ca. 16 nm diameter) by using SA with silver and graphite carbon as electrodes and by carefully controlling the cooling process near the spark by modulating the flow rate of carrier nitrogen gas. The spark-generated ZnO NPs were deposited onto a silicon substrate by thermophoretic deposition and the substrate used for growing a CNT array on ZnO via chemical vapor deposition (CVD) [167]. A similar approach using unagglomerated, equally sized metal NPs was introduced by Zonnevylle in 2009 [168].

13.2.2.3 Fluorescent carbon nanoparticles

Graphene quantum dots (GQDs) and carbon dots (CDs) are quantum-confined fluorescent carbon nanostructures that have been used in biosensing and bioimaging [169] and have a large potential here. These are less toxic, water soluble, chemically inert, environmentally friendly, and photostable compared to organic dyes and conventional semiconductor QDs such as CdX (X = S, Se, Te) and PbS. In addition, such particles have been shown to exhibit drug-loading capability for cancer chemotherapy.

CDs are quasispherical NPs (<10 nm) that are mostly amorphous, although nanocrystalline CDs with sp2 carbons have been reported [170]. These CDs are rich in surface functional groups, including carboxyl, amino, hydroxyl, and other groups, which enables covalent bonding for surface passivation. As SA of carbon electrodes produces particles that exhibit mostly amorphous structures, but sp2 carbons have also been reported [122], we regard it as quite probable that particles produced by SA can be converted to CDs by applying downstream reactions in the gas phase or when brought into the liquid phase. However, this has not been demonstrated yet, it may not be straightforward, and approaches for SA production of GQDs have not been reported either. We encourage considering SA, especially for CD production, but will not further review their applications in the biomedical field to fit into the scope of this book.

13.2.3 Quantum Dots

QDs are fluorescent single-crystal semiconductor NPs with dimensions in the range of 1 to 20 nm (or smaller than their bulk-exciton Bohr radius) and generally made up of elements from groups II (Zn, Cd, Hg) to VI (S, Se, Te), or groups III (In) to VI (Se), or groups IV (Pb, Sn) to VI (S, Se, Te), or transitional metals (Cu, Fe, Mo, W, Ag) to group VI (S, Se). Some of the most commonly investigated QDs are CdSe, InAs, CdS, and PbSe.

They absorb light of energy greater than their bandgap, resulting in the transition of electrons from the valence band to the conduction band, leaving behind a hole. This electron–hole pair (or exciton) recombines and emits a photon [171]. QDs generate a narrow emission profile with full width at half maximum (FWHM) of ca. 15–30 nm, depending on the size due to their quantum-confined properties. The nanocrystals smaller than the Bohr exciton radius of the material exhibit quantized energy states, with energy levels correlating to the QD size [172]. Thus, QD absorption and emission are size dependent, which enables the tunable design of QDs for multicolor labeling for the simultaneous detection and profiling of multiple targets. Larger QDs have smaller energy bandgaps and red-shifted emission. Generally, the quantum yield of QDs is improved by passivation with thin layers of ZnS or CdS, which have a larger energy bandgap than the core, resulting in exciton confinement in the core. This also leads to a reduction in surface trap states, leaching of core materials, and photodegradation [173].

QD absorption is broad with large molar absorption coefficients (100,000–1,000,000 M^{-1} cm^{-1}) compared to organic dyes (25,000–250,000 M^{-1} cm^{-1}), while emission peaks are narrow [47]. QDs also exhibit longer fluorescence lifetimes (>10 ns) compared to organic fluorophores (1–5 ns) [174]. This enables significant improvement in SNRs in biological applications for time-gated fluorescence studies, as autofluorescence of tissues and cells that contributes to a high background signal have shorter lifetimes. Despite these advantageous properties, QDs are toxic to living cells because of leaching of ions from the core of the QDs.

QDs have been used as superior fluorescent labels for visualizing cancer cells and bacteria in in vitro assays without affecting the cells [175]. Jaiswal et al. (2003) used CdSe/ZnS QDs for long-term multicolor imaging of live cells and demonstrated that labeling live cells does not affect their growth and cell signaling [176].

QDs are also reported for their application in in vivo imaging which requires superb biocompatibility (low toxicity), high photostability, and contrast. Yong et al. (2009) showed in vivo tumor targeting and imaging using CdSe/CdS/ZnS quantum nanorods coated with PEGylated phospholipids and conjugated with cyclic RGD peptide in tumor xenografts in mice [177].

SA has been used for the synthesis of transition metal chalcogenide QDs. Poudel et al. (2018) produced Ag_2S QDs in the gas-phase by continuous photoionization (λ = 185 nm) of spark-generated Ag NPs merged within 1-hexanethiol droplets via an aerosol reactor consisting of SA, a collison atomizer, and a UV-irradiation chamber [178].

13.2.4 Photothermal Therapy

In PTT, EM radiation is absorbed by photoabsorbers localized in the cancer cells or tumor interstitium to generate heat sufficient for thermal ablation of cancer cells. Photoabsorbers can be organic dye molecules or NPs. Compared to traditional treatment strategies (surgical resection, radiotherapy, chemotherapy), PTT is minimally invasive, non-ionizing, highly specific, and spatiotemporally selective. Unlike conventional cancer therapies, cancer cell death via the photothermal effect is immediate regardless of resistance to chemicals or radiation.

Nanoparticulate PTT agents are designed to be preferentially internalized by tumor cells either via the enhanced permeability and retention (EPR) effect by virtue of nanoscale size or via active targeting using specific ligands. Accumulation of PTT agents in tumors enables localized heating with minimal adverse effects on adjacent healthy tissue.

Human tissues exhibit strong extinction coefficients for EM radiation in the visible spectrum, that is, tissues are essentially opaque to visible radiation. Biological tissue transmissivity is highest in the NIR I (λ = 700–980 nm) and NIR II (λ = 1000–1400 nm) regions, allowing NIR lasers to penetrate to a depth of up to 10 cm. This is called the biological window [179].

PTT agents are designed to possess large absorption cross sections for laser wavelength in the biological window to allow maximum absorption of radiation. This ensures large light-to-heat conversion efficiency and helps reduce the irradiation intensity and time.

As mentioned in Section 13.2.1.2.1, metallic NPs (Au NP, Ag NP) exhibit LSPR in the visible region. The LSPR of these NPs can be tuned to the NIR range by changing their morphology or composition. The dynamics of plasmonic oscillations is strongly modified in anisotropic Au NPs (rod, shell, cage, hollow shell, nanoaggregates, and star), and Mie theory can no longer accurately predict the optical properties. Among these NPs of different shapes, Au nanorods have garnered a great deal of attention as they allow fine-tuning of LSPR on the basis of the aspect ratio (length:breadth) [180]. Metallic NPs constitute a major portion of PTT agents (see also Chapter 12 for thermoplasmonics). Carbon nanostructures constitute another class of nanostructures with superb PTT effects. GO and single-walled CNTs (SWCNTs) show broad extinction spectra along the whole biological window, while MWCNTs show several absorption peaks that correspond to the transition between van Hove singularities [181–183]. Unlike metallic NPs, the optical properties of carbon nanostructures do not depend on the particular geometry. Cu-based

QDs (CuS and CuSe) and Ge QDs are also reported to display spectral properties similar to noble metal NPs. Cu-based QDs, in particular, have been shown to generate substantial heat under NIR irradiation [184].

Many SA-generated NPs and nanocomposites have been reported to have excellent photothermal effects. In one study, Au-NP-decorated silica NPs (Au-SiO_2 NPs) were reported. A carrier gas with spark-generated Au NPs was passed over the atomizer orifice, where they mixed with the atomized droplets of SiO_2, ibuprofen (IBU), and poly-l-lysine (PLL) to form hybrid droplets. The droplets were dried by passing through a denuder to obtain Au-SiO_2/IBU-PLL NPs. The NPs exhibited a concentration-dependent increment in temperature ($\Delta T > 16°C$) upon NIR ($\lambda = 808$ nm) laser irradiation, which also resulted in augmented IBU release. The NPs showed higher transfection efficiency for plasmid DNA compared to polyethyleneimine. The viability of HeLa cells decreased significantly under NIR laser irradiation (5 mW cm^{-2}) when compared to non-irradiated cells, which might be because of increased IBU release and the photothermal effect [185]. In another report, spark-generated Au NPs were encapsulated in PDMS-PLL droplets from a collision atomizer in the gas phase to serve as a gene vector to HeLa cells with high efficiency and as a PTT agent [185].

Byeon and Kim (2013) synthesized chitosan-conjugated Ag dendritic NPs using a serial system consisting of SA that produced Pd NPs, a polyol cell, and a collison atomizer. The Pd NPs were directed to the ultrasound-assisted polyol cell containing a silver nitrate and chitosan solution, where the Pd NPs acted as seeds and a reductant for Ag growth, forming dendritic Ag NPs. The so-formed chitosan-conjugated dendritic NPs were able to electrostatically load plasmid DNA and transfect HeLa cells with high efficiency. The dendritic NPs showed LSPR in the NIR region. The NPs suspended in agar gel showed localized increase in temperature (>28°C) upon irradiation with an NIR laser of wavelength 830 nm (which was near the LSPR of dendritic Ag NPs). HeLa cells incubated with Ag NPs showed a viability of >85%, even at 100 μg mL^{-1} concentration, indicating excellent biocompatibility. However, cell viability decreased in a dose-dependent manner when exposed to laser irradiation, implying the feasibility of NPs as a PTT agent against cancer [186].

13.2.5 Drug Delivery

NPs are primarily used in cancer chemotherapy, especially owing to a phenomenon called the EPR effect. Rapidly growing tumors (>2 mm^3) initiate local angiogenesis (formation of blood vessels) to accommodate the ever-increasing demand of nutrients and oxygen. The angiogenesis is aberrant and shows morphological abnormalities. The fenestration between the endothelial cells that line the blood vessel wall of the tumor vasculature becomes much larger (from 150 nm to 2 μm, depending on the tumor type) than in normal tissues (ca. 10 nm) [187]. Thus, NPs (<200 nm) can extravasate from the bloodstream via "leaky" vasculature and preferentially accumulate into tumor tissue. In addition, these tumor masses tend to have poor lymphatic drainage, which leads to long retention times of NPs in the tumor. This is called the EPR effect, or passive targeting.

13.2.5.1 Factors affecting in vivo use

The optimum NP-based nanomedicine warrants elucidation of well-defined in vivo behaviors (biodistribution, clearance, toxicological profile, and biocompatibility). The physicochemical properties, such as shape, size, and surface properties, are known to significantly affect the pharmacokinetics and biocompatibility of NPs. The overall effect of shape, size, and surface properties of NPs in determining the pharmacokinetic profiles is the interplay of these interdependent variables and should be understood in relation to each other.

13.2.5.1.1 *Nanoparticle surface*

Surface properties of NPs primarily influence their interactions with cells and tissues in the body. NPs undergo opsonization, a process by which serum proteins adsorb onto the NP surface, enabling the cell surface receptors of resident mononuclear phagocytes in the reticuloendothelial system (RES) organs (e.g., hepatocellular Kupffer cells, alveolar macrophages, lymphatic dendritic cells, B cells in spleen) to recognize and bind to NPs, inducing rapid clearance [188, 189]. Hence, capping agents are used to passivate the NP surface, which reduces nonspecific adsorption of serum proteins and enhances circulation time [190, 191]. Capping agents/ligands for NPs designed for in vivo applications are mostly hydrophilic polymer brushes that minimize protein adsorption. PEG is the most commonly used capping agent [190]. Capping agents also help maintain colloidal stability via the steric effect. It should be noted that shape (curvature) and size also affect opsonization of NPs [192].

13.2.5.1.2 *Nanoparticle shape*

Generally, anisotropic NPs exhibit longer circulation times in the blood, which is possibly due to their decreased probability to permeate endothelial gaps of hepatic fenestrations [47]. Arnida et al. (2008) reported that rod-shaped Au NPs accumulated more efficiently in tumors and had a longer circulation time compared to spherical Au NPs. Rod-shaped Au NPs also showed lesser uptake by macrophages in vitro compared to spherical NPs [193].

13.2.5.1.3 *Nanoparticle size*

For intravenously administered particles, the circulation system has a variety of size-based clearance checkpoints. Large particles are typically trapped in the capillary beds of the lungs [194]. Microparticles (1–5 μm) are mostly trapped in the liver, while NPs larger than 200 nm are mechanically filtered in the spleen [195]. Particles smaller than 200 nm extravasate to tissues through the vascular endothelial lining. Interestingly, particles with hydrodynamic diameters less than 6 nm can be captured by the liver and can pass through the glomerular capillary wall of kidneys for rapid excretion. This reduces tissue accumulation and diminishes in vivo toxicity [196]. Jong et al. (2008) found that the biodistribution of Au NPs was size dependent, with Au NPs ca. 10 nm in size showing the most widespread organ distribution in the heart, lungs, liver, spleen, kidneys, thymus, brain, and reproductive organs compared to larger Au NPs (50, 100, and 250 nm) [197]. A similar finding was reported by Sonavane et al. (2008) [198].

13.2.5.2 Surface modification for in vivo use

NPs produced by SA are clean and can be produced without any adsorbates, capping agents, or ligands. This provides maximum freedom in adding any species the application requires. The high surface-area-to-volume ratio of NPs enables surface loading with various cargoes, including targeting moieties, therapeutic agents, imaging agents, and other functional molecules. Judicious surface modification or ligand coating confers NPs with:

- Enhanced colloidal stability in storage media (water, saline) via electrostatic or steric repulsion
- Enhanced stability in physiological milieu (blood, extracellular fluid)
- Increased biocompatibility and modulation of biodistribution
- Stimuli sensitivity by using polymers that undergo reversible conformational changes as a function of external stimuli (temperature, light, heat, ultrasound, magnetic field, and mechanical strength) or internal stimuli (pH variation, ionic strength, hormone level, glucose level, redox gradient, and enzyme concentration)
- Active targeting to diseased site by using moieties (antibodies, aptamers, and peptides) that recognize and bind to molecules overexpressed at the diseased site via antigen–antibody or ligand–receptor interactions, followed by receptor-mediated endocytosis
- Encapsulation of therapeutic agents in surface ligands via conjugation, complexation, partitioning, or host–guest interaction

13.3 Conclusion

SA provides a simple, cost-effective, and scalable process for the fabrication of ultrapure mono- and multimetallic NPs. It is a versatile, environmentally benign, and continuous process that can be potentially used for the manufacture of biomedical platforms, including biosensing, diagnosis, drug/biomolecule delivery, and therapy. However, reports of its use in biomedical and clinical fields are still scant, probably owing to the novelty of the method. The SA process also requires consecutive steps to make the metallic nanoplatforms biologically compatible. This chapter is an attempt to explore the feasible bioapplicable fields where the SA process can be successfully used.

References

1. Schwyn, S., Garwin, E. and Schmidt-Ott, A. (1988). Aerosol generation by spark discharge, *J. Aerosol Sci.*, **19**(5), pp. 639–642.
2. http://www.buonapart-e.eu/
3. Tabrizi, N. S., Ullmann, M., Vons, V., Lafont, U. and Schmidt-Ott, A. (2009). Generation of nanoparticles by spark discharge, *J. Nanopart. Res.*, **11**(2), p. 315.
4. Vons, V. A., de Smet, L. C., Munao, D., Evirgen, A., Kelder, E. M. and Schmidt-Ott, A. (2011). Silicon nanoparticles produced by spark discharge, *J. Nanopart. Res.*, **13**(10), p. 4867.
5. Feng, J., Biskos, G. and Schmidt-Ott, A. (2015). Toward industrial scale synthesis of ultrapure singlet nanoparticles with controllable sizes in a continuous gas-phase process, *Sci. Rep.*, **5**, p. 15788.
6. Roth, C., Ferron, G. A., Karg, E., Lentner, B., Schumann, G., Takenaka, S. and Heyder, J. (2004). Generation of ultrafine particles by spark discharging, *Aerosol Sci. Technol.*, **38**(3), pp. 228–235.
7. Byeon, J. H. and Kim, J.-W. (2010). Production of carbonaceous nanostructures from a silver-carbon ambient spark, *Appl. Phys. Lett.*, **96**(15), p. 153102.
8. Sanchez, C., Soler-Illia, G. D. A., Ribot, F., Lalot, T., Mayer, C. R. and Cabuil, V. (2001). Designed hybrid organic–inorganic nanocomposites from functional nanobuilding blocks, *Chem. Mater.*, **13**(10), pp. 3061–3083.
9. Byeon, J. H. and Roberts, J. T. (2012). Aerosol-based fabrication of biocompatible organic–inorganic nanocomposites, *ACS Appl. Mater. Interfaces*, **4**(5), pp. 2693–2698.
10. Byeon, J. H. (2016). Aerosol nanoencapsulation: single-pass floating self-assembly of biofunctional hybrid nanoplatforms, *ACS Appl. Mater. Interfaces*, **8**(28), pp. 17757–17762.
11. Rai, M., Deshmukh, S., Ingle, A. and Gade, A. (2012). Silver nanoparticles: the powerful nanoweapon against multidrug-resistant bacteria, *J. Appl. Microbiol.*, **112**(5), pp. 841–852.
12. Malarkodi, C., Rajeshkumar, S., Paulkumar, K., Gnanajobitha, G., Vanaja, M. and Annadurai, G. (2013). Biosynthesis of semiconductor nanoparticles by using sulfur reducing bacteria Serratia nematodiphila, *Adv. Nano Res.*, **1**(2), pp. 83–91.
13. Huh, A. J. and Kwon, Y. J. (2011). "Nanoantibiotics": a new paradigm for treating infectious diseases using nanomaterials in the antibiotics resistant era, *J. Control. Release*, **156**(2), pp. 128–145.

14. Pelgrift, R. Y. and Friedman, A. J. (2013). Nanotechnology as a therapeutic tool to combat microbial resistance, *Adv. Drug Deliv. Rev.*, **65**(13–14), pp. 1803–1815.
15. Silhavy, T. J., Kahne, D. and Walker, S. (2010). The bacterial cell envelope, *Cold Spring Harbour Perspect. Biol.*, **2**(5), p. a000414.
16. Leung, Y. H., Ng, A., Xu, X., Shen, Z., Gethings, L. A., Wong, M. T., Chan, C., Guo, M. Y., Ng, Y. H. and Djurišić, A. B. (2014). Mechanisms of antibacterial activity of MgO: non-ROS mediated toxicity of MgO nanoparticles towards Escherichia coli, *Small*, **10**(6), pp. 1171–1183.
17. Slavin, Y. N., Asnis, J., Häfeli, U. O. and Bach, H. (2017). Metal nanoparticles: understanding the mechanisms behind antibacterial activity, *J. Nanobiotechnol.*, **15**(1), p. 65.
18. Raghunath, A. and Perumal, E. (2017). Metal oxide nanoparticles as antimicrobial agents: a promise for the future, *Int. J. Antimicrob. Agents*, **49**(2), pp. 137–152.
19. Matsumura, Y., Yoshikata, K., Kunisaki, S.-i. and Tsuchido, T. (2003). Mode of bactericidal action of silver zeolite and its comparison with that of silver nitrate, *Appl. Environ. Microbiol.*, **69**(7), pp. 4278–4281.
20. Vatansever, F., de Melo, W. C., Avci, P., Vecchio, D., Sadasivam, M., Gupta, A., Chandran, R., Karimi, M., Parizotto, N. A. and Yin, R. (2013). Antimicrobial strategies centered around reactive oxygen species–bactericidal antibiotics, photodynamic therapy, and beyond, *FEMS Microbiol. Rev.*, **37**(6), pp. 955–989.
21. Bogdan, J., Zarzyńska, J. and Pławińska-Czarnak, J. (2015). Comparison of infectious agents susceptibility to photocatalytic effects of nanosized titanium and zinc oxides: a practical approach, *Nanoscale Res. Lett.*, **10**(1), p. 309.
22. Nel, A., Xia, T., Mädler, L. and Li, N. (2006). Toxic potential of materials at the nanolevel, *Science*, **311**(5761), pp. 622–627.
23. Lill, R. (2009). Function and biogenesis of iron–sulphur proteins, *Nature*, **460**(7257), p. 831.
24. Wigginton, N. S., Titta, A. D., Piccapietra, F., Dobias, J., Nesatyy, V. J., Suter, M. J. and Bernier-Latmani, R. (2010). Binding of silver nanoparticles to bacterial proteins depends on surface modifications and inhibits enzymatic activity, *Environ. Sci. Technol.*, **44**(6), pp. 2163–2168.
25. Feng, Q. L., Wu, J., Chen, G., Cui, F., Kim, T. and Kim, J. (2000). A mechanistic study of the antibacterial effect of silver ions on Escherichia coli and Staphylococcus aureus, *J. Biomed. Mater. Res.*, **52**(4), pp. 662–668.

26. Macomber, L. and Imlay, J. A. (2009). The iron-sulfur clusters of dehydratases are primary intracellular targets of copper toxicity, *Proc. Natl. Acad. Sci. U.S.A.*, **106**(20), pp. 8344–8349.

27. Yamanaka, M., Hara, K. and Kudo, J. (2005). Bactericidal actions of a silver ion solution on Escherichia coli, studied by energy-filtering transmission electron microscopy and proteomic analysis, *Appl. Environ. Microbiol.*, **71**(11), pp. 7589–7593.

28. Azimzadehirani, M., Elahifard, M., Haghighi, S. and Gholami, M. (2013). Highly efficient hydroxyapatite/TiO_2 composites covered by silver halides as E. coli disinfectant under visible light and dark media, *Photochem. Photobiol. Sci.*, **12**(10), pp. 1787–1794.

29. Lonnen, J., Kilvington, S., Kehoe, S., Al-Touati, F. and McGuigan, K. (2005). Solar and photocatalytic disinfection of protozoan, fungal and bacterial microbes in drinking water, *Water Res.*, **39**(5), pp. 877–883.

30. Foster, H. A., Ditta, I. B., Varghese, S. and Steele, A. (2011). Photocatalytic disinfection using titanium dioxide: spectrum and mechanism of antimicrobial activity, *Appl. Microbiol. Biotechnol.*, **90**(6), pp. 1847–1868.

31. Tankhiwale, R. and Bajpai, S. (2012). Preparation, characterization and antibacterial applications of ZnO-nanoparticles coated polyethylene films for food packaging, *Colloids Surf., B*, **90**, pp. 16–20.

32. Joost, U., Juganson, K., Visnapuu, M., Mortimer, M., Kahru, A., Nõmmiste, E., Joost, U., Kisand, V. and Ivask, A. (2015). Photocatalytic antibacterial activity of nano-TiO_2 (anatase)-based thin films: effects on Escherichia coli cells and fatty acids, *J. Photochem. Photobiol., B*, **142**, pp. 178–185.

33. Wypij, M., Czarnecka, J., Świecimska, M., Dahm, H., Rai, M. and Golinska, P. (2018). Synthesis, characterization and evaluation of antimicrobial and cytotoxic activities of biogenic silver nanoparticles synthesized from Streptomyces xinghaiensis OF1 strain, *World J. Microbiol. Biotechnol.*, **34**(2), p. 23.

34. Anasane, N., Golińska, P., Wypij, M., Rathod, D., Dahm, H. and Rai, M. (2016). Acidophilic actinobacteria synthesised silver nanoparticles showed remarkable activity against fungi-causing superficial mycoses in humans, *Mycoses*, **59**(3), pp. 157–166.

35. Birla, S., Tiwari, V., Gade, A., Ingle, A., Yadav, A. and Rai, M. (2009). Fabrication of silver nanoparticles by Phoma glomerata and its combined effect against Escherichia coli, Pseudomonas aeruginosa and Staphylococcus aureus, *Lett. Appl. Microbiol.*, **48**(2), pp. 173–179.

36. Gajbhiye, M., Kesharwani, J., Ingle, A., Gade, A. and Rai, M. (2009). Fungus-mediated synthesis of silver nanoparticles and their activity against pathogenic fungi in combination with fluconazole, *Nanomed. Nanotechnol. Biol. Med.*, **5**(4), pp. 382–386.

37. Shahverdi, A. R., Fakhimi, A., Shahverdi, H. R. and Minaian, S. (2007). Synthesis and effect of silver nanoparticles on the antibacterial activity of different antibiotics against Staphylococcus aureus and Escherichia coli, *Nanomed. Nanotechnol. Biol. Med.*, **3**(2), pp. 168–171.

38. Azam, A., Ahmed, A. S., Oves, M., Khan, M. S., Habib, S. S. and Memic, A. (2012). Antimicrobial activity of metal oxide nanoparticles against gram-positive and gram-negative bacteria: a comparative study, *Int. J. Nanomed.*, **7**, p. 6003.

39. Alzahrani, K. E., Niazy, A. A., Alswieleh, A. M., Wahab, R., El-Toni, A. M. and Alghamdi, H. S. (2018). antibacterial activity of trimetal (cuZnFe) oxide nanoparticles, *Int. J. Nanomed.*, **13**, p. 77.

40. Stankic, S., Suman, S., Haque, F. and Vidic, J. (2016). Pure and multi metal oxide nanoparticles: synthesis, antibacterial and cytotoxic properties, *J. Nanobiotechnol.*, **14**(1), p. 73.

41. Kaweeteerawat, C., Chang, C. H., Roy, K. R., Liu, R., Li, R., Toso, D., Fischer, H., Ivask, A., Ji, Z. and Zink, J. I. (2015). Cu nanoparticles have different impacts in Escherichia coli and Lactobacillus brevis than their microsized and ionic analogues, *ACS Nano*, **9**(7), pp. 7215–7225.

42. Panáček, A., Kvitek, L., Prucek, R., Kolář, M., Večeřová, R., Pizúrová, N., Sharma, V. K., Nevěčná, T. J. and Zbořil, R. (2006). Silver colloid nanoparticles: synthesis, characterization, and their antibacterial activity, *J. Phys. Chem. B*, **110**(33), pp. 16248–16253.

43. Morones, J. R., Elechiguerra, J. L., Camacho, A., Holt, K., Kouri, J. B., Ramírez, J. T. and Yacaman, M. J. (2005). The bactericidal effect of silver nanoparticles, *Nanotechnology*, **16**(10), p. 2346.

44. Pal, S., Tak, Y. K. and Song, J. M. (2007). Does the antibacterial activity of silver nanoparticles depend on the shape of the nanoparticle? A study of the gram-negative bacterium Escherichia coli, *Appl. Environ. Microbiol.*, **73**(6), pp. 1712–1720.

45. Byeon, J. H. (2016). Rapid green assembly of antimicrobial nanobunches, *Sci. Rep.*, **6**, p. 27006.

46. Poudel, B. K., Park, J. H. and Byeon, J. H. (2017). On-demand gas-to-liquid process to fabricate thermoresponsive antimicrobial nanocomposites and coatings, *ACS Appl. Mater. Interfaces*, **9**(18), pp. 15342–15349.

47. Chinen, A. B., Guan, C. M., Ferrer, J. R., Barnaby, S. N., Merkel, T. J. and Mirkin, C. A. (2015). Nanoparticle probes for the detection of cancer biomarkers, cells, and tissues by fluorescence, *Chem. Rev.*, **115**(19), pp. 10530–10574.

48. Huang, X. and El-Sayed, M. A. (2011). Plasmonic photo-thermal therapy (PPTT), *Alex J. Med.*, **47**(1), pp. 1–9.

49. Unser, S., Bruzas, I., He, J. and Sagle, L. (2015). Localized surface plasmon resonance biosensing: current challenges and approaches, *Sensors*, **15**(7), pp. 15684–15716.

50. Huang, X. and El-Sayed, M. A. (2010). Gold nanoparticles: optical properties and implementations in cancer diagnosis and photothermal therapy, *J. Adv. Res.*, **1**(1), pp. 13–28.

51. Jain, P. K., Lee, K. S., El-Sayed, I. H. and El-Sayed, M. A. (2006). Calculated absorption and scattering properties of gold nanoparticles of different size, shape, and composition: applications in biological imaging and biomedicine, *J. Phys. Chem. B*, **110**(14), pp. 7238–7248.

52. Yang, L., Jiang, X., Ruan, W., Zhao, B., Xu, W. and Lombardi, J. R. (2008). Observation of enhanced Raman scattering for molecules adsorbed on TiO_2 nanoparticles: charge-transfer contribution, *J. Phys. Chem. C*, **112**(50), pp. 20095–20098.

53. Hone, D. C., Haines, A. H. and Russell, D. A. (2003). Rapid, quantitative colorimetric detection of a lectin using mannose-stabilized gold nanoparticles, *Langmuir*, **19**(17), pp. 7141–7144.

54. Zhao, Q., Duan, R., Yuan, J., Quan, Y., Yang, H. and Xi, M. (2014). A reusable localized surface plasmon resonance biosensor for quantitative detection of serum squamous cell carcinoma antigen in cervical cancer patients based on silver nanoparticles array, *Int. J. Nanomed.*, **9**, p. 1097.

55. Muehlethaler, C., Leona, M. and Lombardi, J. R. (2015). Review of surface enhanced Raman scattering applications in forensic science, *Anal. Chem.*, **88**(1), pp. 152–169.

56. Bantz, K. C., Meyer, A. F., Wittenberg, N. J., Im, H., Kurtuluş, Ö., Lee, S. H., Lindquist, N. C., Oh, S.-H. and Haynes, C. L. (2011). Recent progress in SERS biosensing, *PCCP*, **13**(24), pp. 11551–11567.

57. Jeanmaire, D. L. and Van Duyne, R. P. (1977). Surface Raman spectroelectrochemistry: Part I. Heterocyclic, aromatic, and aliphatic amines adsorbed on the anodized silver electrode, *J. Electroanal. Chem. Interfacial Electrochem.*, **84**(1), pp. 1–20.

58. Zhao, L. L., Jensen, L. and Schatz, G. C. (2006). Surface-enhanced Raman scattering of pyrazine at the junction between two Ag_{20} nanoclusters, *Nano Lett.*, **6**(6), pp. 1229–1234.

59. Zhang, X., Young, M. A., Lyandres, O. and Van Duyne, R. P. (2005). Rapid detection of an anthrax biomarker by surface-enhanced Raman spectroscopy, *J. Am. Chem. Soc.*, **127**(12), pp. 4484–4489.

60. Huang, G. G., Han, X. X., Hossain, M. K. and Ozaki, Y. (2009). Development of a heat-induced surface-enhanced Raman scattering sensing method for rapid detection of glutathione in aqueous solutions, *Anal. Chem.*, **81**(14), pp. 5881–5888.

61. Lin, C.-C., Yang, Y.-M., Chen, Y.-F., Yang, T.-S. and Chang, H.-C. (2008). A new protein A assay based on Raman reporter labeled immunogold nanoparticles, *Biosens. Bioelectron.*, **24**(2), pp. 178–183.

62. Han, X. X., Cai, L. J., Guo, J., Wang, C. X., Ruan, W. D., Han, W. Y., Xu, W. Q., Zhao, B. and Ozaki, Y. (2008). Fluorescein isothiocyanate linked immunoabsorbent assay based on surface-enhanced resonance Raman scattering, *Anal. Chem.*, **80**(8), pp. 3020–3024.

63. Shin, M. H., Hong, W., Sa, Y., Chen, L., Jung, Y.-J., Wang, X., Zhao, B. and Jung, Y. M. (2014). Multiple detection of proteins by SERS-based immunoassay with core shell magnetic gold nanoparticles, *Vib. Spectrosc.*, **72**, pp. 44–49.

64. Driskell, J. D., Kwarta, K. M., Lipert, R. J., Porter, M. D., Neill, J. D. and Ridpath, J. F. (2005). Low-level detection of viral pathogens by a surface-enhanced Raman scattering based immunoassay, *Anal. Chem.*, **77**(19), pp. 6147–6154.

65. Xu, S., Ji, X., Xu, W., Li, X., Wang, L., Bai, Y., Zhao, B. and Ozaki, Y. (2004). Immunoassay using probe-labelling immunogold nanoparticles with silver staining enhancement via surface-enhanced Raman scattering, *Analyst*, **129**(1), pp. 63–68.

66. Yakes, B. J., Lipert, R. J., Bannantine, J. P. and Porter, M. D. (2008). Detection of Mycobacterium avium subsp. paratuberculosis by a sonicate immunoassay based on surface-enhanced Raman scattering, *Clin. Vaccine Immunol.*, **15**(2), pp. 227–234.

67. Smolsky, J., Kaur, S., Hayashi, C., Batra, S. K. and Krasnoslobodtsev, A. V. (2017). Surface-enhanced raman scattering-based immunoassay technologies for detection of disease biomarkers, *Biosensors*, **7**(1), p. 7.

68. Porter, M. D., Lipert, R. J., Siperko, L. M., Wang, G. and Narayanan, R. (2008). SERS as a bioassay platform: fundamentals, design, and applications, *Chem. Soc. Rev.*, **37**(5), pp. 1001–1011.

69. Byeon, J. H., Park, D. and Kim, J. Y. (2015). An aerosol-based soft lithography to fabricate nanoscale silver dots and rings for spectroscopic applications, *Nanoscale*, **7**(6), pp. 2271–2275.

70. Jung, K., Hahn, J., In, S., Bae, Y., Lee, H., Pikhitsa, P. V., Ahn, K., Ha, K., Lee, J. K., Park, N. and Choi, M. (2014). Hotspot-engineered 3D multipetal flower assemblies for surface-enhanced Raman spectroscopy, *Adv. Mater.*, **26**(34), pp. 5924–5929.

71. Lee, K., Choi, H., Kim, D. S., Jang, M. S. and Choi, M. (2017). Vertical stacking of three-dimensional nanostructures via an aerosol lithography for advanced optical applications, *Nanotechnology*, **28**(47), p. 475302.

72. Jabbour, E. and Kantarjian, H. (2014). Chronic myeloid leukemia: 2014 update on diagnosis, monitoring, and management, *Am. J. Hematol.*, **89**(5), pp. 547–556.

73. Boyd, S. D. (2013). Diagnostic applications of high-throughput DNA sequencing, *Annu. Rev. Pathol.-Mech.*, **8**, pp. 381–410.

74. Tsagarakis, N. J., Kentrou, N. A., Kakiopoulos, G., Androutsos, G., Galanopoulos, A., Michaelidis, C., Rontogianni, D., Tolis, A., Chini, S. and Gortzolidis, G. (2011). Flow cytometry as a diagnostic tool in the early diagnosis of aggressive lymphomas mimicking life-threatening infection, *Case Rep Med.*, **2011**, p. 743817.

75. Singhal, R. and Myneedu, V. P. (2015). Microscopy as a diagnostic tool in pulmonary tuberculosis, *Int. J. Mycobacteriol.*, **4**(1), pp. 1–6.

76. Sieroń, A., Sieroń-Stołtny, K., Kawczyk-Krupka, A., Latos, W., Kwiatek, S., Straszak, D. and Bugaj, A. M. (2013). The role of fluorescence diagnosis in clinical practice, *Onco. Targets Ther.*, **6**, pp. 977–982.

77. Bauch, M., Toma, K., Toma, M., Zhang, Q. and Dostalek, J. (2014). Plasmon-enhanced fluorescence biosensors: a review, *Plasmonics*, **9**(4), pp. 781–799.

78. Darvill, D., Centeno, A. and Xie, F. (2013). Plasmonic fluorescence enhancement by metal nanostructures: shaping the future of bionanotechnology, *PCCP*, **15**(38), pp. 15709–15726.

79. Dong, J., Zhang, Z., Zheng, H. and Sun, M. (2015). Recent progress on plasmon-enhanced fluorescence, *Nanophotonics*, **4**(1), pp. 472–490.

80. Cannone, F., Chirico, G., Bizzarri, A. R. and Cannistraro, S. (2006). Quenching and blinking of fluorescence of a single dye molecule bound to gold nanoparticles, *J. Phys. Chem. B*, **110**(33), pp. 16491–16498.

81. Chen, J., Jin, Y., Fahruddin, N. and Zhao, J. X. (2013). Development of gold nanoparticle-enhanced fluorescent nanocomposites, *Langmuir*, **29**(5), pp. 1584–1591.

82. Capehart, S. L., Coyle, M. P., Glasgow, J. E. and Francis, M. B. (2013). Controlled integration of gold nanoparticles and organic fluorophores using synthetically modified MS2 viral capsids, *J. Am. Chem. Soc.*, **135**(8), pp. 3011–3016.

83. Chen, Y., Munechika, K. and Ginger, D. S. (2007). Dependence of fluorescence intensity on the spectral overlap between fluorophores and plasmon resonant single silver nanoparticles, *Nano Lett.*, **7**(3), pp. 690–696.

84. Liang, J., Li, K., Gurzadyan, G. G., Lu, X. and Liu, B. (2012). Silver nanocube-enhanced far-red/near-infrared fluorescence of conjugated polyelectrolyte for cellular imaging, *Langmuir*, **28**(31), pp. 11302–11309.

85. Sugawa, K., Tamura, T., Tahara, H., Yamaguchi, D., Akiyama, T., Otsuki, J., Kusaka, Y., Fukuda, N. and Ushijima, H. (2013). Metal-enhanced fluorescence platforms based on plasmonic ordered copper arrays: wavelength dependence of quenching and enhancement effects, *ACS Nano*, **7**(11), pp. 9997–10010.

86. Zhang, Y., Aslan, K., Previte, M. J. and Geddes, C. D. (2007). Metal-enhanced fluorescence from copper substrates, *Appl. Phys. Lett.*, **90**(17), p. 173116.

87. Chowdhury, M. H., Ray, K., Gray, S. K., Pond, J. and Lakowicz, J. R. (2009). Aluminum nanoparticles as substrates for metal-enhanced fluorescence in the ultraviolet for the label-free detection of biomolecules, *Anal. Chem.*, **81**(4), pp. 1397–1403.

88. Zhang, Y., Aslan, K., Previte, M. J. and Geddes, C. D. (2007). Metal-enhanced fluorescence from copper substrates, *Appl. Phys. Lett.*, **90**(17), p. 173116.

89. Zhang, Y. and Geddes, C. D. (2010). Metal-enhanced fluorescence from thermally stable rhodium nanodeposits, *J. Mater. Chem.*, **20**(39), pp. 8600–8606.

90. Sabanayagam, C. R. and Lakowicz, J. R. (2006). Increasing the sensitivity of DNA microarrays by metal-enhanced fluorescence using surface-bound silver nanoparticles, *Nucleic Acids Res.*, **35**(2), pp. e13–e13.

91. Wang, Y., Liu, B., Mikhailovsky, A. and Bazan, G. C. (2010). Conjugated polyelectrolyte–metal nanoparticle platforms for optically amplified DNA detection, *Adv. Mater.*, **22**(5), pp. 656–659.

92. Peng, H.-I., Strohsahl, C. M., Leach, K. E., Krauss, T. D. and Miller, B. L. (2009). Label-free DNA detection on nanostructured Ag surfaces, *ACS Nano*, **3**(8), pp. 2265–2273.

93. Aslan, K., Huang, J., Wilson, G. M. and Geddes, C. D. (2006). Metal-enhanced fluorescence-based RNA sensing, *J. Am. Chem. Soc.*, **128**(13), pp. 4206–4207.

94. Saha, A., Basiruddin, S., Sarkar, R., Pradhan, N. and Jana, N. R. (2009). Functionalized plasmonic– fluorescent nanoparticles for imaging and detection, *J. Phys. Chem. C*, **113**(43), pp. 18492–18498.

95. Szmacinski, H., Murtaza, Z. and Lakowicz, J. R. (2009). Time-resolved fluorometric method for one-step immunoassays using plasmonic nanostructures, *J. Phys. Chem. C*, **114**(16), pp. 7236–7241.

96. Lakowicz, J. R., Geddes, C. D., Gryczynski, I., Malicka, J., Gryczynski, Z., Aslan, K., Lukomska, J., Matveeva, E., Zhang, J. and Badugu, R. (2004). Advances in surface-enhanced fluorescence, *J. Fluoresc.*, **14**(4), pp. 425–441.

97. Cui, Q., He, F., Li, L. and Moehwald, H. (2014). Controllable metal-enhanced fluorescence in organized films and colloidal system, *Adv. Colloid Interface Sci.*, **207**, pp. 164–177.

98. Kim, J., Dantelle, G. R., Revaux, A. L., Bérard, M., Huignard, A., Gacoin, T. and Boilot, J.-P. (2010). Plasmon-induced modification of fluorescent thin film emission nearby gold nanoparticle monolayers, *Langmuir*, **26**(11), pp. 8842–8849.

99. Degliangeli, F., Kshirsagar, P., Brunetti, V., Pompa, P. P. and Fiammengo, R. (2014). Absolute and direct microRNA quantification using DNA–gold nanoparticle probes, *J. Am. Chem. Soc.*, **136**(6), pp. 2264–2267.

100. Beni, V., Zewdu, T., Joda, H., Katakis, I. and O'Sullivan, C. K. (2012). Gold nanoparticle fluorescent molecular beacon for low-resolution DQ2 gene HLA typing, *Anal. Bioanal. Chem.*, **402**(3), pp. 1001–1009.

101. Shi, J., Chan, C., Pang, Y., Ye, W., Tian, F., Lyu, J., Zhang, Y. and Yang, M. (2015). A fluorescence resonance energy transfer (FRET) biosensor based on graphene quantum dots (GQDs) and gold nanoparticles (AuNPs) for the detection of mecA gene sequence of Staphylococcus aureus, *Biosens. Bioelectron.*, **67**, pp. 595–600.

102. Zhou, Y., Yao, J. and Wang, L. V. (2016). Tutorial on photoacoustic tomography, *JBO*, **21**(6), p. 061007.

103. Yoon, S. J., Mallidi, S., Tam, J. M., Tam, J. O., Murthy, A., Johnston, K. P., Sokolov, K. V. and Emelianov, S. Y. (2010). Utility of biodegradable

plasmonic nanoclusters in photoacoustic imaging, *Opt. Lett.*, **35**(22), pp. 3751–3753.

104. Li, W. and Chen, X. (2015). Gold nanoparticles for photoacoustic imaging, *Nanomedicine*, **10**(2), pp. 299–320.

105. Viator, J. A., Gupta, S., Goldschmidt, B. S., Bhattacharyya, K., Kannan, R., Shukla, R., Dale, P. S., Boote, E. and Katti, K. (2010). Gold nanoparticle mediated detection of prostate cancer cells using photoacoustic flowmetry with optical reflectance, *J. Biomed. Nanotechnol.*, **6**(2), pp. 187–191.

106. Song, K. H., Kim, C., Cobley, C. M., Xia, Y. and Wang, L. V. (2008). Near-infrared gold nanocages as a new class of tracers for photoacoustic sentinel lymph node mapping on a rat model, *Nano Lett.*, **9**(1), pp. 183–188.

107. Lu, W., Melancon, M. P., Xiong, C., Huang, Q., Elliott, A., Song, S., Zhang, R., Flores, L. G., Gelovani, J. G. and Wang, L. V. (2011). Effects of photoacoustic imaging and photothermal ablation therapy mediated by targeted hollow gold nanospheres in an orthotopic mouse xenograft model of glioma, *Cancer Res.*, **71**(19), pp. 6116–6121.

108. Manohar, S., Ungureanu, C. and Van Leeuwen, T. G. (2011). Gold nanorods as molecular contrast agents in photoacoustic imaging: the promises and the caveats, *Contrast Media Mol. Imaging*, **6**(5), pp. 389–400.

109. Ye, S., Marston, G., McLaughlan, J. R., Sigle, D. O., Ingram, N., Freear, S., Baumberg, J. J., Bushby, R. J., Markham, A. F. and Critchley, K. (2015). Engineering gold nanotubes with controlled length and near-infrared absorption for theranostic applications, *Adv. Funct. Mater.*, **25**(14), pp. 2117–2127.

110. Yoon, S. J., Mallidi, S., Tam, J. M., Tam, J. O., Murthy, A., Johnston, K. P., Sokolov, K. V. and Emelianov, S. Y. (2010). Utility of biodegradable plasmonic nanoclusters in photoacoustic imaging, *Opt. Lett.*, **35**(22), pp. 3751–3753.

111. De La Zerda, A., Zavaleta, C., Keren, S., Vaithilingam, S., Bodapati, S., Liu, Z., Levi, J., Smith, B. R., Ma, T.-J. and Oralkan, O. (2008). Carbon nanotubes as photoacoustic molecular imaging agents in living mice, *Nat. Nanotechnol.*, **3**(9), p. 557.

112. Xie, L., Wang, G., Zhou, H., Zhang, F., Guo, Z., Liu, C., Zhang, X. and Zhu, L. (2016). Functional long circulating single walled carbon nanotubes for fluorescent/photoacoustic imaging-guided enhanced phototherapy, *Biomaterials*, **103**, pp. 219–228.

113. Sheng, Z., Song, L., Zheng, J., Hu, D., He, M., Zheng, M., Gao, G., Gong, P., Zhang, P. and Ma, Y. (2013). Protein-assisted fabrication of nano-reduced graphene oxide for combined in vivo photoacoustic imaging and photothermal therapy, *Biomaterials*, **34**(21), pp. 5236–5243.

114. Moon, H., Kumar, D., Kim, H., Sim, C., Chang, J.-H., Kim, J.-M., Kim, H. and Lim, D.-K. (2015). Amplified photoacoustic performance and enhanced photothermal stability of reduced graphene oxide coated gold nanorods for sensitive photoacoustic imaging, *ACS Nano*, **9**(3), pp. 2711–2719.

115. Weber, J., Beard, P. C. and Bohndiek, S. E. (2016). Contrast agents for molecular photoacoustic imaging, *Nat. Methods*, **13**(8), p. 639.

116. Qian, X., Shen, S., Liu, T., Cheng, L. and Liu, Z. (2015). Two-dimensional TiS 2 nanosheets for in vivo photoacoustic imaging and photothermal cancer therapy, *Nanoscale*, **7**(14), pp. 6380–6387.

117. Yu, J., Yin, W., Zheng, X., Tian, G., Zhang, X., Bao, T., Dong, X., Wang, Z., Gu, Z. and Ma, X. (2015). Smart MoS2/Fe3O4 nanotheranostic for magnetically targeted photothermal therapy guided by magnetic resonance/photoacoustic imaging, *Theranostics*, **5**(9), p. 931.

118. Ku, G., Zhou, M., Song, S., Huang, Q., Hazle, J. and Li, C. (2012). Copper sulfide nanoparticles as a new class of photoacoustic contrast agent for deep tissue imaging at 1064 nm, *ACS Nano*, **6**(8), pp. 7489–7496.

119. Song, X. R., Wang, X., Yu, S. X., Cao, J., Li, S. H., Li, J., Liu, G., Yang, H. H. and Chen, X. (2015). Co9Se8 nanoplates as a new theranostic platform for photoacoustic/magnetic resonance dual-modal-imaging-guided chemo-photothermal combination therapy, *Adv. Mater.*, **27**(21), pp. 3285–3291.

120. Wang, S., Li, X., Chen, Y., Cai, X., Yao, H., Gao, W., Zheng, Y., An, X., Shi, J. and Chen, H. (2015). A facile one-pot synthesis of a two-dimensional MoS2/Bi2S3 composite theranostic nanosystem for multi-modality tumor imaging and therapy, *Adv. Mater.*, **27**(17), pp. 2775–2782.

121. Mou, J., Li, P., Liu, C., Xu, H., Song, L., Wang, J., Zhang, K., Chen, Y., Shi, J. and Chen, H. (2015). Ultrasmall Cu2-xS nanodots for highly efficient photoacoustic imaging-guided photothermal therapy, *Small*, **11**(19), pp. 2275–2283.

122. Knauer, M., Schuster, M. E., Su, D., Schlögl, R., Niessner, R. and Ivleva, N. P. (2009). Soot structure and reactivity analysis by Raman microspectroscopy, temperature-programmed oxidation, and high-resolution transmission electron microscopy, *J. Phys. Chem.*, **113**, pp. 13871–13880.

123. Stephen, Z. R., Kievit, F. M. and Zhang, M. (2011). Magnetite nanoparticles for medical MR imaging, *Mater. Today*, **14**(7–8), pp. 330–338.

124. Grover, V. P., Tognarelli, J. M., Crossey, M. M., Cox, I. J., Taylor-Robinson, S. D. and McPhail, M. J. (2015). Magnetic resonance imaging: principles and techniques: lessons for clinicians, *J. Clin. Exp. Hepatol.*, **5**(3), pp. 246–255.

125. Ni, D., Bu, W., Ehlerding, E. B., Cai, W. and Shi, J. (2017). Engineering of inorganic nanoparticles as magnetic resonance imaging contrast agents, *Chem. Soc. Rev.*, **46**(23), pp. 7438–7468.

126. Estelrich, J., Sánchez-Martín, M. J. and Busquets, M. A. (2015). Nanoparticles in magnetic resonance imaging: from simple to dual contrast agents, *Int. J. Nanomed.*, **10**, p. 1727.

127. Werner, E. J., Datta, A., Jocher, C. J. and Raymond, K. N. (2008). High-relaxivity MRI contrast agents: where coordination chemistry meets medical imaging, *Angew. Chem. Int. Ed.*, **47**(45), pp. 8568–8580.

128. Hu, B., Zeng, M., Chen, J., Zhang, Z., Zhang, X., Fan, Z. and Zhang, X. (2016). External magnetic field-induced targeted delivery of highly sensitive iron oxide nanocubes for MRI of myocardial infarction, *Small*, **12**(34), pp. 4707–4712.

129. Hou, Y., Qiao, R., Fang, F., Wang, X., Dong, C., Liu, K., Liu, C., Liu, Z., Lei, H. and Wang, F. (2012). NaGdF4 nanoparticle-based molecular probes for magnetic resonance imaging of intraperitoneal tumor xenografts in vivo, *ACS Nano*, **7**(1), pp. 330–338.

130. Kimura, Y., Kamisugi, R., Narazaki, M., Matsuda, T., Tabata, Y., Toshimitsu, A. and Kondo, T. (2012). Size-controlled and biocompatible Gd2O3 nanoparticles for dual photoacoustic and MR imaging, *Adv. Healthcare Mater.*, **1**(5), pp. 657–660.

131. Zheng, X.-Y., Sun, L.-D., Zheng, T., Dong, H., Li, Y., Wang, Y.-F. and Yan, C.-H. (2015). PAA-capped GdF3 nanoplates as dual-mode MRI and CT contrast agents, *Sci. Bull.*, **60**(12), pp. 1092–1100.

132. Na, H. B., Lee, J. H., An, K., Park, Y. I., Park, M., Lee, I. S., Nam, D. H., Kim, S. T., Kim, S. H. and Kim, S. W. (2007). Development of a T1 contrast agent for magnetic resonance imaging using MnO nanoparticles, *Angew. Chem.*, **119**(28), pp. 5493–5497.

133. An, K., Park, M., Yu, J. H., Na, H. B., Lee, N., Park, J., Choi, S. H., Song, I. C., Moon, W. K. and Hyeon, T. (2012). Synthesis of uniformly sized manganese oxide nanocrystals with various sizes and shapes and characterization of their T1 magnetic resonance relaxivity, *Eur. J. Inorg. Chem.*, **2012**(12), pp. 2148–2155.

134. Jun, Y.-w., Huh, Y.-M., Choi, J.-s., Lee, J.-H., Song, H.-T., Kim, S., Yoon, S., Kim, K.-S. and Shin, J.-S. (2005). Nanoscale size effect of magnetic nanocrystals and their utilization for cancer diagnosis via magnetic resonance imaging, *J. Am. Chem. Soc.*, **127**(16), pp. 5732–5733.

135. Huang, J., Bu, L., Xie, J., Chen, K., Cheng, Z., Li, X. and Chen, X. (2010). Effects of nanoparticle size on cellular uptake and liver MRI with polyvinylpyrrolidone-coated iron oxide nanoparticles, *ACS Nano*, **4**(12), pp. 7151–7160.

136. Kim, B. H., Lee, N., Kim, H., An, K., Park, Y. I., Choi, Y., Shin, K., Lee, Y., Kwon, S. G. and Na, H. B. (2011). Large-scale synthesis of uniform and extremely small-sized iron oxide nanoparticles for high-resolution T1 magnetic resonance imaging contrast agents, *J. Am. Chem. Soc.*, **133**(32), pp. 12624–12631.

137. Evans, D. E., Harrison, R. M. and Ayres, J. G. (2003). The generation and characterization of metallic and mixed element aerosols for human challenge studies, *Aerosol Sci. Technol.*, **37**(12), pp. 975–987.

138. Byeon, J. H. and Kim, Y. W. (2014). Ambient plasma synthesis of Si-Fe hollow nanoparticles and their biocompatibility and lithium storage capacity, *Adv. Mater. Interfaces*, **1**(5), p. 1300134.

139. Kroto, H. W., Allaf, A. and Balm, S. (1991). C60: Buckminsterfullerene, *Chem. Rev.*, **91**(6), pp. 1213–1235.

140. Iijima, S. (1991). Helical microtubules of graphitic carbon, *Nature*, **354**(6348), p. 56.

141. Novoselov, K. S., Geim, A. K., Morozov, S. V., Jiang, D., Zhang, Y., Dubonos, S. V., Grigorieva, I. V. and Firsov, A. A. (2004). Electric field effect in atomically thin carbon films, *Science*, **306**(5696), pp. 666–669.

142. Nair, R. R., Blake, P., Grigorenko, A. N., Novoselov, K. S., Booth, T. J., Stauber, T., Peres, N. M. and Geim, A. K. (2008). Fine structure constant defines visual transparency of grapheme, *Science*, **320**(5881), pp. 1308–1308.

143. Shao, G., Lu, Y., Wu, F., Yang, C., Zeng, F. and Wu, Q. (2012). Graphene oxide: the mechanisms of oxidation and exfoliation, *J. Mater. Sci.*, **47**(10), pp. 4400–4409.

144. Eigler, S., Enzelberger-Heim, M., Grimm, S., Hofmann, P., Kroener, W., Geworski, A., Dotzer, C., Röckert, M., Xiao, J. and Papp, C. (2013). Wet chemical synthesis of graphene, *Adv. Mater.*, **25**(26), pp. 3583–3587.

145. Kim, J., Park, S.-J. and Min, D.-H. (2016). Emerging approaches for graphene oxide biosensor, *Anal. Chem.*, **89**(1), pp. 232–248.

146. Goenka, S., Sant, V. and Sant, S. (2014). Graphene-based nanomaterials for drug delivery and tissue engineering, *J. Control. Release*, **173**, pp. 75–88.

147. Chang, H., Tang, L., Wang, Y., Jiang, J. and Li, J. (2010). Graphene fluorescence resonance energy transfer aptasensor for the thrombin detection, *Anal. Chem.*, **82**(6), pp. 2341–2346.

148. Lu, C.-H., Zhu, C.-L., Li, J., Liu, J.-J., Chen, X. and Yang, H.-H. (2010). Using graphene to protect DNA from cleavage during cellular delivery, *Chem. Commun.*, **46**(18), pp. 3116–3118.

149. He, S., Song, B., Li, D., Zhu, C., Qi, W., Wen, Y., Wang, L., Song, S., Fang, H. and Fan, C. (2010). A graphene nanoprobe for rapid, sensitive, and multicolor fluorescent DNA analysis, *Adv. Funct. Mater.*, **20**(3), pp. 453–459.

150. Husale, B. S., Sahoo, S., Radenovic, A., Traversi, F., Annibale, P. and Kis, A. (2010). ssDNA binding reveals the atomic structure of graphene, *Langmuir*, **26**(23), pp. 18078–18082.

151. Lu, C. H., Yang, H. H., Zhu, C. L., Chen, X. and Chen, G. N. (2009). A graphene platform for sensing biomolecules, *Angew. Chem.*, **121**(26), pp. 4879–4881.

152. Mohanty, N. and Berry, V. (2008). Graphene-based single-bacterium resolution biodevice and DNA transistor: interfacing graphene derivatives with nanoscale and microscale biocomponents, *Nano Lett.*, **8**(12), pp. 4469–4476.

153. Perreault, F., De Faria, A. F., Nejati, S. and Elimelech, M. (2015). Antimicrobial properties of graphene oxide nanosheets: why size matters, *ACS Nano*, **9**(7), pp. 7226–7236.

154. Byeon, J. H. and Kim, Y.-W. (2013). Continuous gas-phase synthesis of metal oxide–graphene hybrid nanoflakes for the enhancement of lithium storage, *RSC Adv.*, **3**(20), pp. 7259–7264.

155. Thapa, R. K., Byeon, J. H., Ku, S. K., Yong, C. S. and Kim, J. O. (2017). Easy on-demand self-assembly of lateral nanodimensional hybrid graphene oxide flakes for near-infrared-induced chemothermal therapy, *NPG Asia Mater.*, **9**(8), p. e416.

156. Roldo, M. and Fatouros, D. G. (2013). Biomedical applications of carbon nanotubes, *Annu. Rep. Prog. Chem., Sect. C: Phys. Chem.*, **109**, pp. 10–35.

157. Tasis, D., Tagmatarchis, N., Bianco, A. and Prato, M. (2006). Chemistry of carbon nanotubes, *Chem. Rev.*, **106**(3), pp. 1105–1136.

158. Jacobs, C. B., Peairs, M. J. and Venton, B. J. (2010). Carbon nanotube based electrochemical sensors for biomolecules, *Anal. Chim. Acta*, **662**(2), pp. 105–127.

159. Singh, R. and Torti, S. V. (2013). Carbon nanotubes in hyperthermia therapy, *Adv. Drug Deliv. Rev.*, **65**(15), pp. 2045–2060.

160. Bekyarova, E., Ni, Y., Malarkey, E. B., Montana, V., McWilliams, J. L., Haddon, R. C. and Parpura, V. (2005). Applications of carbon nanotubes in biotechnology and biomedicine, *J. Biomed. Nanotechnol.*, **1**(1), pp. 3–17.

161. Zhang, L., Wang, Z., Xu, C., Li, Y., Gao, J., Wang, W. and Liu, Y. (2011). High strength graphene oxide/polyvinyl alcohol composite hydrogels, *J. Mater. Chem.*, **21**(28), pp. 10399–10406.

162. Shin, S. R., Jung, S. M., Zalabany, M., Kim, K., Zorlutuna, P., Kim, S. B., Nikkhah, M., Khabiry, M., Azize, M. and Kong, J. (2013). Carbon-nanotube-embedded hydrogel sheets for engineering cardiac constructs and bioactuators, *ACS Nano*, **7**(3), pp. 2369–2380.

163. Shin, S. R., Bae, H., Cha, J. M., Mun, J. Y., Chen, Y.-C., Tekin, H., Shin, H., Farshchi, S., Dokmeci, M. R. and Tang, S. (2011). Carbon nanotube reinforced hybrid microgels as scaffold materials for cell encapsulation, *ACS Nano*, **6**(1), pp. 362–372.

164. Abarrategi, A., Gutierrez, M. C., Moreno-Vicente, C., Hortigüela, M. J., Ramos, V., Lopez-Lacomba, J. L., Ferrer, M. L. and del Monte, F. (2008). Multiwall carbon nanotube scaffolds for tissue engineering purposes, *Biomaterials*, **29**(1), pp. 94–102.

165. Di Giorgio, M. L., Di Bucchianico, S., Ragnelli, A. M., Aimola, P., Santucci, S. and Poma, A. (2011). Effects of single and multi walled carbon nanotubes on macrophages: cyto and genotoxicity and electron microscopy, *Mutat. Res. Genet. Toxicol. Environ. Mutagen.*, **722**(1), pp. 20–31.

166. Li, X., Peng, Y. and Qu, X. (2006). Carbon nanotubes selective destabilization of duplex and triplex DNA and inducing B–A transition in solution, *Nucleic Acids Res.*, **34**(13), pp. 3670–3676.

167. Byeon, J. H. and Kim, J.-W. (2013). Aerosol assisted fabrication of carbon nanotube/zinc oxide arrays for a field emission device, *J. Colloid Interface Sci.*, **393**, pp. 397–401.

168. Zonnevylle, A. C, Hagen, C. W., Kruit, P., Valenti, M. and Schmidt-Ott, A. (2009). Positioning Pd catalyst particles for carbon nanotube growth using charge patterns created with a scanning electron microscope, *J. Vac. Sci. Technol. B*, **27**, pp. 3048–3050.

169. Wang, J. and Qiu, J. (2016). A review of carbon dots in biological applications, *J. Mater. Sci.*, **51**(10), pp. 4728–4738.

170. Ray, S., Saha, A., Jana, N. R. and Sarkar, R. (2009). Fluorescent carbon nanoparticles: synthesis, characterization, and bioimaging application, *J. Phys. Chem. C*, **113**(43), pp. 18546–18551.

171. Chan, W. C., Maxwell, D. J., Gao, X., Bailey, R. E., Han, M. and Nie, S. (2002). Luminescent quantum dots for multiplexed biological detection and imaging, *Curr. Opin. Biotechnol.*, **13**(1), pp. 40–46.

172. Alivisatos, A. P. (1996). Semiconductor clusters, nanocrystals, and quantum dots, *Science*, **271**(5251), pp. 933–937.

173. Walling, M. A., Novak, J. A. and Shepard, J. R. (2009). Quantum dots for live cell and in vivo imaging, *Int. J. Mol. Sci.*, **10**(2), pp. 441–491.

174. Efros, A. L. and Rosen, M. (2000). The electronic structure of semiconductor nanocrystals, *Annu. Rev. Mater. Sci.*, **30**(1), pp. 475–521.

175. Li, H., Shih, W. Y. and Shih, W.-H. (2007). Synthesis and characterization of aqueous carboxyl-capped CdS quantum dots for bioapplications, *Ind. Eng. Chem. Res.*, **46**(7), pp. 2013–2019.

176. Jaiswal, J. K., Mattoussi, H., Mauro, J. M. and Simon, S. M. (2003). Long-term multiple color imaging of live cells using quantum dot bioconjugates, *Nat. Biotechnol.*, **21**(1), p. 47.

177. Yong, K.-T., Hu, R., Roy, I., Ding, H., Vathy, L. A., Bergey, E. J., Mizuma, M., Maitra, A. and Prasad, P. N. (2009). Tumor targeting and imaging in live animals with functionalized semiconductor quantum rods, *ACS Appl. Mater. Interfaces*, **1**(3), pp. 710–719.

178. Poudel, B. K., Doh, K. O. and Byeon, J. H. (2018). Ag photoionization-induced single-pass assembly of Ag2S nanodots in flowing thiol droplets, *Green Chem.*, **20**(5), pp. 978–983.

179. Smith, A. M., Mancini, M. C. and Nie, S. (2009). Bioimaging: second window for in vivo imaging, *Nat. Nanotechnol.*, **4**(11), p. 710.

180. Jain, P. K., El-Sayed, I. H. and El-Sayed, M. A. (2007). Au nanoparticles target cancer, *Nano Today*, **2**(1), pp. 18–29.

181. Strano, M. S., Doorn, S. K., Haroz, E. H., Kittrell, C., Hauge, R. H. and Smalley, R. E. (2003). Assignment of (n, m) Raman and optical features of metallic single-walled carbon nanotubes, *Nano Lett.*, **3**(8), pp. 1091–1096.

182. Levi-Polyachenko, N. H., Merkel, E. J., Jones, B. T., Carroll, D. L. and Stewart IV, J. H. (2009). Rapid photothermal intracellular drug delivery

using multiwalled carbon nanotubes, *Mol. Pharm.*, **6**(4), pp. 1092–1099.

183. Sun, X., Liu, Z., Welsher, K., Robinson, J. T., Goodwin, A., Zaric, S. and Dai, H. (2008). Nano-graphene oxide for cellular imaging and drug delivery, *Nano Res.*, **1**(3), pp. 203–212.

184. Jaque, D., Maestro, L. M., Del Rosal, B., Haro-Gonzalez, P., Benayas, A., Plaza, J. L., Rodriguez, E. M. and Sole, J. G. (2014). Nanoparticles for photothermal therapies, *Nanoscale*, **6**(16), pp. 9494–9530.

185. Byeon, J. H. and Kim, Y. W. (2014). Aero-self-assembly of ultrafine gold incorporated silica nanobunches for NIR-induced chemo-thermal therapy, *Small*, **10**(12), pp. 2331–2335.

186. Byeon, J. H. and Kim, Y. W. (2014). Chitosan-conjugated dendritic Ag nanopowders for photothermal therapy applications, *ACS Macro Lett.*, **3**(2), pp. 205–210.

187. Byrne, J. D., Betancourt, T. and Brannon-Peppas, L. (2008). Active targeting schemes for nanoparticle systems in cancer therapeutics, *Adv. Drug Deliv. Rev.*, **60**(15), pp. 1615–1626.

188. Lynch, I., Salvati, A. and Dawson, K. A. (2009). Protein-nanoparticle interactions: what does the cell see?, *Nat. Nanotechnol.*, **4**(9), p. 546; Mahmoudi, M., Lynch, I., Ejtehadi, M. R., Monopoli, M. P., Bombelli, F. B. and Laurent, S. (2011). Protein- nanoparticle interactions: opportunities and challenges, *Chem. Rev.*, **111**(9), pp. 5610–5637.

189. Karakoti, A. S., Das, S., Thevuthasan, S. and Seal, S. (2011). PEGylated inorganic nanoparticles, *Angew. Chem. Int. Ed.*, **50**(9), pp. 1980–1994.

190. Amoozgar, Z. and Yeo, Y. (2012). Recent advances in stealth coating of nanoparticle drug delivery systems, *Wiley Interdiscip. Rev. Nanomed. Nanobiotechnol.*, **4**(2), pp. 219–233.

191. Moghimi, S. M., Hunter, A. C. and Murray, J. C. (2001). Long-circulating and target-specific nanoparticles: theory to practice, *Pharmacol. Rev.*, **53**(2), pp. 283–318.

192. Janát-Amsbury, M., Ray, A., Peterson, C. and Ghandehari, H. (2011). Geometry and surface characteristics of gold nanoparticles influence their biodistribution and uptake by macrophages, *Eur. J. Pharm. Biopharm.*, **77**(3), pp. 417–423.

193. Ilium, L., Davis, S., Wilson, C., Thomas, N., Frier, M. and Hardy, J. (1982). Blood clearance and organ deposition of intravenously administered colloidal particles. The effects of particle size, nature and shape, *Int. J. Pharm.*, **12**(2–3), pp. 135–146.

194. Champion, J. A., Katare, Y. K. and Mitragotri, S. (2007). Particle shape: a new design parameter for micro-and nanoscale drug delivery carriers, *J. Control. Release*, **121**(1–2), pp. 3–9.

195. Liu, J., Yu, M., Zhou, C. and Zheng, J. (2013). Renal clearable inorganic nanoparticles: a new frontier of bionanotechnology, *Mater. Today*, **16**(12), pp. 477–486.

196. De Jong, W. H., Hagens, W. I., Krystek, P., Burger, M. C., Sips, A. J. and Geertsma, R. E. (2008). Particle size-dependent organ distribution of gold nanoparticles after intravenous administration, *Biomaterials*, **29**(12), pp. 1912–1919.

197. Sonavane, G., Tomoda, K. and Makino, K. (2008). Biodistribution of colloidal gold nanoparticles after intravenous administration: effect of particle size, *Colloids Surf., B*, **66**(2), pp. 274–280.

198. Love, J. C., Estroff, L. A., Kriebel, J. K., Nuzzo, R. G. and Whitesides, G. M. (2005). Self-assembled monolayers of thiolates on metals as a form of nanotechnology, *Chem. Rev.*, **105**(4), pp. 1103–1170.

Chapter 14

Application of a Spark Discharge Generator for Production of Combustion-Like Aerosols

Heinz Burtscher
Institute for Aerosol and Sensor Technology,
University of Applied Sciences, Northwestern Switzerland,
Klosterzelgstrasse 2, Windisch, 5210, Switzerland
Heinz.burtscher@fhnw.ch

Model soot particles are needed for a variety of applications like instrument calibration, filter testing, inhalation studies, etc. The present chapter deals with spark ablation as an interesting particle production technique for these purposes. Operation and particle characteristics of spark discharge generators are compared to other techniques like sooting flame generators and to real combustion soot (mainly from diesel engines). The comparison shows that significant differences in particle properties exist; however, the simplicity and robustness of spark discharge generators makes them a useful tool in many cases.

Spark Ablation: Building Blocks for Nanotechnology
Edited by Andreas Schmidt-Ott

ISBN 978-981-4800-82-2 (Hardcover), 978-0-367-81709-1 (eBook)
www.jennystanford.com

14.1 Introduction

Soot particles or more generally carbonaceous particles play an important role in atmospheric aerosols. Every combustion process emits them. They are omnipresent in the environment. Particles may have adverse health effects and contribute to global warming [16].

The main sources of these particles are combustion processes where soot is produced by incomplete combustion. The nature of these particles strongly depends on the fuel used and the way of combustion. They are agglomerates consisting of primary particles typically 10–30 nm in diameter, and the mobility diameter of the agglomerates is in the range of 20 to a few hundred nanometers. The particles consist of elemental carbon (EC), a wide variety of hydrocarbons, and, depending on the fuel, inorganic species (ashes), sulfates, and water. In the case of combustion engines, metals are important; for wood combustion, potassium and calcium compounds (oxides and salts).

A lot of work has been done recently to reduce emissions of soot particles, on the one hand, by optimizing the combustion and, on the other, by after treatment devices such as particle traps, electrostatic precipitators, etc. New legal limits have been introduced to consider also very small particles. For combustion engines, a number limit for solid particles >23 nm has been established (PMP protocol, UNECE Regulations No. 83 [36]). This requires also new instrumentation.

New instruments have been developed for this purpose. The "microsoot sensor" by AVL [32], based on photoacoustics, measures the EC concentration. Several systems (based on condensation particle counters [CPCs] or diffusion charging sensors) measure the number concentration according to the PMP protocol.

To study these particles, test and calibrate measurement devices, and test and evaluate after-treatment devices and their regeneration strategies, particle generators, simulating "real" soot particles, are needed. Traceable calibration procedures for a wide range of aerosol instruments are essential. As the instrument response may strongly depend on particle chemistry and morphology, appropriate selection of the calibration aerosol is of crucial importance. Ideally, the calibration aerosol should resemble that of the emission source [26].

The new emission limits also require that soot particles be efficiently removed from diesel and gasoline engine exhaust and in the future eventually also from wood combustion emissions. A wide range of particle-trapping systems and exhaust after-treatment technologies have been developed for this purpose. These filters have to be regenerated, that is, the soot collected has to be burned. The behavior of this regeneration step is strongly influenced by the structure and reactivity of the deposited soot particles. Especially the formation of highly reactive soot would make it possible to perform this regeneration step at relatively low temperatures [22]. To study these processes, again model soot is needed.

Also for environmental studies, such as atmospheric pollution, and investigations concerning health effects of soot particles, there is a need for laboratory production techniques for soot-like carbon nanoparticles.

Several types of generators for these purposes have been developed; one widely used candidate is the spark discharge generator (SDG). In the following section, the characteristics of this generator will be discussed and compared to other techniques. Comparison of the model particles with real soot can be done by comparing size distributions, primary particle size, morphology, composition, surface (e.g., in terms of the Brunauer–Emmett–Teller (BET) surface (determined by the BET method) or the lung deposited surface area [LDSA]), or oxidation properties (important for filter regeneration).

14.2 Model Soot Particle Generators

Carbonaceous particles can be produced by a variety of processes. Frequently used techniques are:

- **Dispersion:** Commercially available carbon black in a suspension (e.g., alcohol) can be nebulized [38]; the size distribution depends on the size of the carbon particles and the droplet size. Vibrating orifice generators [38] can be used instead of nebulizers. Another possibility is the use of powder dispersers. Usually particles produced by these techniques are larger than real soot particles. Collecting ambient particulate matter onto substrates, such as filters, and resuspending the particles in experimental studies by the above-mentioned techniques is also possible.

- **Sooting flames:** Soot particles can also be generated by incomplete combustion as it is done worldwide for soot production on an industrial scale. A large variety of laboratory burners using diffusion flames or premixed flames have been built for this purpose. Very well known are the McKenna-type burners [39]. The Combustion Aerosol Standard (CAST; www.sootgenerator.com) by Jing is a frequently used commercial product based on incomplete combustion. The burner is a coflow, laminar diffusion flame, consisting of a coaxial inner tube containing the fuel gas stream and an outer tube containing the dry, particle-free oxidant air stream. The flame is quenched with dry gas (N_2), which freezes the combustion process and prevents oxidation of the soot particles formed near the flame front. The particle containing gas is then rapidly diluted with dry, particle-free air to inhibit particle coagulation. Soot concentration and properties can be varied by changing the flow rates of fuel, oxidant, N_2, and dilution gases. These settings allow varying the size distribution in a wide range and affect the chemistry of the emitted soot. Depending on settings, black soot (mainly EC) or brown soot (higher organic fraction) can be produced. Moore et al. [28] give a detailed characterization of the frequently used mini CAST.
- **Pyrolysis:** Particles can also be produced by pyrolysis alone (flameless): a mixture of air and a hydrocarbon gas is guided through a tube furnace, and soot is produced by pyrolysis. Again, the choice of fuel and flow rates allows varying size distribution and particle composition [1].
- **Spark discharge**: A discharge between carbon electrodes produces sufficiently high temperatures to evaporate a small fraction of the electrodes, and the carbon atoms condense to particles a few nanometers in size, which then coagulate to form agglomerates in the size range of a few nanometers to some hundred nanometers.

Kim et al. [22] introduced a generator to simulate diesel particulate matter. They tried to reproduce the nucleation mode by H_2SO_4 and benzene saturators and the accumulation mode with particles from an SDG also coupled to a benzene saturator. By varying the operating parameters (flow rates, spark energy) different operating conditions of an engine can be simulated. Whereas this allows reproducing "real" diesel size distributions (also bimodal ones) well, the chemical composition differs, as only H_2SO_4, benzene, and carbon are used.

For calibration often NaCl particles (produced by nebulizing a solution or in tube furnaces) or nebulized emery oil particles are used, even if their properties are far from those of soot particles.

14.3 The Spark Discharge Generator for Production of Soot-Like Particles

The SDG is a versatile device to produce nanoparticle aerosols (see Chapter 2). It can produce aerosol nanoparticles in the range from a few nanometers up to some hundred nanometers. Spark discharge nanoparticle generation is technically simple; it can be used for any conducting or semiconducting materials. Depending on the system used, these nanoparticles can be contamination free and composed of one or more materials. A spark between two electrodes leads to erosion and formation of a vapor plume (Fig. 14.1).

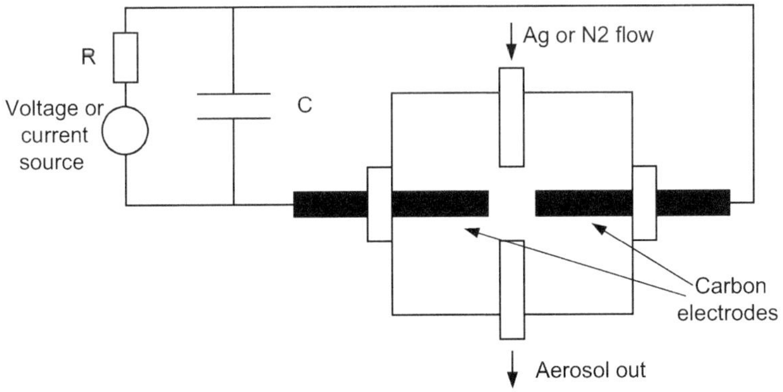

Figure 14.1 Spark discharge generator.

Subsequently particles form by nucleation, condensation, and agglomeration. Electrical discharges have been used since 1907 for vaporizing solids [13], at that time for applications in spectroscopy. Burtscher and Schmidt-Ott produced carbon nanoparticles by spark ablation in 1982 [3], and the same authors [4] used such particles for adsorption experiments. Schwyn et al. [34] give the first detailed description.

By using an SDG with carbon electrodes, agglomerates are formed from primary carbon particles [12]. The chemical composition of the produced particles differs from real soot aerosols, but they are still a good model for the physical properties of combustion products. Real soot particles contain hydrocarbons and contamination from the more complex combustion situation, whereas the SDG particles will be pure carbon with little or no contamination [27].

In the past decades, many scientists investigated and characterized the SDG and presented several designs optimized for distinct characteristics.

Already Schwyn et al. [34] show that by having a fast turbulent flow, agglomeration can be avoided, and primary particles of only a few nanometers in diameter can be emitted. Meuller et al. [27] presented different approaches to the design of SDGs in a review. Chae et al. [6] showed a "wire in-hole configuration," which allows producing mainly unagglomerated carbon particles. Kohut et al. [24] investigated the effects of the electrical circuit used on the generator output. Pfeiffer et al. [30] showed methods to control particle properties like size etc. Kim et al. [21] investigated the influence of electrode distance and spark frequency on the size distribution. Aerosol purity largely depends on the purity of the electrodes used. Standard graphite electrodes contain approximately 20 μg/g ash, and low-ash-content electrodes contain less than 2 μg/g [21].

The method has been standardized in the technical rule VDI 3491 [37]. Palas (PALAS-GfG-1000, [12]) built the first commercial instrument. Horvath and Gangl [15] introduced a triggered version of an SDG. The voltage then can be kept low (typically 200 V), even if the distance between electrodes is large (10 mm). A high-voltage/high-frequency pulse, produced with a Tesla transformer, does the triggering. This setup allows decoupling the spark energy (determined by the capacitance and voltage, but independent of the electrode distance) and the pulse repetition rate (determined by the trigger). The size distribution then becomes mainly independent of the electrode distance and the shape of the electrodes. The mass per pulse can be adjusted by the pulse energy (capacitance and applied voltage). During several 10 hours of operation, no significant change in concentration has been observed. The single scattering albedo of the particles, defined as the ratio of the scattering coefficient and the extinction coefficient, which is very sensitive to changes in size distribution, remained within 0.11 ± 0.003, which demonstrates the stability of particle properties. It is to be noted that this concept is only feasible with an *LC* product (inductance × capacitance) in the circuit that is many orders of magnitude higher compared to the classical circuit (see Chapter 2). This drastically increases the duration of the discharge and lowers the temperature for a given energy per spark. It is therefore not the right choice for mixed materials (see Chapter 5) but well suitable for pure carbon, and the authors of [15] demonstrate the latter. The long discharge time presumably leads to larger primary particles compared to the classical approach, and the similarity with combustion-derived soot may thus be larger. Studies investigating this are yet to be carried out.

14.3.1 The PALAS-GfG-1000

The PALAS-GfG-1000 was for a long time the only commercial instrument and was up to publication of this book the most frequently used and best characterized generator (see Ref. [12]), while Chapter 2 now gives a comprehensive description of the recently introduced G1 from VSParticle. This unit is designed for research purposes for any conducting material, including mixed ones, and includes the option of producing unagglomerated singlet particles. It avoids any outgassing material near the spark zone and therefore allows very clean operation.

To allow long-term operation the GfG-1000 uses a feedback control to keep the distance between the electrodes constant. As a measure for this distance, the breakdown voltage is used. The distance is adjusted by an electric motor that drives spindles connected to the electrodes. The mass production rate is set by changing the spark frequency and ranges from 20 µg/h to 7 mg/h. The primary particle size is around 5 nm, forming agglomerates with a 50 to 200 nm mobility diameter. Production of singlet particles is not possible with this unit. Even if later used with many different electrode materials the generator was initially designed to produce carbon aerosols to be used as "soot like" particles [12], where it delivers by far the largest mass production rate. Earlier generators had very low production rates; the GfG-1000 was the first one delivering a high enough amount of particles per unit time, for example, for filter testing (filtration of diesel soot). Spark frequency is selected by the current charging the capacitor; 0.13–20 mA corresponds to 3–300 Hz. According to Roth [31] it is not possible to generate ultrafine carbon particles without contaminations with this generator. However, the authors were able to decrease these contaminations from 25% to 6% by replacing organic components of the generator by pure inorganic components.

A detailed characterization of the GFG-1000 in terms of size distribution, density, etc., as a function of flows and spark frequency can be found in Ref. [7].

Grob et al. [9] show an application of the GfG-1000 as a particle generator for the calibration system to measure particle number emission from combustion engines (PMP systems).

Meanwhile, the GfG-1000 has been replaced by new models, the DNP digital 2000 and 3000 (the 3000 for higher mass production rates). Whereas the GfG-1000 was designed for argon as quench gas, the new models use nitrogen (www.palas.de/de/product/dnpdigital). The GFG-1000, operated with argon, yields much higher particle numbers compared to operation with nitrogen. This is due to a much lower breakdown threshold under sparking [9].

14.4 Comparison of Particle Properties

If nothing else is mentioned, the results presented for spark generators in the following section refer to the PALAS-GfG-1000. Several literature results are presented, which sometimes differ significantly.

14.4.1 Mass, Size, Density, and Surface

If the spark frequency is not too high, the mass production rate is linearly related to the spark repetition frequency (Fig. 14.2).

However, increasing the spark frequency leads to larger particles due to coagulation. The diameter for some spark frequencies is given in Table 14.1. According to Ref. [12] the maximum number concentration (undiluted) is 2.7×10^8 cm^{-3}. Figure 14.3 shows the particle mass concentration versus pulse energy for a flow rate of 2 L/min and a 25 Hz repetition rate, as found for the design by Horvath and Gangl [15]. The near-exponential increase of ablated mass with spark energy is probably related to the exponential increase of vapor pressure with temperature, because temperature could rise nearly linearly with the energy per spark. As pointed out before, conditions are very much different for the short sparks produced by the classical circuit (Chapter 2), where the spark temperatures are between 10,000 and 20,000 K, and ablation is probably dominated by ion bombardment rather than gas heat conduction (see Chapter 4). Number concentration, average size, and surface are given in Table 14.2. Jeong and Lee [17] found average diameters of 50–70 nm (depending on settings), and some primary particles 15–36 nm in diameter are visible in one micrograph. This is quite puzzling, as it is not consistent with other findings using the same generator. Perhaps contaminant particles explain this single observation. To be able to compare the characteristics of particles, produced by SDGs with "real" soot particles and those from other model soot generators, a review of literature data is given in Table 14.3.

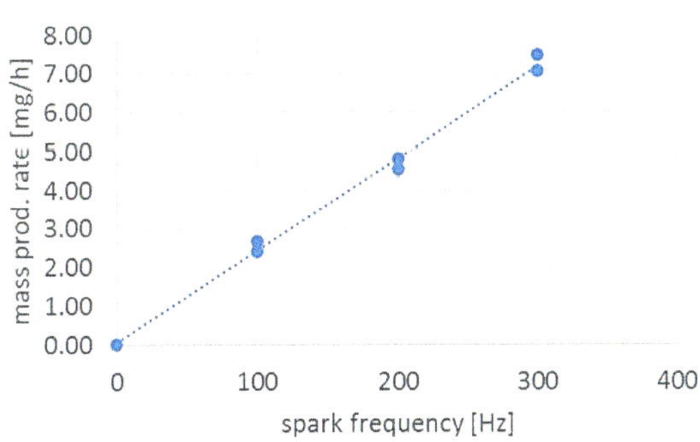

Figure 14.2 Mass production rate as a function of spark frequency (data from [12]).

The effective density of SDG particles is 4–10 times lower compared to real combustion sources [10]. Figure 14.4 shows the effective density as a function of diameter for different current settings. More structure information in terms of fractal parameters (compared to diesel soot) is given by Schnaiter (Table 14.4). Hinot et al. investigated the elemental composition of particles from an SDG, diesel particles, and particles, produced by pyrolysis; results are shown in Table 14.5.

Table 14.1 Geometrical mean diameter versus spark frequency (data from [12])

Spark frequency (Hz)	Average diameter (nm)
60	50
175	67
290	80

Table 14.2 GfG-1000 carbon particle characteristics according to data from Roth [31]

Operating conditions			Particle characteristics			
Spark frequency (Hz)	N_2 flow rate (L/min)	Air flow rate (L/min)	Mode diameter (nm)	Number conc. (cm^{-3})	Mass conc. ($\mu g/cm^3$)	BET surface (m^2/g)
3	17	170	25	10^6	25	750
300	17	125	70	10^7	115	750
300	0	125	150	NA	NA	NA

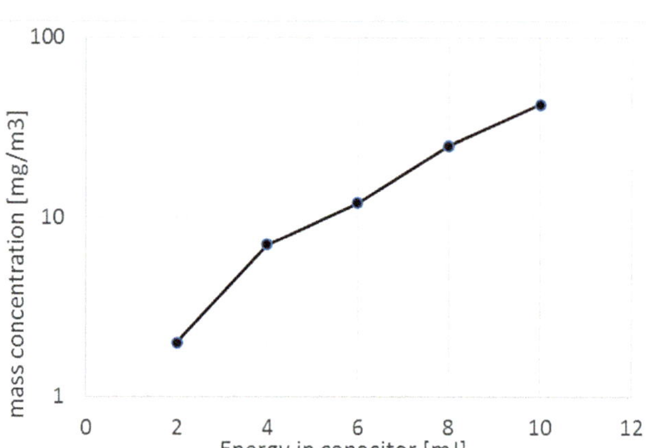

Figure 14.3 Carbon mass concentration versus pulse energy for the triggered version by Gangl and Horvath (data from [15]).

Table 14.3 Summary of particle properties from different references

Source	Effective density (g/cm³)	Primary particle diameter (nm)	Fractal dimension	BET surface (m²/g)	Reference
GfG-1000	0.019 (200 nm), 0.035 (500 nm)	5–10	1.7–2.1		Gysel [10]
	0.12 ... 0.036	15–36	1.78		Jeong and Lee [17]
				639	Byeon [5]
				308 ± 30	Kuznetsov [25]
				300	Kalberer [19]
		5		395	Helsper [12]
		7		700	Hinot [14]
		7	2		Schnaiter [32]
				750	Roth [31]
			1.8		Jurányi [18]
Activated carbon				800–1500	Beyon [5]
Diesel	0.3–1.2				Jeong and Lee [17]
		16		160	Hinot [14]
			2		Schnaiter 32]
Pyrolysis		18		140	Hinot [14]
Combustion soot				48	Kuznetsov [25]
Graphon soot				83	Kuznetsov [25]
CAST	0.3–1.2	20–40			Moore [28]
	0.2–1.4	15–45	2.11–2.23		Durdina [8]

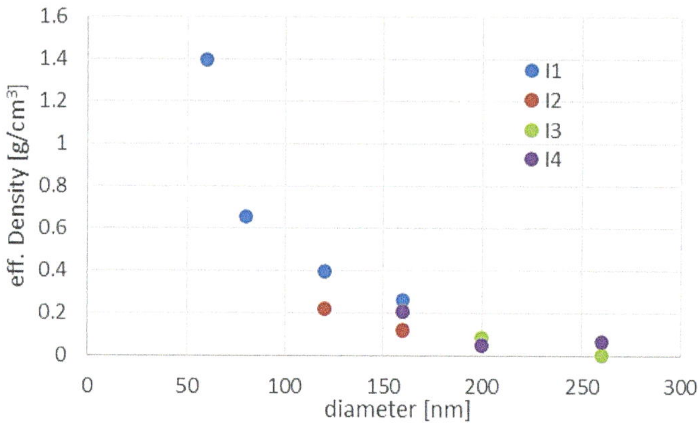

Figure 14.4 Effective density as a function of diameter (data from from [17]). *I*1 corresponds with the lowest current and *I*4 the highest. Obviously, a higher current (higher particle production rate) leads to a lower effective density, due to the higher degree of agglomeration.

Table 14.4 Fractal parameters deduced by modeling the coagulation dynamics (data from [33])

Parameter	GFG-1000 soot	Diesel soot
Fractal dimension Df	2.0	2.0
Prefactor kf	5.6	5.6
Primary particle size dp	7 nm	27 nm
Primary particle density	2.0 g cm^{-3}	1.7 g cm^{-3}

14.4.2 Microstructure

Not only primary the particle size, degree of agglomeration, and resulting density or fractal dimension but also the structure of the primary particle may be of importance, for example, in the context of filter regeneration, where the reactivity is crucial. Knauer et al. [23] made a profound study on particle structure and reactivity using Raman microspectroscopy, temperature-programmed oxidation (TPO), high-resolution transmission electron microscopy (HRTEM), and electron energy loss spectroscopy (EELS). They compared particles from a GfG-1000 with particles in the emissions of EURO IV and EURO VI diesel engines and graphite powder. This study

Table 14.5 Elemental composition and some physical properties (data from [14]). The values for the density here correspond to the volume density.

	Elemental composition (wt%)					Physical properties					
	C	H	N	O	S	Density (g/cm^3)	d_{mob} (nm)	D_{prim} (nm)	σ (nm)	S_{TEM} (m^2/g)	S_{BET} (m^2/g)
SDG	86.7	0.6	8	4.5	0.1	1.97	70	7	2	470	700
Pyrolysis	96.5	1.4	<0.1	1.9	<0.1	2.19	110	16	3	175	140
Diesel	90.9	1.4	0.9	5.6	0.1	2.06	110	18	5	170	160

showed that untreated GfG soot possesses a more heterogeneous and functionalized structure (a highly reactive molecular carbon fraction) than EURO IV soot. In GfG soot the high number of defects and deviation from a perfect graphitic structure results in increased surface functionalization. An increase of structural order and a decrease of chemical heterogeneity of GfG soot are found upon oxidation (TPO up to 770 K). EURO VI soot indicates a much lower degree of chemical heterogeneity and structural disorder compared to GfG soot. The amount of molecular carbon in GfG can be estimated as 20%. The consequences for oxidation properties are shown in Section 14.4.4.

14.4.3 Optical Properties

Schnaiter et al. [33] compared the optical properties of spark generator soot with real diesel soot in the wavelength (λ) ranges from the near-infrared to UV. They found a mass-specific absorption cross section at λ = 450 nm, about a factor of two smaller for the generator than for diesel soot (=10.6 m^2/g). The corresponding data are shown in Tables 14.6 and 14.7.

Table 14.6 Extinction (σ_e), scattering (σ_s), and absorption (σ_a) coefficients for GFG-1000 and diesel soot (data from [33])

	Wavelength λ (nm)	GFG-1000 soot (m^2/g)	Diesel soot (m^2/g)
σ_e		5.7 ± 0.3	10.6 ± 0.5
σ_s	450	1.2 + 0.3–0.2	2.3 + 0.3–0.1
σ_a		4.5 + 0.3–0.4	8.3 + 0.5–0.7
σ_e		3.9 ± 0.4	8.2 ± 0.4
σ_s	550	1.0 + 0.2–0.1	1.6 ± 0.2
σ_a		2.9 + 0.4–0.5	6.6 ± 0.4
σ_e		2.6 ± 0.4	6.2 ± 0.3
σ_s	700	0.8 ± 0.2	1.0 ± 0.1
σ_a		1.8 + 0.4–0.5	5.2 ± 0.3

Jeong and Lee [17] found an absorption coefficient of 6.7 m^2/g fairly independent of the particle size. Particle growth by coagulation does not affect the soot extinction cross section and has only a small influence on scattering properties. This agrees with the findings by Gysel et al. [10], who showed that GFG-1000 soot particles are invisible for laser-induced incandescence (using the single particle soot photometer [SP2] by Droplet Measurement Technologies), although the size of the agglomerates is well above the detection limit. The primary particles are very small (5–10 nm) and due to the low fractal dimension (1.7–2.1) every primary particle acts independently. The primary particles dominate light scattering.

Table 14.7 Complex refractive index $m = n + ik$ (data from [33])

Wavelength	GFG-1000 soot	Diesel soot
450 nm	2.18 + *i*0.94	1.41 + *i*0.64
550 nm	2.65 + *i*1.32	1.49 + *i*0.67
700 nm	–	1.57 + *i*0.73

Juranyi et al. [18] investigated combustion aerosols with a dual-wavelength, multiangle measurement and compared the results to those of carbon particles from a GFG-1000. The fractal dimension D_f found in this study for spark ablation particles is lower ($D_f = 1.8$) compared to particles from a flaming fire ($D_f = 2.12$). Smoldering fires have D_f close to 3.

Thus, concerning the optical properties, spark ablation soot is not a good proxy for atmospheric soot and soot particles from real combustion. This should be considered when calibrating soot sensors based on the extinction or absorption cross sections of soot.

Combustion aerosol consists of EC and hydrocarbons. The condensation of hydrocarbons when the exhaust gas cools down is a growth mechanism of diesel particles in the exhaust pipe. The same can be done in "model" particles by condensing hydrocarbons on particles from an SDG. If the supersaturation is high enough, nucleation can occur in addition to condensation, as this can also occur in engine exhaust. Kim et al. [21] showed that this also works with model particles and allows coming closer to "real" combustion particles.

The coating of spark-generated soot with organic material results in a strong increase of the single scattering albedo. A significant increase of the absorption coefficient at λ = 473 nm during the coating process can be attributed to an enhancement of the specific soot absorption cross section by more than 30%.

14.4.4 Oxidation Behavior

Two classes of measures have reduced emissions from diesel engines: the optimization of the combustion process, to emit as little products of incomplete combustion as possible, and the use of efficient particle filters. These filters must be periodically regenerated by burning the collected soot. This regeneration can be done continuously (continuous regenerating trap [CRT], used for heavy-duty engines, producing enough NO_2 as oxidant) or by periodically increasing the exhaust temperature to temperatures above the ignition temperature of the collected soot (the usual technique for passenger cars). A third option, used, for example, for construction engines, is heating the filter externally (electrically or by burners). This, however, cannot be done during operation.

For the regeneration process the oxidation behavior is of crucial importance; lower ignition temperature allows softer regeneration. The oxidation depends on the soot morphology. Su et al. [35] investigated the microstructure and oxidation behavior of soot from the raw exhaust of a EURO IV heavy-duty diesel engine and compared it to that of spark ablation soot and hexa-benzocoronene (HBC; $C_{42}H_{18}$). They found a microstructure-controlled reactivity toward oxidation of all three samples in 5% O_2 in N_2. The spark ablation soot with its fine primary particles has an onset temperature of 423 K for combustion, while HBC with its well-ordered crystallites has a high onset temperature of 773 K. Due to an improved combustion process in the EURO IV heavy-duty diesel engine, the emitted soot consists of more fullerenoid- or onion-like particles agglomerated in a chain-like secondary structure. The onset temperature of the EURO IV heavy-duty engine soot combustion is 573 K.

To study the regeneration of diesel particle filters Hinot et al. [14] investigated catalytically assisted soot oxidation. They used real soot and several "model soots," among them also particles from an SDG.

In this study, particles from various sources have been compared in TPO experiments. Figure 14.5 shows that oxidation of SDG soot starts at lower temperatures but becomes slower later; already 50% conversion is reached earlier by diesel soot and pyrolysis soot. Soot from pyrolysis is very close to diesel soot. To investigate the catalytically assisted oxidation, soot particles and catalyst particles were produced, deposited in a sinter metal filter, and then measured in a TPO experiment. Several configurations have been tested:

- An SDG is operated with one carbon and one platin electrode.
- Pt and carbon particles are produced independently, and then the two aerosols are mixed prior to deposition.
- Layers of carbon and Pt particles are deposited upon each other.
- Configurations 1 and 2 show a significant reduction in oxidation temperature, indicating a close contact of carbon particles and the catalyst, whereas almost no effect is observed for configuration 3 [14].

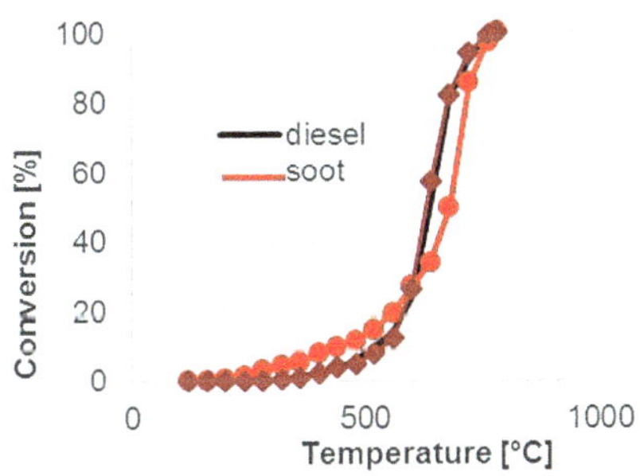

Figure 14.5 Conversion of diesel soot and particles produced by an SDG and pyrolysis (data from [14]).

Knauer et al. [23] also investigated mass conversion of soot upon oxidation. During the TPO, Fourier transform infrared (FTIR) spectroscopy and Raman microspectroscopy and EELS were used to get an overview about the reactivity of the oxidized soot. Figure 14.6 shows the mass conversion over time, which has been calculated by the CO_2 evolution from GfG, EURO VI, EURO IV, and graphite during TPO. The comparison of these curves reveals even larger differences than observed by Hinot et al. [14]. Again, GfG soot was found to be the most reactive. The differences in reactivity are associated by the authors with differences in structure: Starting from 423 K a low mass loss due to desorption of volatile surface functional groups is observed. A rapid oxidation step, indicating the oxidation of highly reactive strongly curved fullerenoidic and molecular carbon structures starts at about 450 K, showing a mass conversion of ca. 65% at 773 K. The lowest reactivity was observed for graphite, which shows a mass conversion of only 2% at 773 K in the range of accuracy of the experiment. The differences in the structure can also well be seen by the fraction of sp2-bonded carbon. The sp2 ratio increases from 56% to 76% for GfG soot if the temperature is increased from 300°C to 800°C, for EURO IV soot from 86% to 90%, and fairly constant at 88% for EURO VI.

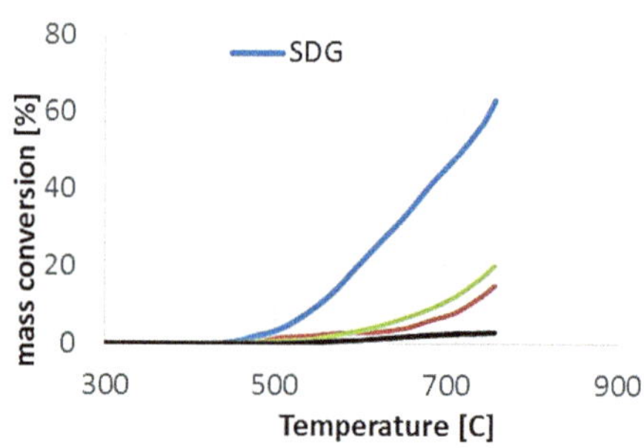

Figure 14.6 Mass conversion in a TPO experiment according to Knauer [23].

Continuous regeneration of diesel filters was investigated by Hackfort et al. [11] using a self-made SDG as a particle source. They built a reactor to test different catalysts and the effect of temperature on oxidation.

14.4.5 Other Properties/Applications

14.4.5.1 Hygroscopicity

One important property of soot in the atmosphere is its ability to act as nuclei for water or ice condensation. Kuznetsov et al. [25] studied the hygroscopic properties of SDG particles. They found that the surface area of GfG-1000 soot available for water molecules (SH_2O)

is 33 m^2/g; this is about one-tenth of the BET surface determined by N_2 adsorption (SN_2).

These authors claim that the internal microstructure of the primary particles of GfG-1000 soot is amorphous without any distinctive micropores. This is different from the onion-like microstructure of combustor soot particles, which determines their internal microporosity. The ratio of the surface areas determined for water and nitrogen adsorption is utilized as a quantitative measure of the surface hydrophilicity. Ratios of SH_2O/SN_2 for soots under investigation and for reference substrates are presented in Table 14.8.

Table 14.8 Surface area, hydrophilicity, and SH_2O/SN_2 ratio

	Surface area (m^2/g)	Relative hydrophilicity	SH_2O/SN_2
Silica gel	330	1	0.38
GfG-1000 soot	310	0.35	0.1
Combustion soot	48	0.6	0.33
Graphon soot	83	0.0037	6.7E-04

14.4.5.2 Radiolabeled "soot like" particles for inhalation studies

Health effects of ultrafine particles are still controversially discussed.

According to Roth [31] the following hypotheses are currently tested using in vitro and in vivo studies:

- Deposition of ultrafine particles in the respiratory tract causes adverse reactions in the respiratory and cardiovascular systems.
- Ultrafine particles deposited in the lungs are harmful because of their generic nature as particles. In other words, they carry a "physical toxicity" independent of their chemical composition.

To test these hypotheses, ultrafine particles simulating the physical properties of ambient particles (size, surface area, number concentration) are required. Roth [31] used a GfG-1000 for this application. In four animal exposure experiments with 30 nm carbon particles over a period of one year, the reproducibility of the modal diameter for fixed operation conditions of the generator (spark frequency and flow rates) was 10%. The operation stability proved spark discharging superior over most of the particle generation techniques. To investigate deposition in the respiratory tract and to follow particle transport to other organs, radioactively labeled particles are used. Brown et al. [2] used a PALAS-GfG-1000 to produce technetium-99-labeled carbon particles. Pertechnetate (TcOy) solution was repeatedly applied to the tips of electrodes until a desired level of activity was achieved. Prior to applying the pertechnetate solution, the tips of the electrodes were wetted with ethyl alcohol. Without the application of alcohol, the solution would form a bead when applied to the electrode, that is, it would not spread out or wet the entire tip of the electrode.

Oberdörster et al. [29] used also a GfG-1000 for inhalation/translocation studies in rats utilizing a 13C-labeled "soot."

14.4.5.3 Simulation of atmospheric aerosol

An important fraction of the atmospheric aerosol is combustion-derived particles; several attempts have been made to produce "model atmospheric particles." One approach is to start with carbon particles, for example, produced by SDGs, and to coat them with organic material. This can be done by directly coating the particles with organic material, as already demonstrated by Burtscher and Schmidt-Ott in 1986 [4]. Another option is to oxidize volatile organic compounds (VOCs) in a micro smog chamber, which reduces the vapor pressure and leads to condensation on existing carbon particles [20]. The production of such synthetic aerosols is investigated in the framework of the EMPIR-program "Aerosol Metrology for Atmospheric Science and Air Quality".[a] Another possibility would be collecting ambient particles and resuspending them for lab experiments.

Adsorption or condensation on particles could also be used to remove VOCs. Byeon et al. [5] tried to use spark-generated carbon particles for removal of VOCs. They used a PALAS-GfG-1000 to produce the particles and mixed them with the HC in a test duct. However, the removal efficiency achieved was poor (4%).

---

[a]https://www.euramet.org/research-innovation/search-research-pro-jects/details/?eurametCtcp_project_show%5Bproject%5D=1486&eurametCtcp_project%5Bback%5D=943&cHash=dd24da2c1ffbc68bde2ae1e6853934b5

14.5 Conclusions

The SDG is a simple-to-use tool to produce carbon particles. Although significant differences exist between the spark ablation-generated aerosols and carbonaceous particles produced by combustion in terms of morphology and composition, its simplicity and stability make it a useful tool to produce model combustion aerosols. This is true, in particular, as long as mainly the physical properties are of concern. Size distributions are comparable.

An important application could be the calibration of measurement systems for combustion engine measurements. Until now, no calibration material is prescribed for calibration in the regulation. Often emery oil or CAST is used. The counting efficiency of emery oil is about 16% for 23 nm and 10% for 41 nm, higher than for CAST particles. This strong discrepancy leads to significant differences in the total number concentration measured for diesel engines. Therefore an inert material is needed that can be delivered anywhere and is standardized. The SDG equipped with carbon electrodes has the advantage that the particles consist of an analytically pure material and is operated only with the noble gas argon as the carrier gas. This allows the production of carbon particles with a highly repeatable morphology. The SDG aerosol leads to almost the same response of the CPC as the CAST aerosol, at least for a particle size of about 23 nm. For 41 nm, the difference is still only about 8%. Therefore, for a CPC calibrated with an SDG, a similar response to diesel soot can be expected when compared to a CPC calibrated with a CAST [9].

Compared to real diesel soot and other carbon particles the SDG produces particles with a higher degree of disorder, leading to low oxidation temperature. Using the SDG for filter regeneration studies or other oxidation studies therefore is critical. There the pyrolysis generator yields the most similar particles to real diesel exhaust, not only as far as oxidation behavior is concerned, but also in primary particles size or BET surface. However, no commercial generator of this kind exists, and the generator may have problems with long term operation due to deposits.

SDG generators may be employed in studies to investigate the human health impacts from model combustion aerosol. The possibility to do this with labeled particles is very useful.

References

1. Baumgartner, Th., Burtscher, H. and Fierz, M. (2005). Soot production by pyrolysis, *ETH Conference on Combustion Generated Nanoparticles*, http://nanoparticles.ch/archive/2005 Baumgartner-_PR.pdf
2. Brown, J. S., Kim, C. S., Reist, P. C., Zeman K. L. and Bennett, W. D. (2000). Generation of radiolabeled "soot-like" ultrafine aerosols suitable for use in human inhalation studies, *Aerosol Sci. Technol.*, **32**, pp. 325–337.
3. Burtscher, H. and Schmidt-Ott, A. (1982). Enormous enhancement of van der Waals forces between small silver particles, *Phys. Rev. Lett.*, **48**, p. 1734.
4. Burtscher, H. and Schmidt-Ott, A. (1986). In situ measurement of adsorption and condensation of a polyaromatic hydrocarbon on ultrafine C particles by means of photoemission, *J. Aerosol Sci.*, **17**, pp. 699–703.
5. Byeon, J. H., Park, J. H., Yoon, K. Y., Ko, B. J.,Ji, J. H. and Hwang, J.. (2006). Removal of volatile organic compounds by spark generated carbon aerosol particles, *Carbon*, **44**, pp. 2106–2109.
6. Chae, S., Lee, D., Kim, M. C., Kim, D. S. and Choi, M. (2015). Wire-in-hole-type spark discharge generator for long-time consistent generation of unagglomerated nanoparticles, *Aerosol Sci. Technol.*, **49**, pp. 463–471.
7. Douglas, E., Evans, D. E., Harrison, R. M. and Ayres, J. G. (2003). The generation and characterisation of elemental carbon aerosols for human challenge studies, *J. Aerosol Sci.*, **34**, pp. 1023–1041.
8. Durdina, L., Lobo, P., Trueblood, M. P. Black, E. A., Achterberg, S., Hagen, D. E., Brem, B. T. and Wang, J. (2016). Response of real-time black carbon mass instruments to mini-CAST soot, *Aerosol Sci. Technol.*, **50**, pp. 906–918.
9. Grob, B., Wolf, J.-C., Kiwull, B. and Niessner R. (2014). Calibration system with an indirect photoelectric charger for legislated vehicle number emission measurement counters in the single counting mode, *J. Aerosol Sci.*, **70**, pp. 50–58.
10. Gysel, M., Laborde, M., Mensah, A. A., Corbin, J. C., Keller, A. Kim, J., Petzold, A. and Sierau B. (2012). Technical Note: The single particle soot photometer fails to reliably detect PALAS soot nanoparticles, *Atmos. Meas. Tech.*, **5**, pp. 3099–3107.
11. Hackfort, H., Hünnekes, E. and Dornseiffer, J. (2000). Fundamental investigation of continuous regeneration of carbon particle traps with oxygen, *SAE Technical Paper*, 2000-01-2984.

12. Helsper C., Mölter, W., Löffler, F., Wadenpohl, C. and Kaufmann, S. (1993). Investigations of a new aerosol generator for the production of carbon aggregate particles, *Atmos. Environ.*, **27A**, pp. 1271–1275.

13. Hemlasech, G. A. and de Watteville, M. C. (1907). Sur une nouvelle méthode de production des spectres de flamme des corps mètalliques, *C.R. Hebd. Seances Acad. Sci.*, **144**, p. 1338.

14. Hinot, K., Burtscher, H., Weber, A. K. and Kasper, G. (2007). The effect of the contact between platinum and soot particles on the catalytic oxidation of soot deposits on a diesel particle filter, *Appl. Catal., B*, **71**, pp. 271–278.

15. Horvath, H. and Gangl, M. (2003). A low-voltage spark generator for production of carbon particles, *J. Aerosol Sci.*, **34**, pp. 1581–1588.

16. Jacobson, M. Z. (2001). Strong radiative heating due to the mixing state of black carbon in atmospheric aerosols, *Nature*, **409**, pp. 695–697.

17. Jeong, B. and Lee, J. (2017). Effective density and light absorption cross section of black carbon generated in a spark discharger, *J. Aerosol Sci.*, **107**, pp. 55–64.

18. Jurányi, Z., Loepfe, M., Nenkov, M. and Burtscher, H. (2017). Multi-angle, dual wavelength scattering measurement chamber for the structural measurement of combustion generated particles, *J. Aerosol Sci.*, **103**, pp. 83–92.

19. Kalberer, M., Ammann, M., Gägeler, H. W. and Baltensperger, U. (1999). Adsorption of NO_2 on carbon aerosol particles in the low ppb range, *Atmos. Environ.*, **33**, pp. 2815–2822.

20. Keller, A. and Burtscher, H. (2012). A continuous photo-oxidation flow reactor for a defined measurement of the SOA formation potential of wood burning emissions, *J. Aerosol Sci.*, **49**, pp. 9–20.

21. Kim, H., Kim, J., Choi, Y., et al. (2006). Generation of model diesel particles by spark discharge and hydrocarbon condensation, *J. Mech. Sci. Technol.*, **20**, pp. 1972–1979.

22. Kim, W., Kim, S. B., Kim, H., Kim, Y. J. and Kim, S. S. (2010). Development of a novel simulated diesel particulate matter generator system, *Fuel*, **89**, pp. 2047–2053.

23. Knauer, M., Schuster, M. E., Su, D., Schlögl, R., Niessner, R. and Ivleva, N. P. (2009). Soot structure and reactivity analysis by Raman microspectroscopy, temperature-programmed oxidation, and high-resolution transmission electron microscopy, *J. Phys. Chem. A*, **113**, pp. 13871–13880.

24. Kohut, A., Villy, L. P., Ajtai, T., Geretovszky, Zs. and Galbács, G. (2018). The effect of circuit resistance on the particle output of a spark discharge nanoparticle generator, *J. Aerosol Sci.*, **118**, pp. 59–63.

25. Kuznetsov, B. V., Rakhmanova, T. A., Popovicheva, O. B. and Shonija N. K. (2003). Water adsorption and energetic properties of spark discharge soot: specific features of hydrophilicity, *J. Aerosol Sci.*, **34**, pp. 1465–1479.

26. Mamakos, A., Khalek, I., Giannelli, R. and Matthe Spears, M. (2013). Characterization of combustion aerosol produced by a mini-CAST and treated in a catalytic stripper, *Aerosol Sci. Technol.*, **47**, pp. 927–936.

27. Meuller, B. O., Messing, M. E., Engberg, D. L. J., Jansson, A. M. Johansson, L. I. M., Norlén, S. M., Tureson, N. and Deppert, K. (2012). Review of spark discharge generators for production of nanoparticle aerosols, *Aerosol Sci. Technol.*, **46**, pp. 1256–1270.

28. Moore, R. H., Ziemba, L. D., Dutcher, D., Beyersdorf, A. J., Chan, k., Crumeyrolle, S, Raymond, T. M., Thornhill, K. L., Winstead, E. L. and Anderson, B. E. (2014). Mapping the operation of the miniature combustion aerosol standard (mini-CAST) soot generator, *Aerosol Sci. Technol.*, **48**, pp. 467–479.

29. Oberdörster, G., Sharp, Z., Atudorei, V., Elder, A., Gelein, R., Lunts, A., Kreyling, W. and Cox, C. (2002). Extrapulmonary translocation of ultrafine carbon particles following whole-body inhalation exposure of rats, *J. Toxicol. Environ. Health Part A*, **65**, pp. 1531–1543.

30. Pfeiffer, T. V., Feng, J. and Schmidt-Ott, A. (2014). New developments in spark production of nanoparticles, *Adv. Powder Technol.*, **25**, pp. 56–70.

31. Roth, C., Ferron, G. A., Karg, E., Lentner, B., Schumann, G., Takenaka, S. and Heyder, J. (2004). Generation of ultrafine particles by spark discharging, *Aerosol Sci. Technol.*, **38**, pp. 228–235.

32. Schindler, W., Haisch, C., Beck, H., Niessner, R., et al. (2004). A photoacoustic sensor system for time resolved quantification of diesel soot emissions, *SAE Technical Paper*, 2004-01-0968, https://doi.org/10.4271/2004-01-0968.

33. Schnaiter, M., Horvath, H., Möhlter, O., Naumann, K.-H., Saafhoff, H. and Schöck, O. W. (2003). UV-VIS-NIR spectral optical properties of soot and soot-containing aerosols, *J. Aerosol Sci.*, **34**, pp. 1421–1444.

34. Schwyn, S., Garwin, E. and Schmidt-Ott, A. (1988). Aerosol generation by spark discharge, *J. Aerosol Sci.*, **19**, pp. 639–642.

35. Su, D. S., Jentoft, R. E., Müller, J. O., Rothe, D., Jacob, E., Simpson, C. D., Tomović, Ž, Müllen, K., Messerer, A., Pöschl, U., Niessner, R. and Schlögl, R. (2004). Microstructure and oxidation behaviour of Euro IV diesel engine soot: a comparative study with synthetic model soot substances, *Catal. Today*, **90**, pp. 127–132.

36. UNECE Regulations No. 83 (Uniform provisions concerning the approval of vehicles with regard to the emission of pollutants according to engine fuel requirements). Available at: http://www.unece.org/trans/main/wp29/wp29regs81-100.html.

37. VDI 3491 Blatt 4:2018-03 Technical rule: measurement of particles, methods for generating test aerosols, condensation methods, Publication date: 2018-03.

38. Hinds, W. C. (2008). *Aerosol Technology: Properties, Behavior, and Measurement of Airborne Particles*, John Wiley & Sons, New York, ISBN: 9780471194101.

39. Lackner, M., Árpád Palotás, Á. and Winter, F. (2013). *Combustion: From Basics to Applications*, Wiley-VCH, ISBN: 9783527667208.

Index